Lecture Notes in Computer Science   8597

Commenced Publication in 1973
Founding and Former Series Editors:
Gerhard Goos, Juris Hartmanis, and Jan van Leeuwen

More information about this series at http://www.springer.com/series/7409

Yueguo Chen · Wolf-Tilo Balke
Jianliang Xu · Wei Xu
Peiquan Jin · Xin Lin
Tiffany Tang · Eenjun Hwang (Eds.)

Web-Age
Information Management

WAIM 2014 International Workshops:
BigEM, HardBD, DaNoS, HRSUNE,
BIDASYS
Macau, China, June 16–18, 2014
Revised Selected Papers

 Springer

Editors

Yueguo Chen
DEKE Laboratory
Renmin University of China
Beijing
China

Wolf-Tilo Balke
Institute for Information Systems
Technical University Braunschweig
Braunschweig
Germany

Jianliang Xu
Department of Computer Science
Hong Kong Baptist University
Kowloon Tong
Hong Kong SAR

Wei Xu
School of Information
Renmin University of China
Beijing
China

Peiquan Jin
School of Computer Science
and Technology
University of Science and Technology
Hefei
China

Xin Lin
Department of Computer Science
East China Normal University
Shanghai
China

Tiffany Tang
Department of Computer Science
Kean University
Wenzhou
China

Eenjun Hwang
School of Electrical Engineering
Korea University
Seoul
Korea, Republic of (South Korea)

ISSN 0302-9743   ISSN 1611-3349 (electronic)
ISBN 978-3-319-11537-5   ISBN 978-3-319-11538-2 (eBook)
DOI 10.1007/978-3-319-11538-2

Library of Congress Control Number: 2014950824

Springer Cham Heidelberg New York Dordrecht London

Printed on acid-free paper

Springer is part of Springer Science+Business Media (www.springer.com)

Preface

Web-Age Information Management (WAIM) is a leading international conference for researchers, practitioners, developers, and users to share and exchange their cutting-edge ideas, results, experiences, techniques, and tools in connection with all aspects of web data management. The conference invites original research papers on the theory, design, and implementation of Web-based information systems. As the 15th event in the increasingly popular series, WAIM 2014 was held in Macau, China, during June 16–18, 2014.

Along with the main conference, WAIM workshops intend to provide international groups of researchers with a forum for the discussion and exchange of research results contributing to the main themes of the WAIM conference. This WAIM 2014 workshop volume contains the papers accepted for the following five workshops that were held in conjunction with WAIM 2014. These five workshops were selected after a public call-for-proposals process, each of which focuses on a specific area that contributes to the main themes of the WAIM conference. The five workshops were as follows:

- The Second International Workshop on Emergency Management in Big Data Age (BigEM 2014).
- The Second International Workshop on Big Data Management on Emerging Hardware (HardBD 2014).
- International Workshop on Data Management for Next-Generation Location-Based Services (DaNoS 2014).
- International Workshop on Human Aspects of Making Recommendations in Social Ubiquitous Networking Environments (HRSUNE 2014).
- International Workshop on Big Data Systems and Services (BIDASYS 2014).

All the organizers of the previous WAIM conferences and workshops have made WAIM a valuable trademark, and we are proud to continue their work. We would like to express our thanks and acknowledgments to all the workshop organizers and Program Committee members who contributed to making the workshop program such a success. They put a tremendous amount of effort into soliciting and selecting research papers with a balance of high quality, novelty, and applications. They also followed a rigorous review process. A total of 38 papers were accepted. Last but not least, we are grateful to the main conference organizers and the local Organizing Committee for their great support and wonderful arrangements.

May 2014

Yueguo Chen
Wolf-Tilo Balke
Jianliang Xu

BigEM 2014 Workshop Organizers' Message

With the advances of emergency management and information communication technologies, to improve the efficiency and accuracy of emergency management systems through modern data processing techniques becomes a crucial research issue. The past decade has witnessed huge technical advances in sensor networks, internet/ web of things, cloud computing, mobile/embedded computing, spatial/temporal data processing, and big data, and these technologies have provided new opportunities and solutions to emergency management.

Data processing/analysis in emergency management is a typical big data scenario. Numerous sensors and monitoring devices continuously sample the states of the physical world, while the web data processing techniques make the internet a huge data repository which can reflect the states of the cyber world and the human world. The efficient processing of these data imposes a big challenge to the data management community. It is important to develop advanced data management and data processing mechanisms to support disaster detection, disaster response and control, rescue resource planning and scheduling, and emergency commanding.

The Second International Workshop on Emergency Management in Big Data Age (BigEM 2014) was held in conjunction with the 15th International Conference on Web-Age Information Management (WAIM 2014) in Macau, China, during June 16–18, 2014. The purpose of BigEM 2014 was to provide a forum for researchers and practitioners to exchange ideas and progress in the related areas of big data management, such as cloud computing, parallel algorithms, internet of things, spatial database, complex event detection, optimization theory, intelligent transportation systems, and social networks. All submissions were reviewed by at least three Program Committee members in order to ensure that high-quality papers were selected. Following a rigorous review process, 15 papers were selected for publication in the workshop proceedings.

The Program Committee of BigEM 2014 consisted of 17 experienced researchers and experts in the area of big data management. We would like to thank all the authors for submitting their papers to the workshop and the valuable contributions of all the Program Committee members during the peer-review process. Also, we would like to thank the WAIM 2014 Workshop Co-chairs for their great support in ensuring the success of BigEM 2014.

May 2014

Xiaofeng Meng
Hui Zhang
Yi Liu
Wei Xu

BigEM 2014 Workshop Organization

General Co-chairs

Xiaofeng Meng   Renmin University of China, China
Hui Zhang   Tsinghua University, China

Program Co-chairs

Yi Liu   Tsinghua University, China
Wei Xu   Renmin University of China, China

Program Committee

Zhidong Cao   Chinese Academy of Sciences, China
Xiaolong Deng   Beijing University of Posts
                          and Telecommunications, China
Zhiming Ding   Chinese Academy of Sciences, China
Danhuai Guo   Chinese Academy of Sciences, China
Hong Huang   Tsinghua University, China
Jianhui Li   Chinese Academy of Sciences, China
Kuien Liu   Chinese Academy of Sciences, China
Xiangfeng Luo   Shanghai University, China
Xiaogang Qiu   National University of Defense Technology, China
Feiyue Wang   Chinese Academy of Sciences, China
Rui Yang   Tsinghua University, China
Dajun Zeng   Chinese Academy of Sciences, China
Lifeng Zhang   Renmin University of China, China

HardBD 2014 Workshop Organizers' Message

Big data management has received much attention from both academia and industries. While many people concentrate on designing new algorithms to mine valuable information from big data, a lot of researches are focused on improving the time efficiency of big data processing and storage. Regarding the time performance issue, one interesting solution is to utilize new and high-performance hardware in big data systems. Nowadays, hardware characteristics, is rapidly changing, imposing new challenges for an efficient utilization of hardware resources. Recent trends include storage-class memory, massive multi-core processing systems, very large main memory systems, fast networking components, big computing clusters, and large data centers that consume massive amounts of energy. It is clear that many aspects of data management have to evolve with these trends. Utilizing new hardware technologies for efficient big data management is of urgent importance.

The Second International Workshop on Big Data Management over Emerging Hardware (HardBD 2014) was held on June 16, 2014 at Macau in conjunction with The 15th International Conference on Web-Age Information Management (WAIM 2014). The overall goal of the workshop is to bring together researchers, practitioners, system administrators, system programmers, and others interested in sharing and presenting their perspectives on the effective management of big data over new hardware platforms, and also to discuss and identify future directions and challenges in this area.

The workshop attracted 12 submissions. All submissions were peer reviewed by two or three Program Committee members to ensure that the high quality of the accepted papers. Based on the reviews, the Program Committee selected four regular papers plus one short paper for inclusion in the workshop proceedings (acceptance rate 42 %). The final program of the workshop also consists of one keynote given by Prof. Shimin Chen who is from the Institute of Computing Technology, Chinese Academy of Sciences.

The Program Committee of the workshop consisted of 15 experienced researchers and experts. We would like to thank the hard work of all the Program Committee members during the peer-review process. Also we would like to acknowledge the WAIM 2014 workshop chairs for their support to HardBD 2014. The workshop is partially supported by the Natural Science Foundation of China (No. 60833005).

May 2014

Xiaofeng Meng
Jianliang Xu
Peiquan Jin

HardBD 2014 Workshop Organization

Workshop Chairs

Xiaofeng Meng   Renmin University of China, China
Jianliang Xu   Hong Kong Baptist University, China
Peiquan Jin   University of Science and Technology
       of China, China

Program Committee Members

Yi Ou   Technical University of Kaiserslautern, Germany
Binsheng He   Nanyang Technological University, Singapore
Luc Bouganim   Inria, France
Bin Cui   Peking University, China
Bin He   IBM Almaden Research, USA
Sang-Wook Kim   Hanyang University, Republic of Korea
Ioannis Koltsidas   IBM Research – Zurich, Switzerland
Ziyu Lin   Xiamen University, China
Vijayan Prabhakaran   Microsoft Research, USA
Xike Xie   Aalborg University, Denmark
Ke Lu   Tencent, China
Qi Zhang   Alibaba, China

HardBD 2014 Workshop Keynote

Effective Data Feeds for Big Data Analysis

Shimin Chen

State Key Laboratory of Computer Architecture
Institute of Computing Technology
Chinese Academy of Sciences, Beijing, China
chensm@ict.ac.cn

Velocity is one of the major challenges for big data processing. While it is natural to link velocity to transaction processing and data streaming, this talk investigates a different aspect of velocity – supporting high-speed data feeds for analytical query processing. A basic requirement is to store incoming data feeds. However, this is often not sufficient. I will discuss two of my previous studies in this respect: (1) achieving efficient online updates in data warehouses and (2) supporting high-speed event ingestion in an event log processing system. The first study exploits solid state disks for caching and organizing data updates while maintaining good query performance in data warehouses. The second study shuffles incoming data feeds to support efficient time window-based join operations. From these studies, I would like to argue that efficient data analysis requires careful handling of incoming data feeds. It is important to organize incoming data feeds in an analysis friendly manner so that data analysis operations can run efficiently.

DaNoS 2014 Workshop Organizers' Message

The proliferation of positioning technologies, such as GPS receivers and Wi-Fi, gives locations good opportunities to meet diversely data/information. Locations are combined with keywords in the spatial web querying that return objects that are near a location argument and are relevant to a text argument. Location-based social networking services allow social networks to connect and coordinate users with local people or events that match their interests. Location-based advertising pinpoint consumers' location and provide location-specific advertisements on their mobile devices. As an example, website Vidcinity (Video+Twitter+Foursquare) allows users to share and discover videos near a location. These various fusions enlarge the scope of location-based services and pose new challenges on the data management.

The First International Workshop on Data Management for Next-Generation Location-Based Services (DaNoS) was held on June 16, 2014, in Macau (China) in conjunction with the 21st International Conference on Web-Age Information Management (WAIM 2014). This workshop aims to facilitate the collaboration between researchers by presenting cutting edge research topics and methodologies on location-based service. The topics of interests in DaNoS 2014 include, but are not limited to: (i) Location-based keyword search, in which both spatial proximity and textual relevance; (ii) Location-based social network services, which match local people with similar interests; (iii) Location-based multimedia services, which aim at tagging spatial information on multimedia materials, e.g., video, photo, etc.; (iv) Location-based advertising services, which provide advertisements to location-specific consumers; (v) Location-based tagging system, which attaches spatial tags on the objects in business review websites; (vi) Privacy issues on location-based service; (vii) Spatial query processing on cloud services; and (viii) Other new type of spatial queries.

The workshop attracted four submissions, which are all peer reviewed by at least three Program Committee members to ensure the high-quality papers were selected. On the basis of the reviews, the Program Committee has selected two full papers for inclusion in the workshop proceedings (acceptance rate 50 %). Additionally, we also invite one paper to be involved in the proceeding. The Program Committee of this workshop consisted of five experienced experts on location-based service. We would like to thank authors of submitted papers of DaNoS 2014, and the Program Committee members for their careful reviews.

May 2014   Xin Lin
                                                                                  Dingming Wu

DaNoS 2014 Workshop Organization

Workshop Chairs

Xin Lin   East China Normal University, China
Dingming Wu   Hong Kong Baptist University, China

Program Committee

Huangliang Sun   Shenyang Jianzhu University, China
Yun Peng   Nanyang Technological University, Singapore
Jiajie Xu   Soochow University, China
Ke Deng   Huawei Noah's Ark Lab, China
Qian Chen   Hong Kong Baptist University, China

HRSUNE 2014 Workshop Organizers' Message

With the huge popular social-rich information environments (e.g., Netflix, Yelp, Facebook, Twitter, Google+) penetrating our daily life, people (and organizations) have become more powerless with the flooding information from which decisions must be made. Fortunately, recommender systems are known to be capable of implicitly or explicitly observing users' online activities, learning their likes and dislikes and making personalized (or group-wise) suggestions accordingly. They have become a well-integrated part of a vast number of web/mobile applications available in the cloud and have been used in a wide variety of application areas such as (digital) entertainment (e.g., news articles, music, movies, books, restaurants, etc.), software engineering (for example, recommending replacement methods for adaptive codes; recommending reusable codes from the internet, etc.), and e-learning contexts (gathering interactions during the learning process both in formal and informal learning scenarios through learning management systems, virtual learning communities, and personal learning environments).

While the majority of earlier research efforts have been focused on the algorithmic understanding of making recommendations, more recent ones have aimed at understanding human and social factors of making suggestions and sharing resources (e.g., content items, people, software widgets, etc.) in existing social ubiquitous networks to answer questions such as, among many others: (1) What types of resources (for example, news articles) are mostly likely to be shared and liked/disliked?; (2) Does human factors matter when rating a resource (and thus, are to be taken into account in the recommendation process) such as the users' mood and emotions or the social ubiquitous environment where the resources is consumed?; (3) What effects do the 'share'/'like'/'follow' buttons have on people's information-seeking behaviors; in other words, should traditional recommendation techniques integrate these non-numeric ratings in making suggestions? If so, how?; and (4) What effects would reviews provided by other users have over the popularity/fall of a resource in a social network and does this effect depends on the context where the review has been made?

The First International Workshop on Human Aspects of Making Recommendations in Social Ubiquitous Networking Environments (HRSUNE) was held on June 16, 2014, in Macau (China) in conjunction with the 21st International Conference on Web-Age Information Management (WAIM 2014). This workshop aims at bringing together researchers and practitioners to explore and share their research results on the human and social aspects of making recommendations in the emerging social and increasingly more and more ubiquitous networking environments. The topics of interests in HRSUNE 2014 include, but are not limited to: (i) The social and human aspects of making recommendations (factors including user mood, emotions, personality, social status, etc.); (ii) The effect on the recommendations of the ubiquitous interactions in the social networks, including geospatial and temporal variability of the user (the same user might prefer different recommendations depending on the physical and temporal context); (iii) Social recommendation in software engineering practices; (iv) Particularities of social and human aspects in making recommendations in e-learning contexts (both formal and informal learning scenarios); (v) Usability of social recommender systems; (vi) Visualizations of recommended resources and of group aspects to made aware to the others in practice; (vii) The psychology and economics of online sharing and recommendations; (viii) Recommending cloud services to support the information needs in social ubiquitous networking environments; and (ix) Any other relevant topic to the theory and application of recommendation system on social activities or cloud services.

The workshop attracted seven submissions. Submissions were peer reviewed by at least three Program Committee members to ensure that high-quality papers were selected. On the basis of the reviews, the Program Committee selected three full papers for inclusion in the workshop proceedings (acceptance rate 43 %). Another three papers were accepted as short papers. The Program Committee of the workshop consisted of 24 experienced researchers and experts. We would like to thank authors of submitted papers for choosing HRSUNE 2014 for the presentation of their research results and the members of the Program Committee for their valuable contributions during the peer-review process. Also, Olga C. Santos would like to acknowledge the support received from the MAMIPEC project (TIN2011-29221-C03-01) funded by the Spanish Ministry of Economy and Competence.

May 2014

Tiffany Tang
Olga C. Santos

HRSUNE 2014 Workshop Organization

General and Program Co-chairs

Tiffany Tang   Kean University, USA
Olga C. Santos   aDeNu Research Group, UNED, Spain

Program Committee

Mária Bieliková   Slovak University of Technology
                            in Bratislava, Slovakia
Ig Ibert Bittencourt   Federal University of Alagoas, Brazil
Jesus G. Boticario   aDeNu Research Group, UNED, Spain
Jüergen Buder   Knowledge Media Research Center, Germany
Rafael Calvo   University of Sydney, Australia
Iván Cantador   Universidad Autónoma de Madrid, Spain
Keith C. Chan   Hong Kong Polytechnic University, Hong Kong
Evandro Costa   Federal University of Alagoas, Brazil
Hendrik Drachsler   Open Universiteit Nederland, The Netherlands
Alexander Felfernig   Graz University of Technology, Austria
Marco de Gemmis   University of Bari, Italy
Guillermo Jiménez Díaz   Universidad Complutense de Madrid, Spain
Pythagoras Karampiperis   National Center of Scientific Research
                            "Demokritos", Greece
Bart Knijnenburg   UC Irvine, USA
Milos Kravcik   RWTH Aachen University, Germany
Estefanía Martín   Universidad Rey Juan Carlos, Spain
Patricia Morreale   Kean University, USA
Woojin Paik   Konkuk University, Republic of Korea
Abelardo Pardo   University of Sydney, Australia
Cristobal Romero   Universidad de Córdoba, Spain
Marko Tkalčič   Johannes Kepler University in Linz, Austria
Nava Tintarev   University of Aberdeen, UK
Pinata Winoto   Kean University, USA
Yuming Zhou   Nanjing University, China

BIDASYS 2014 Workshop Organizers' Message

In the past few years, there has been a rapid growth in big data research and its application technologies. Many novel architectures, algorithms, systems, and applications have been proposed, built, and studied. Compared with traditional databases, big data systems and applications foster many new issues and limitations in actual use and service.

The First International Workshop on Big Data Systems and Services (BIDASYS 2014) was held in conjunction with the 15th International Conference on Web-Age Information Management (WAIM 2014) in Macau, China, on June 16, 2014. The purpose of BIDASYS 2014 was to provide a forum for researchers and practitioners to exchange current issues, challenges, new technologies, and practical experiences, such as cloud/stream computing, software systems to support big data computing, social network search, and big data application/services, etc. The workshop attracted 13 submissions. All submissions were reviewed by at least three Program Committee members in order to ensure that high-quality papers were selected. Following a rigorous review process, nine papers were selected for presentation, covering a wide range of topics and showing interesting experiences.

The Program Committee of BIDASYS 2014 consisted of 15 experienced researchers and experts in the area of data processing and management. We would like to thank all the authors for submitting their papers to the workshop and the valuable contributions of all the Program Committee members during the peer-review process. Also, we would like to thank the WAIM 2014 Workshop Co-chairs for their great support in ensuring the success of BIDASYS 2014.

May 2014

Eenjun Hwang
Yunmook Nah
Xiaofeng Meng
Lizhu Zhou

BIDASYS 2014 Workshop Organization

General Co-chairs

Yunmook Nah   Dankook University, Republic of Korea
Xiaofeng Meng   Renmin University of China, China

Program Co-chairs

Eenjun Hwang   Korea University, Republic of Korea
Lizhu Zhou   Tsinghua University, China

Program Committee

Keun Ho Ryu	Chungbuk National University, Republic of Korea
Zhiyong Peng	Wuhan University, China
Ge Yu	Northeastern University, China
Sanghyun Park	Yonsei University, Republic of Korea
Sang-Wook Kim	Hanyang University, Republic of Korea
Keeyong Han	Polyvore, Inc., USA
Yang-Sae Moon	Kangwon National University, Republic of Korea
Xiaohui Yu	Shandong University, China
Sang-Won Lee	Sungkyunkwan University, Republic of Korea
Young-Koo Lee	Kyung Hee University, Republic of Korea
Jaewoo Kang	Korea University, Republic of Korea
Zhiguo Gong	University of Macau, China
Dugki Min	Konkuk University, Republic of Korea
Jangyoung Kim	University of Suwon, Korea
Lei Chen	Hong Kong University of Science and Technology, China

Contents

**The 2nd International Workshop on Big Data Management
on Emerging Hardware (HardBD 2014)**

**International Workshop on Data Management for Next-Generation
Location-Based Services (DaNoS 2014)**

International Workshop on Human Aspects of Making Recommendations in Social Ubiquitous Networking Environments (HRSUNE 2014)

International Workshop on Big Data Systems and Services (BIDASYS 2014)

The Second International Workshop on Emergency Management in Big Data Age (BigEM 2014)

Predicting Popularity of Microblogs in Emerging Disease Event

Jiaqi Liu[1], Zhidong Cao[1(✉)], and Daniel Zeng[1,2]

[1] The State Key Laboratory of Management and Control for Complex Systems, Institute of Automation, Chinese Academy of Sciences, Beijing, China
{Jiaqi.liu,Zhidong.cao,dajun.zeng}@ia.ac.cn
[2] University of Arizona, Tucson, AZ, USA

Abstract. During emerging disease outbreaks, massive information are disseminated through social network. In China, Sina microblog system as the biggest social network provide a novel way to monitoring the development of emerging disease and public awareness. However, only a small percentage of microblogs could wide spread. Therefore, predict popularity of microblogs timely are meaningful for emergency management. In this paper, a Judgment method for popularity level prediction of microblog is proposed and the temporal pattern between cases number and repost number is verified. Repost number is considered to measure the impact of microblogs. To predict the popularity of microblogs, Granger causality test was used to verify the temporal correlation pattern between development of disease and public concern while an Judgment method based on five classical classification models were proposed. Through analyses, case number of emerging disease are Granger causality of the popularity level of microblogs and the regression model got the best result when lag was three. By Judgment method, more than 86 % microblogs can be classified correctly. The proposed Judgment method based on user, microblog and emerging disease information could analysis the popularity level of microblogs speedily and accurately. This is important and meaningful for monitoring the development of future public health event.

Keywords: Microblogs · Popularity prediction · Granger causality · Classification

1 Introduction

During emerging disease outbreaks, massive information are disseminated not only through online announcements by government agencies but also through lots informal channels [1]. Massive freely available Web-based sources of information make digital disease surveillance possible [2, 3]. Systems based on scanning news media (e.g., GPHIN [4], MedISys [5]), modeling search query (e.g., Google [6], Baidu [7]), and monitoring real-time social media (e.g., Twitter [8], Sina Microblog [9]) were created. In particular, Sina Microblog can be invaluable source of real time individual data in China [10] and millions of users are willing to spread health-related information [11, 12]. However, not all health related messages posted in the microblog system are important that only few of them could attracted public attention and became popular

© Springer International Publishing Switzerland 2014
Y. Chen et al. (Eds.): WAIM 2014, LNCS 8597, pp. 3–13, 2014.
DOI: 10.1007/978-3-319-11538-2_1

through the network [13]. Therefore, concerning and tracking all microblogs would consume a large number of resources and might obtain no achievement. In other words, predict popular microblogs become an urgent problems for situation forecast [14].

Several popularity prediction researches in social media has already obtained some interesting findings. The basic information of user and microblog has strong impact on its spreading scale [15]. More specifically, large repost number tweets would spread quickly in users and exert big influence on users [16] as well as the impact of the temporal [17–19] and spatial [20, 21] characteristics of microblogs are also investigated comprehensively [22]. In addition, some previous research indicated that machine learning methods (e.g., Bayesian Network, Support Vector Machine) have strong classification ability to identify the microblog with high popularity through some basic microblog features [23, 24] and the posting number of tweets and cases number of emerging disease are highly correlated in time [25]. In this paper, the repost number we used is the most common way to evaluate the popularity level of microblogs in information diffusion area [14].

As we consider the popularity level prediction of microblogs about emerging disease event is classification problem, the main work of this paper is proposing a Judgment method by integrating results of five classic classification algorithms. Emerging disease event of human infection with influenza A(H7N9) provided a golden opportunity to test our Judgment Method. Similar with previous studies, several quantitatively features of users and related microblogs are considered in our method. In addition, whether and how the disease situation will influence the popularity of the related microblogs is also an interesting question which is not been studied. In this paper, Granger causality test was used to verify the temporal correlation between the popularity level of microblogs and the development of emerging disease. Based on these analyses, some basic information about the emerging disease were also used in our method. Through analyses, the proposed Judgment method offer an new way to predict the popularity level of microblogs about emerging disease and have important implications for the future digital disease surveillance.

The rest of this paper is organized as follows. Section 2 present the methods we used. Section 3 describes the experimental results. Finally, Sect. 4 concludes the paper and outlines the directions for future work.

2   Methods

2.1   Data Collection and Datasets

H7N9 Case Number. Human infection with influenza A(H7N9) was notified by National Health and Family Planning Commission of the People's Republic of China(NHFPC) on Mach 31, 2013 [26]. It is the first time that human infection with this type of avian virus has been identified. 130 cases were laboratory-confirmed in 8 provinces and 2 munici-palities up to May, 14 [27]. After official announcement of three human H7N9 cases on March 31, detailed disease information was daily updated on NHFPC's website till April 24 and then H7N9 related health information was weekly reported. We manually collected these detail message of each human H7N9 cases. Only location, date and number were used in this paper and no personal information is involved (see Fig. 1).

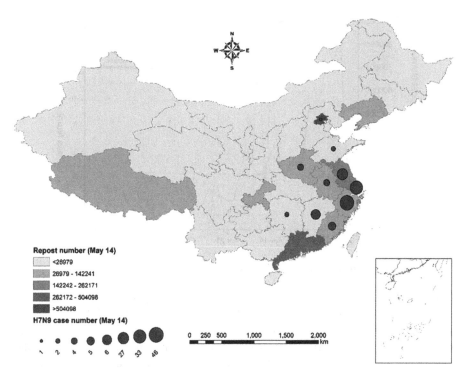

Fig. 1. The geographic distribution of repost and case

Microblogging Data. Sina microblogs (Weibo), which has more than 500 million users and 46.2 million daily active users, is the biggest microblog system ever used in China [28]. After H7N9 event happened, lots of related microblogs were posted and reposted. Hot word index(HWI) (Fig. 2) is calculated according to massive daily microblog data by Sina microblogging system [29]. It reflects the mentioned number of a keyword, includes original and repost, in the whole microblog system. We developed an Internet Epidemic Surveillance Platform(IWSP) which continuously collected textual information from Weibo and other websites or BBS related to public health. Since Weibo has lots of users and a large number of microblogs were created over time [30], we cannot get all of these messages. This platform pay more attention to Internet celebrities, opinion leaders, news agencies and user or account which related to public health. We filter out microblogs which contain keyword or hashtag "H7N9". Unlike Twitter, Weibo count number of repost and comment for each microblog, we can easily evaluate the influence scope of each original microblog. Hence, we excluded repost microblogs. Between March 31 and May 14, 2013, we archived 146,684 original microblogs from 105,721 users. In addition, we updated users' information, such as followers, friends and statuses number and registered province, in August 7.

2.2 Temporal Correlation Analysis

After H7N9 event happened, lots of people concerned about the development of the emerging disease and some of them posted or reposted related microblogs. To verify

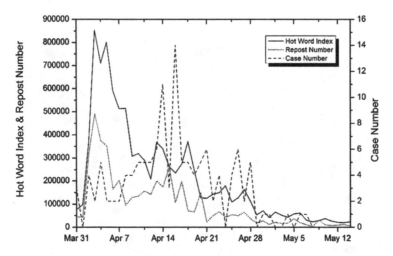

Fig. 2. The time series of hot index, repost number and case number

whether these two aspects have causal relationship, especially temporal correlation pattern, Granger causality test were used. The Granger causality test [31] is a statistical hypothesis test for determining whether one time series is useful in forecasting the another. When values in series X provide statistically significant information about the value or the future values in the other series Y, the first is said to be the Granger causality of the second series.

$$y_t = \sum_q^{i=1} \alpha_i x_{t-i} + \sum_q^{j=1} \beta_j y_{t-i} + u_t \tag{1}$$

In addition, the stationary of two time series is a preconditions for Granger causality test, which means that the joint probability distribution of the series does not change when shifted in time, otherwise spurious regression problems could occur. Thus, unit root test need to be done on both series in advance. The model for unit root test in this paper is Augmented Dickey-Fuller(ADF) test with zero lag length. Due to different trend of the two time series, the ADF test for case number series is just include intercept while trend and intercept are both considered for reposting number series.

2.3 Popularity Level Prediction Model

Predict popularity level of microblogs is a typical classification problems and several models have been proposed. Five classical classification models in WEKA [32] were used in our experiments, which is logistic regression model, J48 decision tree model, Bayes net classification model, support vector machines and adaptive boosting model. Logistic regression model is a type of probabilistic statistical classification model which usually use one or more continuous variables to make the prediction and success to handle the threshold question [33]. Decision tree model uses an attribute selection measure to determine which attribute has high distinguish [34]. Bayes net is based on Bayes' theorem of posterior probability [35]. Sequential minimization algorithm(SMO) is an important application of support vector machine which have yielded excellent generalization performance on lots problems [36]. And adaptive boosting model, as voting classification algorithm, has been shown successful in improving the accuracy of certain classifiers [37]. However, all these classification models have their advantages and disadvantages. Hence, we proposed the Judgment method to integrate all results from the five models by regarding the majority result as the final popularity level. Several common evaluation index, such as accuracy, precision, recall and F-value, were used to measure the classification result.

Table 1. Microblog features

Features	Description	Type
Follower	# of users who follows the author of the microblog	Numerical
Following	# of users the author is following	Numerical
Status	# of microblogs posted by the author since the creation of the account	Numerical
Location	The reported location or register location	Nominal
Lag of Days	# of days after March 31 when the microblog posted	Numerical
Local Lag of Days	# of days after the province been infected when the microblog posted	Numerical
Case	# of humans infected by H7N9 avian influenza	Numerical
URL	Whether the microblog contain URL	Nominal
Tag	Whether the microblog tag other user with '@'	Nominal

After deciding the classification models and the judgment method we use, the input and output of these models need to be determined. The features we selected to make the prediction are shown in Table 1. The first four features are user's basic information, which we can direct get through Sina Application Programming Interface(API). The province is nominal variable while others are numerical. The following two numerical features are time information about microblog posted. Microblog posted in different day got varying concerns and local news agencies also play different role after the area got first attacked. H7N9 case number is also considered as numerical. The last two features are two typical and easy accessible nominal features. Microblog which contain URL [17] or tag [18] other user would be marked. For the popularity level, we sep-arated microblogs into two categories according to whether it had been reposted more than or equal to ten, which is a balance between event description and resource utili-zation. 6.77 % of the microblogs were marked as high reposted microblog (positive tuples) as this criteria. Since class-imbalanced dataset will significant influence the result of most classify model, we used under sampling process to decrease the number of negative tuples. We kept all 9932 microblogs which reposted more than or equal to ten, and then random chose microblogs in negative tuples to constitute a training set with 20000 microblogs.

3   Results

3.1   Temporal Correlation Analysis

The results of ADF unit root test on case and repost number series are shown in Table 2. Both results reject the null hypothesis ($P_{case} < 0.001$, $P_{RTNum} < 0.01$), which means these series don't have unit root. The coefficients of the regression model are all statistical significant ($P < 0.01$). The Akaike information criterion and Schwarz criterion is small enough while Durbin-Watson statistic indicated autocorrelation. According to these results, case and repost number series are both stationary time series.

Table 2.   ADF unit root test result

CASE				
	Coefficient	Std. Error	t-Statistic	Prob.
ADF test statistic	-	-	-4.119420	0.0023
Variable: CASE(-1)	-0.586431	0.142358	-4.119420	0.0002
Variable: C	1.664457	0.594751	2.798575	0.0077
Statistic	Value	Statistic		Value
R-squared	0.287769	Akaike info criterion		4.933848
F-statistic	16.96962	Schwarz criterion		5.014948
Prob(F-statistic)	0.000174	Durbin-Watson stat		2.445098
Repost Number				
	Coefficient	Std. Error	t-Statistic	Prob.
ADF test statistic	-	-	-4.334661	0.0067
Variable: RTNum(-1)	-0.547483	0.126304	-4.334661	0.0001
Variable: C	143347.4	36778.24	3.897613	0.0004
Variable:@TREND(1)	-3910.022	1110.155	-3.522051	0.0011
Statistic	Value	Statistic		Value
R-squared	0.318543	Akaike info criterion		25.08405
F-statistic	9.582602	Schwarz criterion		25.20570
Prob(F-statistic)	0.000385	Durbin-Watson stat		1.984863

Table 3 shows that the case series is Granger causality of the repost series. With lag number of days varying from one to five, the results of statistical tests are all significant in different level. The regression model got the best result when lag was three which means the sum of first three lags of case series can predict the most similar results with repost series.

From the results obtained so far, several conclusions can be made. First, new case number will affect users' attention of health event while case series is causality of repost series. Second, the effect of new case number would last for a few days while user would swing their attention on the health event base on the changing trend of the emerging disease. Third, the repost behavior have hysteresis that some users does not log in frequently and might repost the microblog a few days later. Overall, new case number will affect people's attention and the reposting scale of public health event to a certain extent. However, as long as no burst point of the event, people's attention will gradually decay after reaching the peak. The decreasing trend will not fundamentally changed even sporadic small-scale events.

Table 3. Granger causality result

Lags	Observations	F-Statistic	Prob.
1	44	2.63601	0.1121
2	43	3.51495	0.0398
3	42	4.79999	0.0066
4	41	2.74538	0.0453
5	40	2.14205	0.0885

3.2   Popularity Level Prediction Model

When finished modeling, we tested our models on training dataset first. The results of five classical classification models are shown in Fig. 3. All models achieved good results. Since the training dataset is balanced data, we assigned the β of F-value equal to one which means same weight to precision and recall was given. The F_1-value of models are within the scope of 0.72 (SMO) to 0.90 (J48 decision tree). Moreover, J48 decision tree model obtained the best result in all four measures. In other words, the result indicated that the attributes we selected are effective and information gain ratio showed good performances. Logistic model and adaptive boosting models are good at precision rate (both 0.88) which implies that the high popularity level microblogs they choice out is more likely to be real high popularity level microblogs. In addition, Bayes net and adaptive boosting model got relative high recall (both 0.82) and accuracy rate (Bayes 0.84, Adaboost 0.86) that most of the high popularity level microblogs were found. Majority rules were used to combine the results and the accuracy of Judgment method result is 0.86 while the F_1-value is 0.85. Even if this result is not the highest, all four measures of Judgment method were 3 % higher than the average. Overall, most of the high popularity level microblogs are identified in balanced training dataset.

Then, we test our prediction model on the whole dataset. As class-imbalanced dataset it is, the recall rate is the most important measure for models [38]. Therefore, we assigned $\beta = 2$ which weights recall twice as much as precision. The results of models testing on whole dataset are shown in Fig. 4. The highest F_2-value is 0.69 (J48 decision tree) while the Judgment method got 0.65. For the recall rate, J48 decision tree model obtained the highest value (0.89) while Bayes net, adaptive boosting and Judgment method also got acceptable value above 0.80. The majority of important microblogs were recalled. The Judgment method, with the second highest F_2-value (0.65), achieved the highest precision rate (0.37). Because the dataset is highly imbalanced, even though the precision is relative low, lots of monitoring costs are still save. The density of important microblog was improved from 6.77 % to more than one-third. All models performed very well on accuracy within a range from 0.86 (Bayes net) to 0.89 (Judgment method). Most of the massive low popularity level microblogs are identified and filtered out.

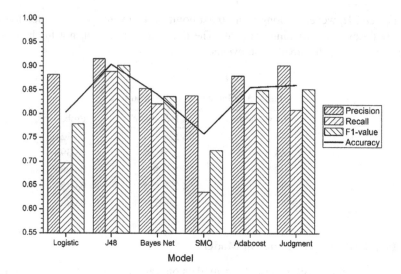

Fig. 3. Classification results on training dataset

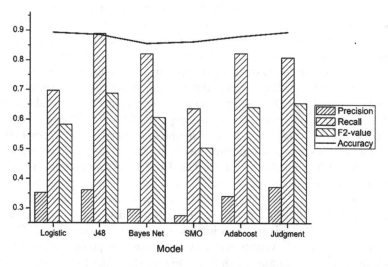

Fig. 4. Classification results on whole dataset

The results obtained by our proposed model proved that the potential popularity of microblogs can be identified exactly through some features of microblogs and disease information. Judgment can be made as soon as the microblog crawled by our platform. Microblogs with high popularity level would be preserved and used for further analysis immediately while low popularity level microblogs would be discarded to save resources.

4 Discussion

With the development of social network and digital disease surveillance, a novel monitoring approach on public health have been proposed and the further intervention would more tangible. The results of our study indicate that the new case number of emerging infectious diseases is the Granger causality of the repost number of microblogs and the new case number will significant affect the influence of microblogs. Therefore, more attention would pay when large number of emerging disease occur to avoid false information or rumor occupy public opinion. Either health related guided approach or warning messages about the emerging disease can widely spread. Moreover, since the sum of first three lags of case series is most predictable for predicting daily microblogs level through the analyses, the control measures should have continuity in order to get enduring influence.

In addition, potential high popularity microblogs have big impact on public awareness on emerging disease event and should be pay attention at any time. The proposed popularity level prediction model which based on the user, microblog and emerging disease information gets high accuracy and F-value on predicting popularity level of microblogs. Using this model, the influence of microblogs can be identified quickly at the time it posted and only potential high popularity level microblogs would be preserved and kept watching. In this way, not only lots storage space and monitoring cost would be saved, but the effect and trend of emerging disease event on Internet would be detected timely. The policymakers could also make judgments based on monitoring results of classification while some interventions could be implemented on both online and offline before the event causing uncontrollable consequences. Several Internet monitoring measures, such as highlight, recommend or delete, can make a hug difference.

Although we got some promising results, there are still several weaknesses to this study. Firstly, the dataset we used is a subset of whole H7N9 related data in Sina microblog system. Although we had tested the representation of our data, some further results might achieve with more comprehensive data. Secondly, the relationships between users might have some impacts on their repost behavior. Once someone in a small group with closely contacted comment on emerging infectious disease event, others have more possibilities to join the discussion. However, these network features were not consider in our research this time since they would significantly increase the complexity of the model. Thirdly, limitation also exist in content analysis that some burst events or words might influence on the popularity level of microblogs, but we didn't include these factors since we think that their impact had already contained in statistical results and wouldn't increase the accuracy visibly. These weaknesses will be our important direction for future research. Although limitations exist, this research has already verified the temporal correlation between emerging disease case number and repost number of microblogs. The proposed prediction model has sufficiently high accuracy and will be deployed in our IWSP platform as soon as possible.

Acknowledgments. This study was funded by National Natural Science Foundation of China (Nos.90924302, 91224008, 91024030, 91324007) and Important National Science & Technology Specific Projects (Nos.2012ZX10004801, 2013ZX10004218).

References

1. Brownstein, J.S., Freifeld, C.C., Mado, L.C.: Digital disease detection-harnessing the web for public health surveillance. New Engl. J. Med. **360**, 2153–2157 (2009)
2. Wilson, K., Brownstein, J.S.: Early detection of disease outbreaks using the internet. Can. Med. Assoc. J. **180**, 829–831 (2009)
3. Lampos, V., De Bie, T., Cristianini, N.: Flu detector - tracking epidemics on twitter. In: Balcázar, J.L., Bonchi, F., Gionis, A., Sebag, M. (eds.) ECML PKDD 2010, Part III. LNCS, vol. 6323, pp. 599–602. Springer, Heidelberg (2010)
4. The global public health intelligence network (GPHIN). http://www.who.int/csr/alertresponse/epidemicintelligence/en/
5. The medical information system (medisys). http://medusa.jrc.it/medisys/
6. Ginsberg, J., Mohebbi, M.H., Patel, R.S., Brammer, L., Smolinski, M.S., et al.: Detecting influenza epidemics using search engine query data. Nature **457**, 1012–1014 (2008)
7. Yuan, Q., Nsoesie, E.O., Lv, B., Peng, G., Chunara, R., et al.: Monitoring influenza epidemics in china with search query from baidu. PLoS ONE **8**, e64323 (2013)
8. Naveed, N., Gottron, T., Kunegis, J., Alhadi, A.C.: Bad news travel fast: A content-based analysis of interestingness on twitter. In: Proceedings of the 3rd International Web Science Conference (2011)
9. Salathé, M., Freifeld, C.C., Mekaru, S.R., Tomasulo, A.F., Brownstein, J.S.: Influenza A (H7N9) and the importance of digital epidemiology. New Engl. J. Med. **369**, 401–404 (2013)
10. Yang, J., Counts, S.: Comparing information diffusion structure in weblogs and microblogs. In: ICWSM (2010)
11. Fernandez-Luque, L., Karlsen, R., Bonander, J.: Review of extracting information from the social web for health personalization. J. Med. Internet Res. **13**, e15 (2011)
12. Kostkova, P.: A roadmap to integrated digital public health surveillance: the vision and the challenges. In: Proceedings of the 22nd International Conference on World Wide Web Companion, International World Wide Web Conferences Steering Committee, pp. 687–694 (2013)
13. Yang, J., Counts, S.: Predicting the speed, scale, and range of information diffusion in twitter. ICWSM **10**, 355–358 (2010)
14. Hong, L., Dan, O., Davison, B.D.: Predicting popular messages in twitter. In: Proceedings of the 20th International Conference Companion on World Wide Web, pp. 57–58. ACM (2011)
15. Jenders, M., Kasneci, G., Naumann, F.: Analyzing and predicting viral tweets. In: Proceedings of the 22nd International Conference on World Wide Web Companion, International World Wide Web Conferences Steering Committee, pp. 657–664 (2013)
16. Starbird, K., Palen, L.: (how) will the revolution be retweeted? information diffusion and the 2011 egyptian uprising. In: Proceedings of the ACM 2012 Conference on Computer Supported Cooperative Work, pp. 7–16. ACM (2012)
17. Chew, C., Eysenbach, G.: Pandemics in the age of twitter: content analysis of tweets during the 2009 H1N1 outbreak. PLoS ONE **5**, e14118 (2010)
18. Yang, J., Leskovec, J.: Patterns of temporal variation in online media. In: Proceedings of the Fourth ACM INTERNATIONAL Conference on Web Search and Data Mining, pp. 177–186. ACM (2011)
19. Lee, C., Kwak, H., Park, H., Moon, S.: Finding influentials based on the temporal order of information adoption in twitter. In: Proceedings of the 19th International Conference on World Wide Web, pp. 1137–1138. ACM (2010)

20. Hay, S.I., George, D.B., Moyes, C.L., Brownstein, J.S.: Big data opportunities for global infectious disease surveillance. PLoS medicine **10**, e1001413 (2013)
21. Paul, M.J., Dredze, M.: You are what you tweet: Analyzing twitter for public health. In: ICWSM (2011)
22. Kwak, H., Lee, C., Park, H., Moon, S.: What is twitter, a social network or a news media? In: Proceedings of the 19th International Conference on World Wide Web, pp. 591–600. ACM (2010)
23. Signorini, A., Segre, A.M., Polgreen, P.M.: The use of twitter to track levels of disease activity and public concern in the us during the influenza A H1N1 pandemic. PLoS ONE **6**, e19467 (2011)
24. Collier, N., Son, N.T., Ngoc, M.N.T.: Omg u got u? analysis of shared health messages for bio-surveillance. In: Semantic Mining in Biomedicine (2010)
25. Chunara, R., Andrews, J.R., Brownstein, J.S., et al.: Social and news media enable estimation of epidemiological patterns early in the 2010 haitian cholera outbreak. Am. J. Trop. Med. Hyg. **86**, 39 (2012)
26. Human infection with influenza A(H7N9) virus in china. http://www.who.int
27. Human infection with influenza A(H7N9) virus in china. http://www.chinapop.gov.cn/yjb/s3578/201305/67d505cd37eb4a419f17518bdbe05b54.shtml
28. Guo, Z., Li, Z., Tu, H.: Sina microblog: an information-driven online social network. In: 2011 International Conference on Cyberworlds (CW), pp. 160–167. IEEE (2011)
29. The hot index of hotword in sina microblog system. http://data.weibo.com/index/hotword
30. The registered users of sina microblog system exceeded 300 million and more than 100 million microblogs a day. http://news.xinhuanet.com/tech/2012-02/29/c_122769084.htm
31. Granger, C.W.: Investigating causal relations by econometric models and cross-spectral methods. Econom. J. Econom. Soc. **37**, 424–438 (1969)
32. Hall, M., Frank, E., Holmes, G., Pfahringer, B., Reutemann, P., et al.: The weka data mining software: an update. ACM SIGKDD Explor. Newsl. **11**, 10–18 (2009)
33. Bishop, C.M., et al.: Pattern Recognition and Machine Learning, vol. 1. Springer, New York (2006)
34. Lim, T.S., Loh, W.Y., Shih, Y.S.: A comparison of prediction accuracy, complexity, and training time of thirty-three old and new classification algorithms. Mach. Learn. **40**, 203–228 (2000)
35. Han, J., Kamber, M., Pei, J.: Data Mining: Concepts and Techniques. Morgan kaufmann, San Francisco (2006)
36. Keerthi, S.S., Shevade, S.K., Bhattacharyya, C., Murthy, K.R.K.: Improvements to platt's SMO algorithm for SVM classifier design. Neural Comput. **13**, 637–649 (2001)
37. Margineantu, D.D., Dietterich, T.G.: Pruning adaptive boosting. In: ICML. Citeseer, volume 97, pp. 211–218 (1997)
38. Barboza, P., Vaillant, L., Mawudeku, A., Nelson, N.P., Hartley, D.M., et al.: Evaluation of epidemic intelligence systems integrated in the early alerting and reporting project for the detection of A/H5N1 influenza events. PLoS ONE **8**, e57252 (2013)

A Cross-Simulation Method for Large-Scale Traffic Evacuation with Big Data

Shengcheng Yuan, Yi Liu[(⊠)], Gangqiao Wang, and Hui Zhang

Department of Engineering Physics, Institute of Public Safety Research,
Tsinghua University, Beijing, China
liuyi@tsinghua.edu.cn

Abstract. Microscopic traffic simulation is one of the effective tools for transportation forecast and decision support. It is a challenge task to make reasonable prediction of traffic scenarios during emergency. Big data technology provides a new solution for this issue. This paper proposes a cross-simulation method to apply the mass data collected in normal situations into large-scale traffic evacuations to provide better supporting information for emergency decision. The method consists of three processes: Acquisition, Analysis and Adaptation. It captures the dynamic distance-speed relation of every vehicles on the real roads and build a database of driving behaviors according to the existing car-following models. After calibration and analysis, various driving behaviors can be identified. During emergency, the distribution of driving behaviors will be refactored to adapt the fast-changing situation automatically so that the simulation system gains the adaptive ability in emergency situations. An experimental result on a real road preliminarily validates the practicability of the method and shows the supporting information which it can provide. The new method will make contributions on enhancing the predictive ability of traffic simulation systems in emergency situations.

Keywords: Transportation system · Large-scale evacuation · Cross simulation · Big data

1 Introduction

Large-scale evacuation is a significant topic in emergency management. A thorough contingency plan which neglects to consider transportation management might eventually collapse. In situations concerning emergency management, making quick decisions on traffic evacuation is usually required. Therefore it is essential for people who take part in emergency management to make quick judgments and reasonable estimations concerning transportation events.

Microscopic traffic simulations are one of the effective tools for the estimation and decisions of traffic situations. By simulating each vehicle individually, this tool is able to express the diversity of driving behavior. However, it is still challenging to reasonably estimate transportation during emergency due to the following reasons:

1. Large amount of data is required. Simulation systems require information drawn from real transportation systems for correction. Yet during emergency such information is often lacking.

© Springer International Publishing Switzerland 2014
Y. Chen et al. (Eds.): WAIM 2014, LNCS 8597, pp. 14–21, 2014.
DOI: 10.1007/978-3-319-11538-2_2

2. High uncertainty. During emergency the driving behavior is often various, rendering the traffic evacuation system highly unstable. The available models cannot cope with the estimation of the changing transportation system.
3. Wide range. Due to high density of population and over-crowdedness, simulation models are required to cover a range of a couple million of vehicles.

Existing research on diverse driving behavior was often achieved through traditional experimentation. For example, a typical research was aimed at the choice of route that a driver makes in order to get to a certain location. Dia et al. presented the belief-desire-intention agent architecture, describing whether or not the driver would give up behavioral choices of route when receiving real-time information [1]. Tawfik et al. researched the diversity in a driver's ability to perceive different route data and their decisions [2]. The interaction between transportation environment and the psychological influences of drivers is an important factor in causing intra-variability [3] in drivers. Parker et al. surveyed the drivers from Finland, UK, and the Netherlands, studying the extend to which drivers react to unreasonable driving behaviors, enabling them to react radically [4]. Kaysi et al. used a U-turn intersection to study the impact of radical behaviors on the transportation environment.

The big data technology has provided an excellent opportunity to solve this issue. It can provide better supporting information for the decision-making process. For example, Yuan et al. used trajectory data of taxies to provide route recommendations, and the results of which are notable [5]. However, in times of emergency the driver's psychological state is unstable, the behavior therefore differs highly from the usual state, which has an impact on transportation systems. How to use normal transportation data gaining useful information in times of emergency has become a challenging issue.

This paper provides a solution to converging the big data technology in a standard situation and adapting it to be used in a large-scale evacuation. This solution is separated into three processes: Acquisition, Analysis, and Adaptation. It uses video cameras in transportation systems and other sensory equipment to collect data, generating the probability distribution of driving behaviors, and analyzing similar behavior patterns. This paper adapted a road into a calculation experiment, proving primarily the practicality of this method and its ability to provide supporting information for the decision-making process.

This method is helpful in improving the estimation capability in a simulation transportation system during an emergency. The decision-making system could distribute a certain evacuation decision to a validation system for multiple meticulous simulation, evaluating the robustness of the decision. Because the driving behavior data comes from a standard scenario, the use of this method could achieve to form a self-adapting simulation evacuation system using big data.

2 System Design

In general, traffic simulation system needs to obtain real information on driving behavior in order to achieve accuracy. In a standard situation, this process would be simple and efficient. However, these methods can hardly be applied to emergencies, because there is limited time for collecting mass data from the emergency system.

Therefore, it is necessary for an emergency system to conditionally use the data collected in a normal state. Using traffic flow data (i.e. the flow rate, quantity and density on motorways) might risk the loss of accuracy, because the relationship between flow-rate and velocity differs greatly in emergencies [6]. Relatively speaking, using micro transportation data (i.e. the acceleration, moderation, and driving route) attains more effective information, because the function of the vehicle does not alter with psychological changes of the driver. Through attaining micro information in normal states, a large range of diverse driving behavior is covered, obtaining driving behavior information in emergencies.

This research captures diverse driving behaviors through real-time video data, providing models for simulation systems. This method consists of three processes: Acquisition, Analysis and Adaption. This method captures diverse driving behaviors through the mass data of normal traffic environment, and adapts them reasonably in situations of emergencies, providing effective data support for emergency evacuation models.

2.1 Acquisition

In the process of Acquisition, the time t of the video data is used to analyze the relationship between the distance between two vehicles and the state of speed:

$$s_{n,t} = \{(v_n, v_{n-1}, g_n)_t\}, \tag{1}$$

where n and $n-1$ represent the identity number of two vehicles traveling in succession while occupying the same lane, $n-1$ is the former, and n is the latter, v_n is the velocity of the latter vehicle, v_{n-1} is the velocity of the latter, g_n is the distance between two vehicles, and t is the time that the data is obtained.

For a single vehicle, the camera needs to obtain a set of data concerning the relationship between distance and velocity:

$$\mathbf{S_{n,t}} = \{s_{n,t_i}\}, \tag{2}$$

$$t_i = t_0 + i\tau, i = 0, 1, 2, \ldots, k. \tag{3}$$

where τ represents the reflective time of the driver (usually 0.33 s [7]), and t_0 is the starting observation time.

In the car-following model, $\mathbf{S_n}$ describes the vehicle's motion in a micro state. This paper is based on the Gipps' model [8], whose form is as follows:

$$v_n^* = \min\left[v_n + 2.5A_n\tau\left(1 - \frac{v_n}{V_n^M}\right)^{1/2}, -B_n\left(\frac{\tau}{2} + \theta\right),\right.$$
$$\left. + \left(B_n^2\left(\frac{\tau}{2} + \theta\right)^2 + B_n\left(2g_n - \tau v_n + \frac{v_{n-1}}{\widehat{B}_{n-1}}\right)\right)^{1/2}\right] \tag{4}$$

where τ is the reaction time, θ is the safety margin time, V_n^M is the speed limit of vehicle n, A_n is the largest acceleration of vehicle n, B_n is the actual braking, and $\widehat{B}_{n-1}$ is the perceived braking.

Through the data in $s_{n,t}$, the Gipps' model would be able to calculate the new velocity of the car v_n^* in the next time step $s_{n,t+\tau}$. Therefore the system could tag the state of these vehicles. Because τ and θ have a small range of changes, and V_n^M is usually only affected by the road, the only parameters that needs to be calibrated are A_n, B_n and $\widehat{B}_{n-1}$. These are the three parameters that form a vector and is used to describe driving behavior:

$$d_{n,t} = \{A_n, B_n, \hat{B}_{n-1}\},\tag{5}$$

Because the driving behavior could be very diverse, in normal states the data set $d_{n,t}$ is not exclusive. The set of driving behaviors that could arise in the standard state is defined as $\mathbf{D_N} = \{d_{n,t}\}$. The set needs to input all of the data sets of driving behaviors which appeared in the normal situation, forming a spatiotemporal database for normal driving behaviors, as Table 1 has shown. This mass database will become the foundation for calculation in the process to come.

Table 1. Table structure of the spatiotemporal database for normal driving behaviors

Name	ID	Time	Road ID	A_n	B_n	$\widehat{B}_{n-1}$
Type	int	datetime	int	float	float	float

2.2 Analysis

In the process of Analysis, the system needs to analyze the spatiotemporal data for normal driving behavior, in order to calculate the distribution of these data, and whether they would form in an emergency scenario.

Apart from the normal driving behavior set $\mathbf{D_N}$, the set of behavior that may occur to all of the vehicles on a route is defined as $\mathbf{D_V}$, and the set which occurs during an emergency is defined as $\mathbf{D_E}$. The probability distribution functions of normal and emergency driving behaviors are set as p_N and p_E.

If the following hypothesis are correct, it can be assumed that the behaviors in standard situations could also appear in situations of emergency:

1. The performance of vehicles and road characteristics in an emergency does not differ from the normal scenario.
2. The amount of data is large enough. In this condition, all of the driving behaviors that could be expressed in all vehicles could be observed in a standard situation:

$$\mathbf{D_E} \subseteq \mathbf{D_V} = \mathbf{D_N}\tag{6}$$

3. In emergencies the distribution of driving behaviors are different from those in normal situations.

If the above stands correct, the outcome of the process of Acquisition D_N could be adapted into the Analysis process, which is divided into three steps:

Step 1: Categorize the spatiotemporal data of driving behaviors using time. For example, using thirty minutes as an interval to categorize the data.

Step 2: Set tolerance rate $\Delta d = \{\Delta A_n, \Delta B_n, \Delta \hat{B}_{n-1}\}$, divide the space of driving behavior by Δd into multiple blocks $\Delta_{m,t}$. For each time interval, calculate the probability of normal driving behavior (i.e. p_N) landing in each block.

Step 3: Using month as a unit, gather individually the information of workdays and weekends, getting the patterns of driving behavior in normal situations, as Table 2 states below.

Table 2. Table structure of database of the patterns for normal driving behaviors

Name	ID	Time	Road ID	Weekend	A_n	B_n	$\hat{B}_{n-1}$	p_N
Type	int	datetime	int	boolean	float	float	float	float

2.3 Adaptation

Finally, in the process of Adaptation, the system adjusts the ratios of each behavioral pattern in the simulation system based on the present scenario, developing a cross-simulation system which surpasses the normal and emergency driving behaviors. The system follows the process of Analysis, calculating the driving behaviors probability of the last time period p^*, using the following formula to determine the outcome of p_E:

$$p_E = kp_N + (1 - k)p^*, \tag{7}$$

where $k \in [0, 1]$ is called the adjustment ratio. It is defined as the probability of normal driving behaviors in the present emergency circumstances. Especially, if $k = 1$, the system remains normal, not considering the impact of recent driving behavior probability. If $k = 0$, the system ignores all of the data in normal situations, only consider the driving behavior information of the last time period. This way, the system could then continuously alternate between normal and emergency situations.

Because all of the processes have been claimed through experimental data, the self-adapting method would not create new driving behaviors, which would ensure the actual meaning in each of the individual driving behaviors generated from the simulation system.

3 Experimental Data

This research uses a main road in a city as an experiment subject. The object road captured by cameras is as Fig. 1 has shown. The road parameters is shown in Table 3. There are no traffic lights for 3 km in front or behind the road so that the impact of driving behavior due to traffic lights could be overlooked.

Fig. 1. Video image of the experimental section

Table 3. Key parameters of the experimental section

Length	Lanes	Lane width	Speed limit	Height limit
84.2 m	3	3.6 m	60 km/h	4 m

During the gathering process, 700 sets of normal data and 175 sets of "emergency data" are recorded. The "emergency data" here uses a different set of data than the normal data, which has been sampled at different times. In this paper it is only used to prove the effectiveness of the procedure. In real operation, it would just be required to transfer man-made data into real data. Since each $s_{n,t}$ is a three-dimensional vector, the figure is drawn as a 3D graph whose colors represent the next action of each vehicle, as shown in Fig. 2. The points in Fig. 2 refer to gap and the speed difference respectively, where speed difference refers to speed of vehicle n−1, and the relative speed between vehicle n and n−1, identifying the position of the points, stand for the micro state of vehicles. The color between red and green shows the next action of each vehicle, i.e. acceleration (shown as green) or deceleration (shown as red). Through the process of Acquisition, all of the data is collected into the database in the spatio-temporal database for normal driving behaviors as shown in Table 1.

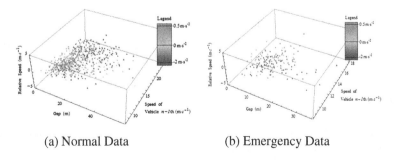

(a) Normal Data   (b) Emergency Data

Fig. 2. Micro traffic state of the normal and emergency data (Color figure online)

Based on traditional statistic method, the data could be fit with the Gipps' model (Eq. 3). The calibrated parameters are listed in Table 4. Statistically, the mean and standard deviation of the difference between calculating results and experimental data are -0.07 m•s^{-2} and 1.05 m•s^{-2} respectively.

Table 4. Calibration parameters of the Gipps' model

θ	τ	A_n	B_n	$\widehat{B}_{n-1}$
0.33 s	0.67 s	2 m•s^{-2}	2 m•s^{-2}	1.8 m•s^{-2}

In the process of Analysis, normal and emergency data has been summarized individually into the probability density graph, as shown in Fig. 3. The horizontal axis is the deviation between the actual vehicle acceleration and the acceleration calculated by the Gipps' model based on the calibration parameters in Table 4. The positive value represents the vehicle has a slower acceleration rate than the model estimation, and negative value represents that the acceleration rate is higher than the estimation. As seen, in the two situations, driving behavior does not fit the Gauss distribution. Multiple peaks suggest the existence of multiple models of driving behavior in research subjects.

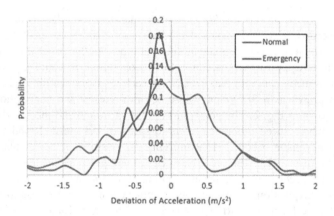

Fig. 3. Probability distribution graph in the process of Analysis

Lastly, in the process of Adaptation, using Eq. 6, the system could estimate recent states of driving behavior distribution patterns. It would be used as supporting data in simulation systems to implement microscopic evacuation.

4   Conclusion

This paper proposes a cross-simulation method to apply the mass data collected in normal situations into large-scale traffic evacuations to provide better supporting information for emergency decision. This method captures micro driving behaviors data from normal traffic environment, generates a series of probability distribution, and adapt the distribution for emergency situations. The paper uses a road as an example for experiments, primarily proving the practicality of this method and its ability to provide supporting decision-making information. Using this method, it is able to achieve the self-adapting simulation in different situations based on big data technologies.

In the further work, the system will continue to collect mass data, implementing the complete process, and providing effective evacuation decision-making support.

Acknowledgement. This work was partially supported by the National Natural Science Foundation of China (No. 91324022, No.91224008, No.91024032, No.70601015, No.70833003).

References

1. Dia, H.: An agent-based approach to modelling driver route choice behaviour under the influence of real-time information. Transp. Res. Part C Emer. Technol. **10**(5), 331–349 (2002)
2. Tawfik, A.M., Rakha, H.A., Miller, S.D.: Driver route choice behavior: Experiences, perceptions, and choices. In: Intelligent Vehicles Symposium (IV), 2010 IEEE, pp. 1195–1200. IEEE (2010)
3. Treiber, M., Helbing, D.: Memory effects in microscopic traffic models and wide scattering in flow-density data. Phys. Rev. E **68**(4), 046119 (2003)
4. Parker, D., Lajunen, T., Summala, H.: Anger and aggression among drivers in three European countries. Accid. Anal. Prev. **34**(2), 229–235 (2002)
5. Yuan, J., Zheng, Y., Zhang, C., Xie, W., Xie, X., Sun, G., Huang, Y.: T-drive: driving directions based on taxi trajectories. In: Proceedings of the 18th SIGSPATIAL International Conference on Advances in Geographic Information Systems, pp. 99–108. ACM (2010)
6. Tu, H., Tamminga, G., Drolenga, H., de Wit, J., van der Berg, W.: Evacuation plan of the city of Almere: simulating the impact of driving behavior on evacuation clearance time. Procedia Eng. **3**, 67–75 (2010)
7. Brackstone, M., McDonald, M.: Car-following: a historical review. Transp. Res. Part F Traffic Psychol. Behav. **2**(4), 181–196 (1999)
8. Gipps, Parker, D., Lajunen, T., Summala, H., 2002. Anger and aggression among drivers in three European countries. Accident Analysis & Prevention, **34**(2), 229-235

Detecting Fake Reviews of Hype About Restaurants by Sentiment Analysis

Run Yu Chen[1(✉)], Jin Yi Guo[1], and Xiao Long Deng[2]

[1] International School, Beijing University of Posts and Telecommunications,
10 Xitucheng Road, Beijing 100876, China
{zjsxcry, 2011213244}@bupt.edu.cn
[2] Key Laboratory of Trustworthy Distributed Computing and Service,
Beijing University of Posts and Telecommunications, 10 Xitucheng Road,
Beijing 100876, China
shannondeng@bupt.edu.cn

Abstract. Fake reviews in e-commerce website can greatly affect the choice of consumers. By categorizing the set of fake reviews, we find that fake reviews of hype make up the largest part, and this type of review is most misleading as well. We analyzed all the characteristics of fake reviews of hype and find the most evident one is in the text of the review. Usually, hype is completely positive or negative. So this article purposes an algorithm to detect fake reviews of hype about restaurants based on sentiment analysis. Reviews are considered in four dimensions: taste, environment, service and overall attitude, if the analysis result of the four dimensions are consistent, the review will be categorized as hype. The testing result shows that the accuracy of the algorithm is about 74 %. The method can also be applied to other areas, such as the sentiment analysis in emergency management.

Keywords: Reviews of hype · Sentiment analysis · Multi-dimension · Bayes

1 Introduction

Most consumers have the habit of scanning the reviews before purchasing. Products with many positive reviews often leaves customers good impression. Therefore, for merchants of e-commerce, it is important to maintain the reputation of their products online. However, some of the merchants, in order to enhance the popularity of their products, hire groups of people called internet "Water Army" to post positive reviews, which greatly affect the choice of consumers.

We have established collaborative relationship with dianping.com and obtained fake review set collected by the company. By observation and analysis, we conclude that most of the fake reviews are hypes. And they are very similar to authentic reviews in many aspects including length, tone and wording etc., that is why they are most misleading.

We did further investigation on fake reviews of hype we sort out. This kind of reviews is most distinctive in its text, i.e. totally positive or negative. Some reviews of this kind, although not suspicious according to analysis on other attributes, possess little reference value because of extreme sentiment.

© Springer International Publishing Switzerland 2014
Y. Chen et al. (Eds.): WAIM 2014, LNCS 8597, pp. 22–30, 2014.
DOI: 10.1007/978-3-319-11538-2_3

Aiming at the problem described above, we apply sentiment analysis to the detection of such kind of fake reviews. To obtain a desirable accuracy, we first find out all the subject words by matching the sentiment words we extracted from all the fake reviews data set with the corresponding features. We manually judged the subject words and divided them into four dimensions: taste, environment, service, overall attitude. To evaluate the authenticity of a review about restaurant, our algorithm will conduct sentiment analysis from the four dimensions of it based on Bayes classifier. If the analysis results of the four dimensions are the same (all positive or negative), the review is hype. By inspection, the algorithm has an accuracy of about 74 %.

2 Related Work

Relevant research of spam detection developed in the spam [1] and rubbish website [2]. Recent years, researchers started to identify spam reviews.

Researchers concentrate on some different characteristics of spam reviews. Jindal etc., proposed the algorithm concerns unusual score [3, 4]. Wang, Xie etc., pay more attention on store reviews to detect spam [5, 6]. Arjun etc., considered a number of indicators to find fake reviewer groups [7]. Some mainly considered relevant, Song etc., defined the relevance of their basic characteristics and features among other comments [8]. Myle etc., proposed the combination of language features SVM modeling [9].

The innovation of this paper is to classify the content itself from the marked spam reviews, for which account the largest number and most influential speculation reviews. Then we draw the appropriate model program. When using sentiment analysis in more specialized areas, the introduction of multi-dimensional analysis, making the identification of such spam reviews greater accuracy.

Adding some traditional methods to identify other kinds of spam reviews, the accuracy of the algorithm is basically similar to previous studies. In comparison, our algorithm requires much more manual handling, but the recall rate reflects better performance.

3 Sentiment Analysis

3.1 Bayes Classifier

Through investigation on the possibility of a specific event happening in the past, Bayes Theorem can calculate the approximate the probability of the event happening in the future. The theorem only require a few parameters for estimation and is not sensitive to the lost data. Algorithms that apply this theorem run fast due to simplicity [10].

Bayes classifier is the application of Bayes Theorem to classification of text. By calculating the probability of each category, the algorithm categorize the text to the one with largest probability.

Bayes equation:

$$P(c_j \mid x_1, x_2, \ldots, x_n) = \frac{P(x_1, x_2, \ldots, x_n \mid c_j)}{P(c_1, c_2, \ldots, c_n)} \tag{1}$$

Since each condition is independent and identically distributed, we have:

$$c_{NB} = P(c_j) \prod_i P(x_i \mid c_j) \tag{2}$$

Because there is no apparent feature in the labeled text, we apply multi-event model based on word frequency here: [11]

$$P(t_i \mid c = \text{spam}) = \frac{1 + Count(t_i)}{n + Counts\left(\sum_1^n t_i\right)} \tag{3}$$

3.2   Obtain Training Set

Calculating Bayes prior probability requires training data, so we manually selected 500 totally positive and negative reviews respectively. We have proved that the training set is enough since by enlarging the size from 300 to 500 we did not find noticeable change in the result. When selecting the sample of training set, we followed the principle that the review is as comprehensive as possible (Table 1).

Table 1. Training set data

Positive	Negative
三人行骨头锅太给力了，冬天啃绝对过瘾，料很足，好多肉，人气爆满，装修美观，去晚了就要排队，钵钵鸡，个人觉得不错，喜欢吃辣的朋友可以尝试下，味道十足。	服务态度极差 提前一周打去说只能提前两天预订 等到提前两天打又没位子了 还说没人接过电话说要提前两天预订 领班态度也差 还挂客人电话。
昨天请朋友来吃饭，听说这里的蟹晏菜挺有名的，于是就点了几只大闸蟹，点了个蟹粉豆腐。还有几道特色菜，片皮鸭，5A牛柳粒，古法焖鱼。朋友都说不错我很高兴。这里环境也挺舒服的，服务态度很好，注重细节，我们想要的在未开口之前服务员都想到了。	第一次去就被咖喱牛腩的牛腩震惊了，那么老的牛腩，都快赶上牛肉干了，牛腩也完全不入味...询问服务员还说他们的牛腩一直是这样的，那请问前面点评里面说牛腩嫩、烂的吃的是什么...强烈不推荐！服务员态度很不好，下次绝对不会去了。蟹粉小笼十分一般，根本不值这个价格。
同事们商量着一起去吃点东西，大家一致赞同，冬天里吃火锅应该是再好不过的选择了吧，就这样我们一伙人就过来了，店里的生意很好的，但还是井然有序的，我们找了个地坐下，点了个火锅，一会的功夫就上来了，菜很 丰富，话不多说，大家就开动了，有菜有肉，真是一顿营养大餐呢，汤鲜味美呀，大家都吃的很高兴，一边聊着天，一 边吃着饭，真是享受呢，	锅贴的品种不多，也没什么可挑选的，贡丸汤一份，偌大一只碗，里面晃悠着两个贡丸；中午时分也没有坐满，三两个服务员在休班闲聊，毫不避讳的高谈阔论着；我们点的锅贴好了，得自己去台面上找，就几个单就有点混乱了，服务水平可见一斑。
这才是生活啊，以后有时间还要常来啊，朋友们有空也可以过来尝一下，相信不会让你失望 的。	
...	

3.3 Evaluation of Sentiment

To find out the sentiment of a review, we use the method of multiplying the conditional probability of each word. Whether a review is positive or negative depends on the value of the result of multiplying process. We multiply the probability of each word with 10 first in case the result is too small.

4 Multidimensional Discrimination

4.1 Establishment of Sentiment Word Library

We sort out 56483 reviews about restaurants from our original data set. With the help of ICTCLA50 [12], we divided all the Chinese content into words. We extract all the sentiment words and set up our sentiment word library. Meanwhile, we calculate the frequency of each adjective, which enables us to get rid of some undesirable words with low frequencies. The final version of our sentiment library contains 1590 adjectives (Table 2).

Table 2. Emotional word library

Words/Frequencies		Words/Frequencies	
很好/a	12.389356882973637	实在/a	7.453503647044743
好吃/a	11.14669772771281	舒适/a	7.406438314516372
不错/a	11.052315097990974	丰富/a	7.3550685280099195
实惠/a	9.742061863915502	亲切/a	7.350785294189377
干净/a	9.651343822312056	安静/a	7.289251208561696
值得/a	9.466860974306062	确实/a	7.262434541770648
周到/a	9.096459212424548	可口/a	7.18616605729332
特别/a	9.000203638575638	贴心/a	7.141585285545526
舒服/a	8.995655552806497	独特/a	7.044336876237392
便宜/a	8.929070356945344	合理/a	7.03207042683553
最好/a	8.40933391245834	一样/a	6.988856149668119
优惠/a	8.213934707582403	美味/a	6.957181112804157
优雅/a	8.193907482484725	重要/a	6.9435164471041055
温馨/a	8.078017307503831	合适/a	6.795440558011954
地道/a	8.023967758424863	有味/a	6.752448672847059
划算/a	8.008506960447109	适中/a	6.7334198176058955
一般/a	7.982753540338287	整洁/a	6.654735396670123
方便/a	7.9191048016013506	一流/a	6.63665330359303
热情/a	7.905370323918099	主动/a	6.636631314797421
鲜美/a	7.887444208296432	过瘾/a	6.61751253180918
地方/a	7.806704408558014	失望/a	6.612146670170864
入味/a	7.788232816199374	随便/a	6.533682077002941
精致/a	7.6268743931156555	耐心/a	6.46625147932886
其实/a	7.513344086828133	开心/a	6.452121735931351
...		...	

4.2   Feature Finding

Feature refers to the corresponding none phrase of existing sentiment words. We manage to find them by searching words before the adjectives that match those in our library, the nearest noun phrase is the matching result. Finally, we obtained 2303 subject words of reviews about restaurants.

4.3   Classification of Subject Words

After finding all the subject words, we analyzed the dimension each of them belongs to and finally limited the dimension into four types: taste, environment, service and overall. Then we classified these subject words. There are 1314 words belong to dimension "taste", including those that describe the look, flavor and taste of foods. 222 words describing the environment, geographical conditions and traffic conditions of the restaurants are categorized into "environment". 286 words are categorized into "service", most of them describe the quality of service, prices. And there are 151 words classified as "overall", including name of restaurants, names of places and holistic description etc (Table 3).

Table 3. Main body word library

Taste	Service	Environment	Overall
口味	阿姨	摆设	安溪
阿拉斯加蟹	按摩	包房	安阳
爱尔兰咖啡	按摩师	包房环境	澳门豆捞
安格斯牛肉	包装	包房装饰	澳门街
安徽菜	包装盒	包间	澳洲
安排的菜品	保安	包间风格	八佰伴
八宝辣酱	杯子	包间环境	巴贝拉
八宝年糕	菜的价格	包厢	巴蜀风
八宝养生茶	菜价	包厢环境	斑鱼府
八宝鱼	菜价格	背景音乐	北京烤鸭
霸王蛙	菜肴的服务	布局	餐厅
霸王鱼头	菜肴的价位	布置	大酒店
白菜	菜肴的质量	餐馆生意	店
白菜粉丝	菜肴服务	餐厅	店气氛
白肉丝	餐厅的厨师	餐厅的布置	店人气
白水鱼	餐厅的服务人员	餐厅的环境	店生意
白汤	餐厅的老板	餐厅的设计	店味道
白汤牛奶	操作过程	餐厅印象	东西总体
白糖	茶水	厕所	饭店
…	…	…	…

5   Sensationalize Identification

5.1   Determination of Dimension

By observing the sample data we find that spans of reviews are usually very large, thus using period to divide the sentences brings many errors. To avoid this, we use comma as monitoring sign to divide the sentences first.

When categorizing each sentences, we first set the default dimension as "overall". Then we matching each subject word in the sentence with our library established before, the sentence will be categorized into the dimension with most matching words.

5.2   Detection of Hype

In our process of detecting reviews of hype, our model analyzed the sentiment of all sub sentences of each dimension. If the results of four dimensions are same (all positive or negative), we will label the review as hype. In the following statistics, we find that positive reviews of hype are far more than negative reviews. Thus we assume that the default sentiment of a review is positive when there is information loss in 1 or 2 dimensions. Result shows that this preprocessing brings error less than 0.1 %.

6   Experiment Result

6.1   Algorithm Difficulty Analysis

Through the analysis and category on the original data provided by dianping.com, we obtain 17681 fake reviews of hype about restaurants, and we applied our model to test these data. With the increase of reviews being tested, the result accuracy remained stable around 68 % (Fig. 1).

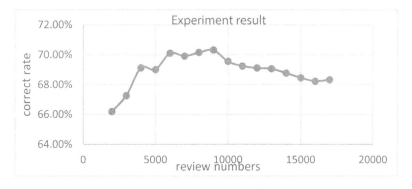

Fig. 1. Experiment result on raw data

Afterwards, we randomly sampled 2000 reviews to analyze. We found some error in labeling among the original data by checking the content of each review. Possible reasons are summarized below: (1) reviews are not about restaurants; (2) reviews are apparently not hype. After correcting the label of the 2000 reviews one by one, we witnessed the accuracy increased from 67 % to 73 %. Therefore we can estimate that same labeling errors exist in other parts of original data set, and the accuracy of our algorithm could reach around 74 % (Table 4).

Table 4. A random sample of 2,000 reviews

Review numbers	Original correct rate	Non-restaurant reviews	Available reviews	Error mark	Revised correct rate
1–500	68.4 %	34	466	7	**73.2 %**
501–1000	62.8 %	29	471	11	**70.3 %**
1001–1500	65.6 %	38	462	9	**71.4 %**
1501–2000	71.8 %	21	479	5	**76.2 %**
All	67.15 %	122	1878	32	**72.79 %**

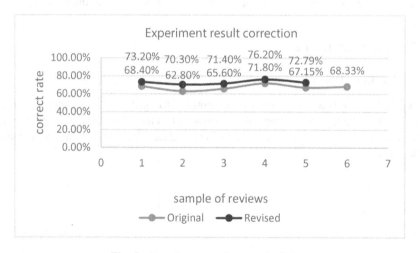

Fig. 2. Experiment result on revised data

In Fig. 2, sample 1 to 6 is respectively stands for review numbers 1–500, 501–1000,1001–1500,1501–2000,1–2000 and 1–17681. It is hard for us to clean all the error marked data, but the ultimate accuracy can be predicted about 74 %.

6.2   Algorithm Difficulty Analysis

During the error analyzing process, we found that nearly 60 % of errors occur for mainly two reasons. (1) Using comma as delimiter can cause incomplete sentence-breaking.

(2) Some neutral reviews can be sorted wrongly by different training set, and it is difficult to judge whether they are hype or not.

7 Application of Emergency Management

The method proposed in this paper can also be used in emergency management for sentiment analysis. We can divide comments into different dimensions to get more specific sentiment analysis. We can also base on individuals, considering some of its comments together, to discover some associations with highly consistent in the emotion.

8 Conclusion

In this paper, we proposed a method to detect fake reviews of hype based on sentiment analysis. Reviews of hype take large part of fake reviews and are influential. The model we established can detect this type of review with an accuracy of around 74 %. We set up our own sentiment word and multi-dimensional subject word library focus on reviews about restaurants as well. This method can also be applied to field other than reviews about restaurants.

Acknowledgement. Thanks to the support of National Natural Science Foundation of China (NNSF) (Grants No.90924029), National Culture Support Foundation Project of China (2013BAH43F01), and National 973 Program Foundation Project of China (2013CB329600).

References

1. Drucker, H., Wu, D., Vapnik, V.N.: Support vector machines for spam categorization. IEEE Trans. Neural Netw. **10**(5), 1048–1054 (2002)
2. Ntoulas, A., Najork, M., Manasse, M., et al.: Detecting spam web pages through content analysis. In: Proceedings of the 15th International World Wide Web Conference (WWW'06), Edinburgh, Scotland, pp. 83–92. ACM, New York, 23–26 May 2006
3. Jindal, N., Liu, B., Lim, E.: Finding unusual review patterns using unexpected rules. In: The 19th ACM International Conference on Information and Knowledge Management (CIKM-2010), Toronto, Canada, pp. 25–628, 26–30 Oct 2010
4. Jindal, N, Liu, B.: Opinion spam and analysis. In: Proceedings of the 1st ACM International Conference on Web Search and Data Mining (WSDM'08), California, USA, pp. 137–142. ACM, New York, 11–12 Feb 2008
5. Wang, G., Xie, S., Liu, B., Yu, P.S.: Review graph based online store review spammer detection. In: 2011 IEEE 11th International Conference on Data Mining (ICDM), Vancouver, BC, pp. 1242–1247, 11–14 Dec 2011
6. Xie, S., Wang, G., Lin, S., Yu, P.S.: Review spam detection via temporal pattern discovery. In: Proceedings of the ACM SIGKDD International Conference on Knowledge Discovery and Data Mining, Beijing, China, pp. 823–831, 12–16 Aug 2012

7. Mukherjee, A., Liu, B., Wang, J., Glance, N., Jindal, N.: Detecting group review spam. In: Proceedings of the 20th International Conference Companion on World Wide Web (WWW '11), New York, NY, USA, pp. 93–94 (2011)
8. Haixia, S., Xin, Y., Zhengtao, Y., et al.: Detection of fake reviews based on adaptive clustering. J. Nanjing Univ. Nat. Sci. **49**(4), 38–43 (2013)
9. Ott, M., Choi, Y., Cardie, C., Hancock, J.T.: Finding deceptive opinion spam by any stretch of the imagination. In: Proceedings of the 49th Annual Meeting of the Association for Computational Linguistics: Human Language Technologies (ACL HLT), Portland, OR, United States, pp. 309–319, 19–24 June 2011
10. Wang, J., Wang, L., Gao, W., Yu, J.: A research on the keywords extraction of naive Bayes. Comput. Appl. Softw. **31**(2), 174–181 (2014)
11. Zhang, F., Wu, Z., Yao, F.: Research of spam filter based on Bayes. J. Yanshan Univ. **33**(1), 47–52 (2009)
12. Xia, T., Fan, X., Liu, L.: Implementation of ICTCLAS system based on JNI. Comput. Appl. **24**(z2), 945–950 (2004). Deng, J.: Control problems of grey system. Syst. Control Lett., **1**, 288–294 (1982)

Where to Find Help When You Are in an Emergency?

HelpMe: A Guidance System for Self-rescue

Jiahai Wu[1,3,4]([⊠]), Huangfu Yang[1,2], Xiaowei Gao[1,2], Juncai Ma[4],
and Kuien Liu[2]

[1] University of Chinese Academy of Sciences, Beijing 100049, China
jiahai.wu@foxmail.com,
{huangfuyang,xiaowei}@nfs.iscas.ac.cn
[2] Institute of Software, Chinese Academy of Sciences (CAS),
Beijing 100190, China
kuien@iscas.ac.cn
[3] Computer Network Information Center, CAS, Beijing 100190, China
[4] Network and Information Center of Institute of Microbiology, CAS,
Beijing 100101, China
ma@im.ac.cn

Abstract. A variety of emergency situations may occur around us at any time in our daily life, especially in an unfamiliar place. For example, a tourist may seek medical treatment for an accidental wound or encounter a problem like running out of gasoline, raising the urgent requirements where to find specific services nearby for help. In this paper, we develop a guidance system, HelpMe, benefiting the mobile-phone user with self-rescue instructions when needed. HelpMe provides reasonable instructions by leveraging the urban transportation patterns discovered from massive historical trajectories of vehicles and an efficient route planning algorithm based on these patterns. Within the process of route planning, HelpMe introduces a novel reach-time-cost model to sort out the candidates of emergency service spots, such as hospital, police station and gas station and so on, as well as the fastest way to the destination. Capturing the current locations from a mobile phone, HelpMe can suggest the phone user where to find secured transport assurance with the highest probability and automatically guide you to the destination by the fastest route. Our experiments have revealed satisfactory conclusions and demonstrated the effectiveness of our system.

Keywords: Location-based services · Emergency management · Self-rescue

1 Introduction

The occurrence of urban emergent issues around individuals is more and more frequent with the process of urbanization and undermines the personal security. These emergencies are often unpredictable and even life-threatening. At present, there are lacks of universally effective solutions for individual users, especially when they are in an unfamiliar place. It is crucial to assist a person suffering an emergency to answer the

© Springer International Publishing Switzerland 2014
Y. Chen et al. (Eds.): WAIM 2014, LNCS 8597, pp. 31–43, 2014.
DOI: 10.1007/978-3-319-11538-2_4

following three questions: (1) "Is there any possible place for help?", (2) "How to get to the destination as soon as possible?" and (3) "which is the best destination for the current situation?", thereby helping find help and avoid the worst. Imagine that your friend is wounded accidently when you two are trapped in an unfamiliar place while few people can help. You are eager to find a hospital nearby and transmit your friend there. Another scenario, when you are in a dangerous area by PIDA (People in Danger Algorithm) [14], you need looking for a rescue station nearby.

There are many other individual emergency situations in our daily life. Thus, trying the self-rescue is often the case. If people are not sure where to go and how to go, they may use a map app to help themselves. It does help, but not always. Does the shortest path to the nearest assistance spot mean the effective, accessible and fast way? It may not work in practice, because it neither considers the opportunities to finding a currently available way of transportation, nor evaluates the time consumed in the whole process traversing three steps: run to intermediate spot where can find a passing vehicle, wait on the spot, and shift to the assistance destination. These are also the key differences between this paper and related work in the traditional emergency evacuation [1, 2, 6, 7, 9] and route recommendation [3–5, 8, 12, 13].

Unlike these conventional guidance systems, which mostly perform path-finding algorithms (e.g., A*) based on some carefully selected metrics, the proposed solution in this paper is a typical application of big data scenarios. One of its advantages is that it tends to find out the "most reasonable" self-rescue routes for users based on massive trajectory data in urban life rather than the "shortest" path on digital maps. In this sense, it is a kind of empiricist approaches which is widely adopted in the age of big data. It highly depends on the deep understanding of massive historical datasets. In other words, if we know little about the transportation patterns of urban infrastructure when developing a guidance system of self-rescue, it is difficult for us to offer practical solutions to the users.

Based on the above considerations, we develop HelpMe, a big-data based guidance system for self-rescue. The system can help you find the possible places providing specific help and search for the fastest routes available to those places, and then the system can navigate you to the destination with compact and efficient instructions in both voice and visual ways. Take the wounded case as an example as shown in Fig. 1, the time to be spent on a candidate route consists of three parts: t_w (time to run to a possible intermediate spot), t_Δ (time to wait for transportation at the intermediate spot) and t_{tc} (time to transmit the wounded to a hospital).

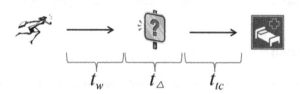

Fig. 1. An example of self-rescue for wounded man.

To summarize, this paper makes the following contributions:

- We analyze the massive historical trajectories of vehicles in Beijing and export the traffic flow model of Beijing downtown road network.
- Based on the traffic flow model, we design a route recommendation algorithm with a novel reach-time-cost (RTC) model to plan the global optimized route to emergency service spot (ESS), such as hospital, police station etc.
- We implement HelpMe, an individual-oriented guidance system for self-rescue, and verify the efficiency of our system by a group of experimental studies in real scenarios.

2   Related Work

Emergency Evacuation. Existing works on evacuation mostly focus on massive population evacuation inside a building [2, 7], or on the traffic network [1, 6, 9]. Kongjin et al. [2] proposed the evacuation strategies through using stairs and elevators in high-rise buildings in situation of both small and large crowds. Shi et al. [7] used an agent-based model to simulate the evacuation process in different fire cases to study the key factors influencing the safe evacuation of metros and support the safe evacuation design. Yu et al. [1] developed a traffic simulation model of downtown San Jose transportation network to devise strategies for evacuation of evacuees and emergency response. Kaisar et al. [6] employed the current public transit system to conduct different evacuation strategies for populations with special needs from large urban areas, e.g., to find the optimum location for evacuation bus stops. Stepanov et al. [9] utilized an integer programming model to compute and advise an optimal routing policy for the population from the affected areas. In another point of view, individual-oriented evacuation technologies to support the guidance are lack of care.

Route Recommendation. Route recommendation is widely investigated in [3–5, 8] with massive historical data. Liu et al. [3] developed a trip recommendation system based on the massive historical trajectory data shared on Web to recommend the satisfactory route meeting user preferences. Sun et al. [4] employed the massive geo-tagged photos on social media to discover tourists' preferences on landmarks and trajectories of tour and recommend the best travel routes between landmarks. Wei et al. [5] proposed a route inference framework based on collective knowledge to construct the popular routes from uncertain historical trajectories. Zheng et al. [8] performed a travel recommendation devoted to offering a user with top interesting locations and travel routes in a given geospatial region in consideration of both the number of users and their knowledge. These works mentioned above are more in the context of tour planning, thus less consideration are taken for the emergency scenarios.

3   Problem Definition

In this section, we first define several concepts mentioned throughout this paper.

Definition 1 (Road Network). A road network G is a directed graph made up by edges E and vertices V, each directed edge e is associated with an id $e.eid$, a length *e.len*, a road type *e.kind*, a starting point *e.beg* and an ending point *e.end*, while each vertex *v* representing a geometric contains two float values, longitude and latitude. All of the edges in our road network are one-way road without branches. An edge contains a set of connected road segments represented as a sequence of GPS points.

Definition 2 (Trajectory). A *trajectory* contains a group of GPS point sequences $p_1 \rightarrow p_2 \rightarrow p_3 \rightarrow ... \rightarrow p_n$ logged by a working taxicab, each GPS point was associated with a time stamps *p.time* and an operating status *p.sta*. In this paper, we mainly consider two operating status for working taxicabs: *occupied* and *vacant*.

Definition 3 (Trip). A trip *Tr* is a special trajectory with only one operating status, either *occupied* or *vacant*. The *trajectory* of each day consists of a set of *trips*.

Definition 4 (Route). Given two points p_i, p_j in a road network *G*, a *route* is a set of connected road edges that start at p_i, and end at p_j.

Definition 5 (Time slice). We divide 24 h (a day) into a number of *time slices* with a fixed time interval to facilitate the statistical learning. As the average sampling interval of the GPS points in our data set is about 35 s (most of the *trajectories* are between 15 and 150 s), we set a time slice μ to 60 s by default.

Definition 6 (Origin Point). An *Origin Point (OP)* represents the location where the user sends the emergency request, e.g., point *op* in Fig. 2.

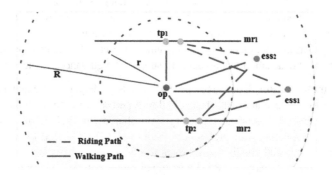

Fig. 2. An example of self-rescue.

Definition 7 (Main Road). *Main Road (MR)* represents the roads around a user's current location, where this user can take a taxi with high probability. In Fig. 2, mr_1 and mr_2 are two main roads around the *op*.

Definition 8 (Emergency Service Spot). An *Emergency Service Spot (ESS)* is the place where user can find specific help. An *ESS* can be considered as a destination of a planned route. In Fig. 2, ess_1 and ess_2 are two possible destinations in that scenario.

Definition 9 (Transit Point). *Transit Point (TP)* is one of the end-points of a main road, and it is the point for waiting a taxi. For example, point tp_1 and tp_2 are two transit points on road mr_1 and mr_2 respectively in Fig. 2.

The **self-rescue problem** then can be defined.

As: given a road network *G*, an origin point *op* and a request time *t*, find out a best route go to an Emergency service spot *ESS*, and guide the user to follow in real-time.

4   Framework

The framework of the system is illustrated in Fig. 3. We develop a back-end server to support the online recommendation and the offline mining. The back-end server consists of three major components: *route planner, optimal route builder and traffic flow miner*. *Route planner* accepts the user's requests, each of which consists of a start location and a request type. And then the *route planner* searches for the candidate routes. The *traffic flow miner* builds the traffic flow model on the massive historical GPS trajectories data and road network of Beijing City. We deduce the traffic information by map-matching trajectories on the underlying road network and calculating traffic flow for each road in ways proposed in our preliminary work [10, 13]. We divide the trajectories into trips and perform statistic study on different operating status. Based on this model, *optimal route builder* selects a global optimal route with the minimum time cost. Finally, the mobile device receives the optimal route from back-end server and provides real-time navigation for the user.

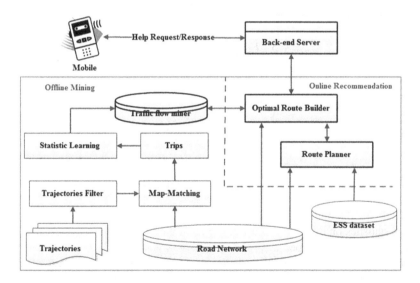

Fig. 3. System overview

5 Approach

5.1 Traffic Flow Model

This section details traffic flow modeling which contains a vacant flow density function φ and a mean occupied speed function ω.

We divide travelling trajectories of taxicabs into trips by operating status, and group these trips into two categories: *vacant* and *occupied*. Given a trip Tr: $e_1 \rightarrow e_2 \rightarrow ... \rightarrow e_n$, we calculate the trip distance L_{Tr} and the travelling time T_{Tr} on this trip. Then we figure out the average speed $S_{Tr} = L_{Tr}/T_{Tr}$. We use $F_{Tr}(e_i)$ (default as 1) to denote the traffic flow contribution of this trip to edge e_i. In addition, we use $S_{Tr}(e_i)$ to denote the average speed on edge e_i and $\tau_{Tr}(e_i)$ to denote the time passing the edge e_i. Suppose within a certain time slice denoted as n_t, and an edge e occurs in k trips. The mean traffic flow on edge e on the given time slice is formulated as:

$$\overline{f_e} = \frac{1}{k}\sum_{i=1}^{k} F_{Tr_i}(e) \tag{1}$$

Actually, the accurate traffic flow contribution of Tr to edge e depends on $\tau_{Tr}(e)$ and the time slice μ. If the taxicab passes the edge e within one time slice, namely $\tau_{Tr}(e) < \mu$, the traffic flow contribution is just 1. Otherwise, its contribution to this edge is supposed less than 1. In this paper, we use $\Psi(\mu/\tau)$ to calculate the accurate traffic flow contribution. Therefore, the mean traffic flow on edge e is devised as:

$$\overline{f_e} = \frac{1}{k}\sum_{i=1}^{k} F_{Tr_i}(e) = \frac{1}{k}\sum_{i=1}^{k} \psi\left(\frac{\mu}{\tau_{Tr}(e)}\right),$$
$$\psi\left(\frac{\mu}{\tau}\right) = \begin{cases} \mu/\tau, & \mu < \tau, \\ 1, & \mu \geq \tau \end{cases} \tag{2}$$

The micro-average speed of edge e on the given time slice is formulated as:

$$\overline{s(e)} = \frac{1}{k}\sum_{i=1}^{k} S_{Tr_i}(e) = \frac{1}{k}\sum_{i=1}^{k} \frac{L_{Tr_i}}{t_{Tr_i}} \tag{3}$$

Considering the traffic flow is influenced by the work and rest schedules of people, we make statistics on every day of a week to learn the traffic flow characteristics. Each day of a week is a date type.

In our system, we calculate the mean vacant traffic flow φ of a certain edge on a given time slice to estimate the waiting time for taking a taxicab, and calculate the mean traffic speed of occupied taxicabs ω of a certain edge on a given time slice to estimate the travelling time by a taxicab. Given a date type n_d, an edge e, and an time slice n_t, the mean traffic flow of e on the n_t of n_d is given as:

$$\varphi(n_d, n_t, e) = \frac{1}{q}\sum_{j=1}^{q}\overline{f_e}, \ (p.sta = vacant) \tag{4}$$

Where q represents the counted number of the date type. Likewise, the mean traffic flow speed is given as:

$$\omega(n_d, n_t, e) = \frac{1}{q}\sum_{j=1}^{q}\overline{s_e}, \ (p.sta = occupied) \tag{5}$$

Because the amount of taxicabs in our training set is less than that in actual situation, we set a global multiplying factor σ. Then, the revised mean flow traffic is given as:

$$\varphi(n_d, n_t, e) = \frac{\sigma}{q}\sum_{j=1}^{q}\overline{f_{e_i}}, \ (p.sta = vacant) \tag{6}$$

5.2　The RTC Model

Time consuming is the key point throughout the process of emergency rescue service. This section introduces the reach-time-cost model and details how it works. Our model is made up of three parts as below.

Walking Time. Given *origin point* P_c and *emergency service spot* P_d, the requester should get the shortest path to a candidate *main road* to take a taxicab. We find out all candidate main roads around P_c and denote them as E_{mr}. For each main road *mr* in candidate set E_{mr}, we can calculate the length L_w of the shortest path from P_c. Suppose we have learned the walking speed s_w of the requester, we can compute the walking time t_w to get to the possible *main road*.

$$t_w = L_w/s_w \tag{7}$$

Suppose the moment that HelpMe client receives the response from the back-end server is T_0 and the moment when the user reaches the transit point P_w is T_1. Then T_1 can be formulated as:

$$T_1(T_0; P_c \rightarrow P_w) = T_0 + t_w = T_0 + L_w/s_w \tag{8}$$

Waiting Time. Obviously, the waiting time is uncertain for the user since he/she knows little about the traffic on the *main road* at a specific time. The waiting time for an empty taxicab can vary from road to road and from time to time. That's the very part we benefit from the traffic flow model. Through statistic learning on massive historical data, we can estimate a mean waiting time on a road at a specific time.

Based on the vacant flow density function $\varphi(n_d, n_p, e)$, we can compute the value of traffic flow density to get the traffic flow using the given location $P(e)$ in the road network, the request time $T(n_d, n_t)$ and a time-span t_Δ. This is formulated as:

$$Fe_{(P(e), T(n_d, n_t))}(x) = \int_{x=n_t}^{x=n_t+t_\Delta} \varphi(n_d, x, e) dx \tag{9}$$

Assuming that there is an empty taxi available when the integral of the traffic flow density increased by one. We got:

$$\Phi_1(t_\Delta) = Fe(n_t + t_\Delta) - Fe(n_t) - 1 = 0 \tag{10}$$

The time-span t_Δ can be worked out. If there are multiple requests at one location $P(e)$ at the same time, there will be a queuing delay.

Suppose the queuing number is N, then the waiting time can be computed as:

$$\Phi_N(t_\Delta) = Fe(n_t + t_\Delta) - Fe(n_t) - N - 1 = 0 \tag{11}$$

Time-span t_Δ satisfies the Eq. (10). The time a taxi arriving at the transit point T2 is formulated as:

$$T_2(T_0; P_c \rightarrow P_w) = T_0 + t_w + t_\Delta = T_0 + L_w/s_w + \Phi_N^{-1}(\Phi_N) \tag{12}$$

Obviously, this queuing number is oversimplified in the equation above. For a mature time model, we need to evaluate the time difference between each other and the possibility of cancelling the operation prompted by the user. So the count number N can be a floating-point number that revised by other factors. The queuing number and the count number N are proportional.

Riding Time. From the candidate main roads E_{mr}, we apply the classic Dijkstra algorithm [11] to find out the shortest path $e_1 \rightarrow e_2 \rightarrow e_3 \rightarrow ... \rightarrow e_m$ to *emergency service spot*. As all of the edges are one-way, without branches, we can figure out the road route $R_y: p_1 \rightarrow p_2 \rightarrow p_3 \rightarrow ... \rightarrow p_{2m}$. Every two points represents a certain edge or part of the edge, and the points order represents the direction. Then we figure out the travel length of each edge $l_1 \rightarrow l_2 \rightarrow l_3 \rightarrow ... \rightarrow l_m$. Based on the mean occupied speed function $\omega(n_d, n_p, e)$, we can estimate the speed $s_1 \rightarrow s_2 \rightarrow s_3 \rightarrow ... \rightarrow s_m$ with the given edge e_i and the request time $T(n_d, n_t)$. Then the travel time by taxicab t_{tc} is given as below:

$$t_{tc} = \sum_{i=1}^{m} \frac{l_i}{s_i} = \sum_{i=1}^{m} \frac{l_i}{\omega\left(n_d, \left(T_2 + \sum_{j=1}^{i-1} \frac{l_i}{s_j}\right), e\right)} \tag{13}$$

The *emergency service spot* reach time T_3 is formulated as:

$$T_3(T_0; P_c \rightarrow P_w \rightarrow P_d) = T_0 + t_w + t_\Delta + t_{tc} = T_0 + L_w/s_w + \Phi_N^{-1}(\Phi_N) + \sum_{i=1}^{m}\left(\frac{l_i}{s_i}\right) \tag{14}$$

If the *emergency service point* is very near, walking to the *emergency service spot* may be the best way rather than taking a taxi. In these cases, we need to figure out the shortest walking time $t_{wk} = L_{wk}/s_{wk}$. L_{wk} is the travel length of the shortest walking route r_{wk} and s_{wk} is the walking speed of the user. Let R_v represent the candidate route set. The minimum time-consuming route can be calculated as:

$$R_{\min} = \arg\min\{r_{wk}|t_{wk}, R_v|(t_w + t_\Delta + t_{tc})\} \tag{15}$$

The expected minimum time to get to the *emergency service spot* is:

$$T_{\min} = \min\{T_0 + t_{wk}, T_3(T_0; P_w \to P_\Delta \to P_d)\} \tag{16}$$

HelpMe client will receive the optimal route R_{min} and the minimum time cost T_{min}.

5.3   Optimal Routing Generation

As the reach-time-cost model introduced above, the best routing planning algorithm includes three steps: (1) searching for candidate main roads and destinations, (2) searching for all candidate routes and (3) selecting the best candidate route, which answer the three questions presented in Sect. 1.

Question 1: Is there any possible places for help?
In order to query all possible roads and *emergency service spots* as quick as possible, we constructed spatial grid indexes on the road network and ESS dataset. The roads and *emergency service spots* are mapped into cells consisting of the spatial grid. Those cells fully or partly covered by the circles with the radiuses R and r ($R > r$), are considered as the candidate cells for *emergency service spots* and candidate roads. The request location is the center point the circle. Those *emergency service spots* should conform to the requested emergency service type (e.g., hospital, police etc.).

Question 2: How to get to the destination as soon as possible?
The path from origin point to each main road ignores the road direction. Therefore, we use an undirected graph based on Dijkstra algorithm [11] to accomplish path searching. The searching process ends when the algorithm finds an edge on the target main road. The end-point of the last edge is considered as a transit point.

The path from the transit point to the destination strictly obeys the road direction. The path searching relies on the Dijkstra algorithm based on a directed graph.

Question 3: Which destination is the best for current issue?
We apply the RTC model to evaluate all candidate routes and pick out the optimal one with the minimum time consumed. The overview of the optimal routing selection algorithm is given in Algorithm 1.

Algorithm 1. Optimal Routing Selection

Input: Road network G, ESS data set D, user location GPS gps, ESS type $type$, $rpath$
means from origin point to main road, $tpath$ means from transit point to destination
Output: optimal route $Route_{optimal}$

1: ESS ← **queryCandidateDestinations**($D, gps, type$)
2: MR ← **queryCandidateMainRoads**(G, gps)
3: CRoutes ← ∅
4: **Foreach** $ess \in ESS$ **do**
5:    rpath(gps, ess) ← *searchRunningPath*(gps, ess)
6:    **Foreach** $mr \in MR$ **do**
7:       rpath(gps, tp) ← *searchRunningPath*(gps, mr)
8:       t_{cost} ← *predictWaitTime*(tp)
9:       **If** t_{cost} + rpath(gps, tp). time > rpath(gps, ess). $time$ **then**
10:          CRoutes ← CRoutes ∪ **CandRoute**(rpath(gps, ess))
11:       **Else**
12:          tpath(tp, ess) ← *searchTaxiPath*(tp, ess)
13:          CRoutes ← CRoutes ∪ **CandRoute**(rpath(gps, tp), tp, tpath(tp, ess))
14: $Route_{optimal}$ ← **selectBestRoute**(CRoutes)
15: **return** $Route_{optimal}$

---

6   Experimental Results

Evaluation of Waiting Time. To evaluate the quality of the guide routings adequately
is quite difficult. Making a full test on the guide routings is really expensive. It needs to
be tested by time and practice. Because of the classic algorithm for path-finding is
credible, we focus on partial test instead of full test. We collected data manually in
some streets of Beijing downtown at different times. Then we figured out a series of
average waiting times, to verify the effectiveness of waiting time prediction. As shown
in Fig. 4, the predicted waiting times derived from the traffic flow model are smoother,
and it is close to the sample times, except a sharp peak. In an actual scenario, a specific
fluctuation takes place sometimes because of a sort of influence factors.

As the amount of GPS trajectories we used is not sufficient and the data is not up-
to-date. The predicted times cannot keep consistent with the actual situations. The
mean absolute error (MAE) values on the samples associated with the Fig. 4 are 49.625
and 0.729. It indicates that the traffic flow model is reasonable in most cases, and the
guide routing generated by our system should be somehow effective and helpful.

Implementation of Application. We have developed an android prototype as shown in
Fig. 5. When the user launches the app, there are three options of ESS service to choose
from, i.e., the gas station, the hospital and the police. When the network is not in a good
condition, the user can open the mute mode.

Once one service button is clicked, the app gathers the user location and sends it to
the back-end server automatically. When the app receives guide routing information
from back-end server, it draws the route on the map and then starts navigation. The app

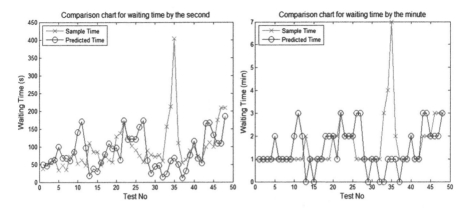

Fig. 4. Comparison chart for waiting time by second and minute

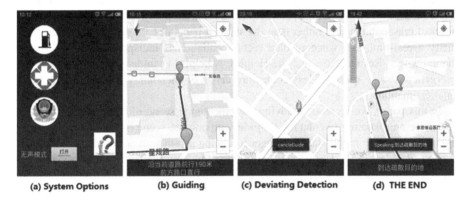

| (a) System Options | (b) Guiding | (c) Deviating Detection | (d) THE END |

Fig. 5. An implement of HelpMe on android platform

displays a route highlighted in blue on the map, and gives voice instructions (e.g., northeast, 200 m, turn left etc.) timely. We can see three types of markers on the map. The blue marks and yellow marks are key-points in the route. The blue markers stand for "by walking" and the yellow markers stand for "by taxi". The red marker represents the current location of the user. It is updated by GPS locator continuously. If the user turns aside from the navigation, the app will cancel the current guide routing and recall the back-end server. When the user reaches the destination, the navigation is finished and the app will send a complete report to the back-end server.

Performance on Recommendation. In order to measure the performance of HelpMe, we randomly select 50 places in Beijing downtown to submit the requests. We capture the candidate places and record the response times as shown in Fig. 6.

It is easy to see that the average responding time is about 0.2819 s. Considering the experiments are running on a personal microcomputer, there is reason to believe our proposed methods scale well.

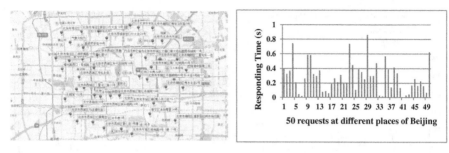

Fig. 6. Distribution of request and the response time

7 Conclusion and Future Work

In this paper, we investigate the problem of self-rescue for individual emergency situations. By leverage big data collected from urban public transportations, a reach-time-cost model and a dynamic routing planning are proposed to guide the users with the most reasonable routes rather than the shortest paths in conventional methods, and finally a real-time and voice-enabled recommendation is indispensable to the smart phone owners. Results show that the HelpMe is helpful in finding the optimal way to *emergency service spots*, of less conjecture and less uncertainty compared with personal estimation.

In this paper, we focus on studying the emergency scene of "go-get-help". In the future, we are planning to enhance HelpMe with another scene of "run-from-danger", i.e., rapid escape from the place in danger following a path with accumulating potential damage. Besides, more datasets will be imported to improve the applicability of proposed system in next steps.

Acknowledgments. This work was supported by the National Natural Science Foundation of China (No. 61202064), the National High Technology Research and Development Program of China (863 Program) (No. 2013AA01A603), and the Strategic Priority Research Program of the Chinese Academy of Sciences (No. XDA06010600).

References

1. Yu, J., Pande, A., Ali, N., Dixit, V., Edwards, F.: Routing strategies for emergency management decision support systems during evacuation. J. Transp. Saf. Secur. (2013)
2. Kongjin, Z., Lizhong, Y.: Comparative study of evacuation efficiency using stairs and elevators in high-rise buildings. J. Appl. Fire Sci. **23**, 105–113 (2013)
3. Shang, S., Ding, R., Yuan, B., Xie, K., Zheng, K., Kalnis, P.: User oriented trajectory search for trip recommendation. In: Proceedings of the 15th International Conference on Extending Database Technology, pp. 156–167. ACM (2012)
4. Sun, Y., Fan, H., Bakillah, M., Zipf, A: Road-based travel recommendation using geo-tagged images. Comput. Environ. Urban Syst. (2013)

5. Wei, L.-Y., Zheng, Y., Peng, W.-C.: Constructing popular routes from uncertain trajectories. In: Proceedings of the 18th ACM SIGKDD International Conference on Knowledge Discovery and Data Mining, pp. 195–203. ACM (2012)

6. Kaisar, E.I., Hess, L.: An emergency evacuation planning model for special needs populations using public transit systems. J. Public Transp. **15**(2), 45–69 (2012)

7. Shi, C., Zhong, M., Nong, X., He, L., Shi, J., Feng, G.: Modeling and safety strategy of passenger evacuation in a metro station in China. Saf. Sci. **50**, 1319–1332 (2012)

8. Zheng, Y., Xie, X.: Learning travel recommendations from user-generated gps traces. ACM Trans. Intell. Syst. Technol. (TIST) **2**, 2:1–2:29 (2011)

9. Stepanov, A., Smith, J.M.: Multi-objective evacuation routing in transportation networks. Eur. J. Oper. Res. **198**, 435–446 (2009)

10. Liu, K., Li, Y., He, F., Xu, J., Ding, Z.: Effective map-matching on the most simplified road network. In: Proceedings of the 20th International Conference on Advances in Geographic Information Systems, pp. 609–612. ACM (2012)

11. Dijkstra, E.W.: A note on two problems in connexion with graphs. Numer. Math. **1**, 269–271 (1959)

12. Yuan, N.J., Zheng, Y., Zhang, L., Xie, X.: T-Finder: a recommender system for finding passengers and vacant taxis. IEEE Trans. TKDE **25**, 2390–2403 (2013)

13. Liu, K., Deng, K., Ding, Z., Li, M., Zhou, X.: MOIR/MT: monitoring large-scale road network traffic in real-time. VLDB **2**(2), 1538–1541 (2009)

14. Zhang, H., Liu, K., Wang, X., Guo, L., Ding, Z.: An effective method to detect people in danger. In: Proceeding of the 2nd International Workshop on Emergency Management in Big Data Age (BIGEM 2014), Macao, China, 16 June 2014

Keyword-Based Semantic Analysis of Microblog for Public Opinion Study in Online Collective Behaviors

Yefeng Ma[1], Qing Deng[1], Xinzhi Wang[2], Jiaqi Liu[3],
and Hui Zhang[1(✉)]

[1] Institute of Public Safety Research, Tsinghua University, Beijing, China
Zhhui@tsinghua.edu.cn
[2] School of Computer Engineering and Science, Shanghai University,
Shanghai, China
[3] Institute of Automation, Chinese Academy of Sciences, Beijing, China

Abstract. This study employs text mining and semantic analysis methods to analyze the change of public opinion in the formation of online collective behaviors in crisis management. A case that related to law enforcement violence is investigated to study the communication between government and the public. The development of event is framed based on news data. The paper conducts a semantic analysis on microblog data and extracts high-frequency keywords and co-occurrence words at each stage of the event. By comparing key words at different stages, the paper proposes a new way to gain insight into the requirement of the public and predict the change of public opinion.

Keywords: Semantic · Keywords · Public opinion · Online collective behaviors · Co-occurrence

1 Introduction

Public opinion has great influence on government response to contingencies. The communication between government and the public in crisis situation is quite different from the normal times. In crisis situation, government faces much more pressure on information release and decision-making. On one hand, it is challenging to identify valuable and trustworthy information from a large volume of information sources with limited time. On the other hand, it is challenging to predict how public opinion will take shape and change. People's appeals and expectations may change quickly with the development of situation. And the rise of social media makes it more difficult to identify the needs of audiences since people are free to express opinions at any time and any place. Interaction between individuals is enhanced greatly on social networking websites. Negative opinions and emotions can easily spread among Internet users and sometimes cause online collective behaviors.

Public opinion study on social media (microblog, forum, discussion board, etc.) has drawn much academic attention [1]. Existing studies mainly focus on three aspects: (1) information spreading network, for instance how rumor spreads, (2) public sentiment,

© Springer International Publishing Switzerland 2014
Y. Chen et al. (Eds.): WAIM 2014, LNCS 8597, pp. 44–55, 2014.
DOI: 10.1007/978-3-319-11538-2_5

specifically public's attitudes and feelings towards a crisis, (3) crisis communication, which reflects the interaction between crisis managers and the public. Much work have been done to investigate information spreading process on social media and develop information diffusion models for simulation and prediction [2, 3]. Sentiment analysis of public opinion is an increasing research hotspot [4, 5]. The sentiment found within comments and critiques can provide useful indicators for crisis managers to perceive the change of public opinion. It is usually combined with text mining, natural language processing and semantic analysis in practice use [6, 7]. Public opinion plays an important role in government decision-making, and at the same time response from government has great impact on shaping public opinion. Many researchers look into the issue of how social media influence crisis communication. For the public side, research focuses on the novelties of people's perception and reaction to crisis communication via social media [8, 9]. And for the government side, research focuses on assessing government's strategy and performance when communicate to the public [10]. These studies and investigations mainly play emphasis on public participation or the way that government communicates to the public. Few studies look at the particular communication problem in dealing with big conflicts such as large-scale online collective behaviors.

Online collective behaviors of Internet users often emerge spontaneously as a reaction to disappointing response from government or other organizations responsible for emergency management. For instance, in the case of BP oil spill incident in 2010, a great number of people condemned BP's behavior of disabling the comments feature on social media sites and made BP sink into widespread criticism [11]. Online collective behaviors can be nonviolent or violent, which are often referred as soft actions (such as massive online discussion, petition or voting) and hard actions (such as human flesh search) [12]. Large-scale online collective behaviors can damage public trust in short term and cause negative influence on government's image in long term, especially for countries that the government does not have good communication mechanism with the public. China is one of the countries deeply troubled with this problem.

In China, online collective behaviors become prevalent after microblog becomes popular. Most online collective behaviors are soft actions, primarily are large-scale online discussions and petitions. Public attention often breaks out suddenly and soon gets the situation out of control. For instance, in a train crash accident happened in 2011, the tardy response of railway administration led to massive online discussions of the possible cover-up of the causes and inadequate efforts on rescue. Discussions on microblog evolved to an online petition for thorough investigation into the cause of the crash and business management of railway administration. There have been many similar cases that Chinese government is criticized for inadequate response or injustice when managing crises, which finally triggers online collective actions. Studies on similar cases imply that the lack of proper channels for public participation in crisis management and expression of grievance could be a contributor to online collective behaviors. But the fact that government has little experience in communicating to the public through social media is a lead cause. From the prospective of government, it is necessary to improve the ability of recognizing public's opinions quickly and responding to the changing demands in time.

The understanding of the change of public opinion and government response is the key to indicate the trend of online collective behaviors and possible outcome in a crisis.

This study employs text mining and semantic analysis methods to analyze the change of public opinion in the formation of online collective behaviors. The paper looks into a case that related to law enforcement violence and collects related information from main news websites and microblog. The development of event is framed based on news data. The paper conducts a semantic analysis on microblog data and extracts high-frequency keywords and co-occurrence words at each stage of the event. By comparing key words at different stages, the paper proposes a new way to gain insight into the requirement of the public and predict the change of public opinion. The proposed method has potential to improve event evolvement and support decision making.

2 Methodology

2.1 Text Mining

Since text is the most common format for information storage, a large amount of textual data may potentially contain a great wealth of knowledge. Text mining aims at finding valuable patterns or rules in text in order to gain insight or knowledge about specific topics [13]. Tan [14] defined text mining as "the process of extracting interesting and non-trivial patterns or knowledge from text documents". The main purpose of text refining is to represent the contents of textual data in particular pattern in order to apply further analysis (e.g. statistical analysis).The process of text mining generally consists of two parts: text refining and knowledge distillation [13]. Textual data is converted to an intermediate form after refining and ready for knowledge distillation, such as for categorization, visualization, modeling or associative discovery. Various text mining technologies have been developed for different purposes, such as information retrieval technology, clustering, classification, Natural Language Processing (NLP), data mining and semantic analysis, etc. [14]. Information retrieval technology is mainly used for searching documents and extracting keywords. Clustering and classification technologies are used for organizing documents and analyzing keyword distribution in order to get overview of topics. NLP, data mining and semantic technologies are used for extracting interesting information from content, or known as knowledge discovery. These technologies have been adapted to all kinds of text type, including long text and short text (specifically twitter-text or microblog text) [15].

2.2 Semantic Network Analysis

Semantic network analysis is a method for content analysis which focuses on infer relevant aspects of what a message means in its content and the relation between messages [16]. Different from traditional content analysis methods that directly use coding to answer the research question, semantic network analysis tries to represent the content of the messages as a network of objects and answer the question by querying the network representation [17]. Semantic network analysis focuses on association between words and the meaning shared by them. It has been widely used in media analysis, such as research on hot topics detection and public opinion analysis [18, 19]. One challenge of semantic network analysis is expensive human coding, so how to realize automatic semantic network analysis has drawn much academic attention. Information technology makes great contribution to save human coding effort. There have been much research focuses on developing information systems and tools for automatic semantic network analysis, such as for automatic labeling of semantic roles [20] and for analyzing co-occurrence words [21].

3 Research Design

3.1 Data Collection

Online collective behaviors could be driven by the motivation to restore justice, which is related to social, political or moral issues, or by the emotions to address discontent [12]. In this study, a case that related to law enforcement violence was investigated to study online collective behaviors and communication between government and the public.

Conflicts between urban management officials (known as ChengGuan in Chinese) and peddlers (known as Xiaofan in Chinese) are typical examples of law enforcement violence in China and have drawn wide attention from the public. In February, 2014, CCTV (China Central Tele Vision) news released several urban management officials (or "Chengguan" in Chinese) engaged in violence when performing duty. The officials were serving at Urban Management Bureau of Huangzhong county in Qinghai province. A pregnant woman was beaten by several officials and the violence resulted in a miscarriage. The report rapidly circulated online and stirred outrage among netizens. The incident led to nationwide discussions and online collective activism. Related news and microblog posts were collected to investigate the development of the event.

The Words 'Qinghai', 'Chengguan' and 'Yunfu (pregnant woman)' were set as keywords for extracting data from news source and microblog. Specifically, news data was collected from Baidu news search engine, the most popular news search site in mainland China. News reports were collected at least two keywords are mentioned in headlines. Microblog data was collected from Sina weibo (the most popular microblog in China) by using weibo API and web page crawling method. The microblog data included content, author, public time, URL, repost count and comment count. The time range of collection started from the first day of event outbreak and ended at the public attention transferred to other news (February 14–20th in this case). About 200 pieces of news were collected and 52 of them were original news and selected for analysis. More than 4000 pieces of microblog data were collected from Sina weibo. A classifier was developed to filter noise data. Firstly, about 1000 weibos were randomly selected and classified into related and unrelated classes manually. Then a decision-tree classification model was developed and trained with several commonly used features, such as repost counts, comment counts, the number of followers, etc. [22]. Every piece of unrelated microblog posts was double-checked in case of wrong selection. Finally, 3100 pieces of microblog data were selected for analysis after filtering.

3.2   Case Study with News Data

52 items of news reports were sorted into a time-ordered dataset. The important information that was supportive to indicate event progress was identified and analyzed by two coders. The result was discussed by coders and 15 key sub-events were coded to demonstrate the development of the Qinghai event, as shown in Fig. 1. The conflict was between Li's family and several urban management officials (UMO) who served at Duoba urban management office of Huangzhong county, Qinghai Province. A team of urban management officials from Duoba town were required to demolish three illegal constructions of Li's family and met with Li's resistance. Li was a pregnant mother and the conflict resulted in her miscarriage. Figure 1 shows the time-line of the event.

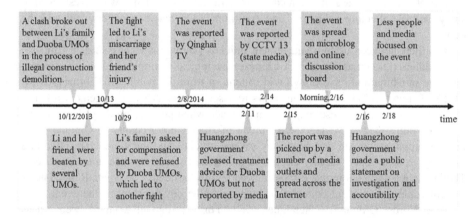

Fig. 1. Time-line of event progress

3.3   Microblog Data Analysis

3.3.1   Data Pre-processing

The development of online collective actions often has four stages: initial stage, growing stage, prosperous stage (burst stage) and degressive stage. In this study, the target case was divided into four stages according to the increasing rate of microblog posts per hour. From the beginning of the event, the period that the increasing rate is less than 10 items per hour is defined as initial stage (stage 1). The following period is defined as growing stage (stage 2) when the increasing rate is between 10 to 200 items/hour. The period that the increasing rate is more than 200 items/hour is defined as prosperous stage (stage 3). And the degressive stage (stage 4) is from the end of prosperous stage to the end of public attention. Figure 2 shows the stage division of Qinghai case.

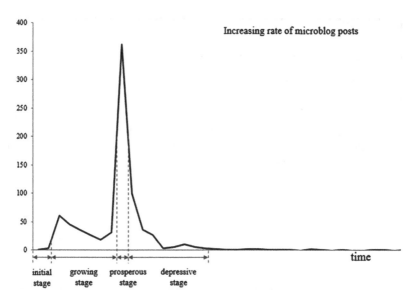

Fig. 2. Four stages of Qinghai case.

3.3.2 Keyword-Based Semantic Analysis

Semantic analysis includes four steps: (1) words segmentation, (2) keywords extraction, (3) document frequency calculation and (4) co-occurrence words statistics. English words can be easily segmented by space sign, while words segmentation for Chinese characters are much more complicated. In this paper, we use a popular Chinese segmenting system ICTCLAS to process Chinese words segmentation and POS (part-of-speech) tagging. The second step is to remove noisy words and extract keywords. We defined a set of stop words, which don't have actual meaning in the context, such as qualifiers (a, an, the, that, those, etc.) and prepositions (of, over, under, above, etc.). We remove all the stop words and leave nouns, verbs and adjectives as keywords. Then we calculate the document frequency (DF) of key nouns, verbs and adjectives in microblog posts at different stages and sort them respectively from the highest frequency to lowest frequency. DF refers to the number of times that a word appears in a document [23]. In this study, we define DF as the number of a keyword appears in the given microblog set. For instance, if the word "happy" appears 100 times in all microblog posts of stage 1, the DF of "happy" at stage 1 is 100. The value of DF can imply the importance of a keyword. The last step is to identify co-occurrence words of the keywords. Co-occurrence refers to "concurrence/coincidence or the above-chance frequent occurrence of two terms from a text corpus alongside each other in certain order" (Wikipedia definition). The high-frequency keywords extracted from the third step were input to a classifier which was developed with pattern-mining algorithm [24] in Matlab to classify co-occurrence words in the original microblog data set, including co-occurrence nous verbs and adjectives. By comparing the DFs of the same type keywords, we can get a clue of the change of people's concern and attitude, which is helpful to predict the trend of the event and public opinions.

4   Results and Discussions

4.1   Co-occurrence Words Analysis for High-Frequency Nouns

The most frequent nouns mentioned by people may give a clue of who are involved in the event and what people concern most. Based on the prior analysis, high frequency nouns at each stage are identified according to occurrence frequency. Figure 3 shows the nouns with highest occurrence frequency and co-occurrence words of each key noun from stage 1 to stage 4. The original words are Chinese words and are translated into English for demonstration.

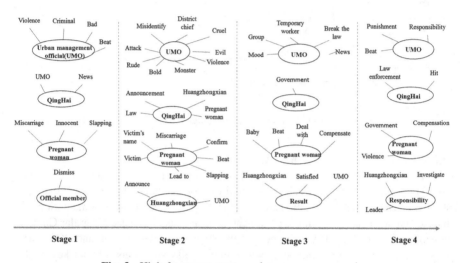

Fig. 3. High frequency nouns and co-occurrence words

At stage 1, 'Chengguan (urban management official)','Qinghai', 'Yunfu (pregnant woman)' and 'Duiyuan (official member)' are the most frequent words that mentioned by people on microblog. Looking at the co-occurrence words of above high frequency nouns, we find that people primarily focus on the details of event at this stage, specifically, people show concern to the female victim who was pregnant and miscarried as a result of beating by several urban management officials. At stage 2, the site of incident and local government become new public focus. People's attention begins to shift to government response. Local government made an announcement that the officials involved were temporary staffs and did not belong to urban management system. This 'Linshigong (temporary staff)' explanation draw a lot of attention, and the co-occurrence words imply that people don't buy it. At stage 3, a bunch of words that related to result and punishment come out, such as 'Chulijieguo (result)', 'Chufen (discipline)', and 'Peichangjin (compensation)', which indicates that people's attention shift to accident investigation and accountability. According to co-occurrence words of each noun, it is shown that people are unhappy with the handling of accident. Most people think that the punishment for urban management officials are too light and the compensation for victims are not enough. The controversy on 'temporary staff explanation' continues and many people blame Qinghai urban management office for passing the buck to temporary staffs. The timeline in Fig. 1 shows that Qinghai government (provincial level) responded again to public demands and promised to deal with the case seriously. High frequency nouns and corresponding co-occurrence words at stage 4 show that less people pay attention to the event, but people are still not satisfied with Qinghai government response and suggest to dismiss the chef of urban management team in Huangzhong county.

4.2   Co-occurrence Words Analysis for High-Frequency Verbs

The most frequent verbs appeared in microblog discussions are helpful to understand the basic information of event and how the situation develops. Figure 4 presents high frequency verbs and corresponding co-occurrence words at each stage of the event. Key messages released from verbs at stage 1 include 'Da (hit)' and 'Rencuo (mis-identify)'. Urban management office stated that officials who involved in violence didn't mean to beat the pregnant woman and had mistaken her for another person. This explanation is questioned by many people and draw a lot of attention at the first stage. People shift attention to the victim's situation and the punishment for brutal officials at stage 2. From the co-occurrence words, we can see that the public are outraged by the report of the woman's miscarriage and show sympathy to the female victim. People appeal to government to give severe punishment to officials. At stage 3, words 'Gongbu (announce)', 'Chezhi (dismiss)' and 'Chuli (deal with)' imply that government or high-level urban management office released the punishment for brutal officials. The co-occurrence words show that public concerns shift to accountability and the result. The high frequency verbs at stage 4 show that people still care about the punishment for those officials and suggest government to pay more attention to grassroots and administrative management.

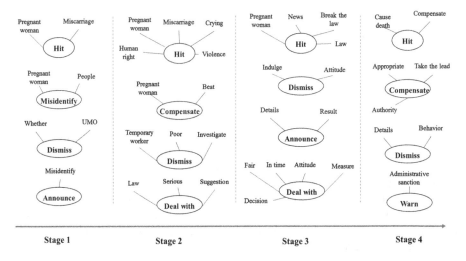

Fig. 4. High frequency verbs and co-occurrence words

4.3   Co-occurrence Words Analysis for Key Adjectives

The most frequent adjectives may imply the development of public sentiment or people's attitudes towards government response. Figure 5 shows the most frequency adjectives and corresponding co-occurrence words at four stages respectively. High frequency adjectives at stage 1 indicate people's negative attitude to urban management officials and the act of violence. According to words 'Chijing (shocked)', 'Wuchi (shameless)' and 'Kewu (hateful)', it seems that resentment to urban management officials is increasingly accelerated. Besides, people start to doubt whether there is corruption involved. High frequency words at stage 3 releases the fact that people are disappointed with government's behavior of shifting responsibility to temporary staff and covering up the truth. They also show concern on social injustice and law enforcement. At stage 4, people's attention turns to rights and justice.

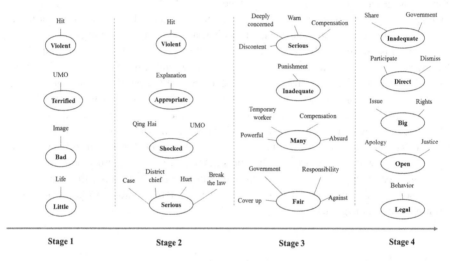

Fig. 5. High frequency adjectives and co-occurrence words

4.4   Discussions

By analyzing the change of high frequency words and the corresponding co-occurrence words, this paper looks into the evolvement of public concerns and key responses from government at different stages, as shown in Fig. 6. At the initial stage, the public are outraged by violent behaviors of urban management officials. They show concern to the pregnant woman and scream for details and apologies. But the response from government (local urban management office) is very disappointed. The government states that the officials mistaken her for another person and don't mean to it. At the growing stage, the public lost trust on local government and shift their concerns on response from higher level. The urban management office at city level explains that the involved officials are not registered as formal staff, which causes more complaints from public and drives online collective petitions to a climax. At the following stage, public concerns move to investigation, accountability, punishment and compensation. The government releases more information on investigation and accountability. However, the public are still unhappy with the punishment and the result. However, there is no further response from government side.

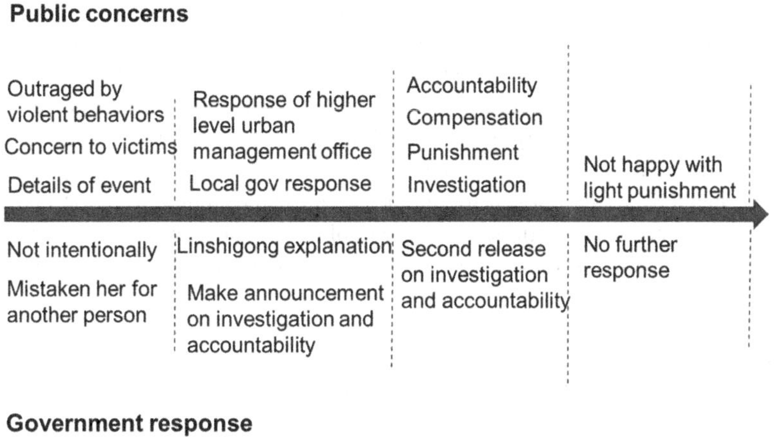

Public concerns

Government response

Fig. 6. Summary of public concerns and government response

Looking at the whole response process, both local government and higher level government have very passive performance, specifically, do not want to take responsibility. The government doesn't make sincere apology to the victim and the public in the early stage. In face of wide-spread complaints and discussions, they attempt to cover up and ignore public appeals, which are the leading reasons of unsuccessful response.

5   Conclusion

The ability of the government to respond to the preferences of public is very important to manage a crisis. Under crisis situation, it is difficult to predict how public opinion will take shape and change. And the rise of social media makes it more difficult. This study employed text mining and semantic analysis methods to analyze the change of public opinion in the formation of online collective behaviors in crisis management. A case that related to law enforcement violence of urban management officials was investigated to study public opinions and government response at different stages of event. There are two parts for research design: case study with news data and public opinion study with microblog data. The data were collected from the most popular news site and microblog site. A web crawler and two classifiers were developed to filter data. A time-line of event was developed from news data analysis, which is helpful to trace the key information of the event. Analysis of microblog data, including analysis on high frequency keywords (nouns, verbs and adjectives) and the co-occurrence words, presents the change of public opinions and sentiments towards government response with the development of event. At initial stage, public attention is on the details of incident. The public show concerns to victims and condemn urban management officials' act of violence. People's attention begins to shift to government response at growing stage, especially response from local government. At prosperous stage, the public mainly focus on investigation and accountability. People are unhappy with the handling of accident at local government level, specifically government's behavior of shifting responsibility to temporary staff and covering up the truth. At depressive stage, people still care about the punishment for those officials and suggest the government pay more attention to grassroots and administrative management. This keyword-based semantic analysis of microblog data is helpful to investigate the evolvement of online collective behaviors (discussions, feelings and appeals) in a crisis. The method can be applied to similar cases to study public opinion and crisis communication between government and the public. The future work will improve the ability of government response assessment and combine assessment with decision-making.

Acknowledgement. This work was partially supported by the National Natural Science Foundation of China (Grant No.91024032, No.91224008, No.70833003). National Basic Research Program of China (973 Program No: 2012CB719705).

References

1. Sobkowicz, P., Kaschesky, M., Bouchard, G.: Opinion mining in social media: Modeling, simulating, and forecasting political opinions in the web. Gov. Inf. Q. **29**(4), 470–479 (2012)
2. Xiong, F., Liu, Y., Zhang, Z.J., Zhu, J., Zhang, Y.: An information diffusion model based on retweeting mechanism for online social media. Phys. Lett. A **376**(30), 2103–2108 (2012)
3. Liu, C., Zhang, Z.K.: Information spreading on dynamic social networks. Commun. Nonlinear Sci. Numer. Simul. **19**(4), 896–904 (2014)
4. Prabowo, R., Thelwall, M.: Sentiment analysis: A combined approach. J. Informetr. **3**(2), 143–157 (2009)
5. Godbole, N., Srinivasaiah, M., Skiena, S.: Large-scale sentiment analysis for news and blogs. In: ICWSM 7 (2007)
6. Li, N., Wu, D.D.: Using text mining and sentiment analysis for online forums hotspot detection and forecast. Decis. Support Syst. **48**(2), 354–368 (2010)
7. Yates, D., Paquette, S.: Emergency knowledge management and social media technologies: A case study of the 2010 Haitian earthquake. Int. J. Inf. Manag. **31**(1), 6–13 (2011)
8. Schultz, F., Utz, S., Göritz, A.: Is the medium the message? Perceptions of and reactions to crisis communication via twitter, blogs and traditional media. Pub. Relat. Rev. **37**(1), 20–27 (2011)
9. Utz, S., Schultz, F., Glocka, S.: Crisis communication online: How medium, crisis type and emotions affected public reactions in the Fukushima Daiichi nuclear disaster. Pub. Relat. Rev. **39**(1), 40–46 (2013)
10. Palttala, P., Vos, M.: Quality indicators for crisis communication to support emergency management by public authorities. J. Contingencies Crisis Manag. **20**(1), 39–51 (2012)
11. Nickel, E.D.: Novato Fire District (2009)

12. Qiu, L., Lin, H., Chiu, C.-Y., Liu, P.: Online collective behaviors in China: dimensions and motivations. Analyses Soc. Issues Pub. Policy (2014). doi:10.1111/asap.12049

13. Nasukawa, T., Nagano, T.: Text analysis and knowledge mining system. IBM Syst. J. **40**(4), 967–984 (2001)

14. Tan, A.H.: Text mining: The state of the art and the challenges. In: Proceedings of the PAKDD 1999 Workshop on Knowledge Discovery from Advanced Databases, pp. 65–70 (1999)

15. Guan, W., Gao, H., Yang, M., Li, Y., Ma, H., Qian, W., Yang, X.: Analyzing user behavior of the micro-blogging website Sina Weibo during hot social events. Phys. A Stat. Mech. Appl. **395**, 340–351 (2014)

16. Krippendorff, K.: Reliability in content analysis. Hum. Commun. Res. **30**, 411–433 (2004). doi:10.1111/j.1468-2958.2004.tb00738.x

17. Van Atteveldt, W.: Semantic Network Analysis: Techniques for Extracting, Representing, and Querying Media Content. Vrije Universiteit, Amsterdam (2008)

18. Yu, H.O.N.G., Yu, Z., Liu, T., Li, S.: Topic detection and tracking review. J. Chin. Inf. Process. **21**(6), 71–87 (2007)

19. Kim, S.Y., Choi, M.I., Reber, B.H., Kim, D.: Tracking public relations scholarship trends: Using semantic network analysis on PR Journals from 1975 to 2011. Public Relat. Rev. **40**(1), 116–118 (2013)

20. Gildea, D., Jurafsky, D.: Automatic labeling of semantic roles. Comput. Linguist. **28**(3), 245–288 (2002)

21. Matsuo, Y., Ishizuka, M.: Keyword extraction from a single document using word co-occurrence statistical information. Int. J. Artif. Intell. Tools **13**(01), 157–169 (2004)

22. Liu, J., Cao, Z., Cui, K., Xie, F.: Identifying important users in sina microblog. In: 2012 Fourth International Conference on Multimedia Information Networking and Security (MINES), pp. 839–842. IEEE (2012)

23. Luo, X., Zhang, J., Ye, F., Wang, P., Cai, C.: Power Series Representation Model of Text Knowledge Based on Human Concept Learning. IEEE Trans. Syst. Man Cybern. Syst. **44**(1), 86–102 (2014)

24. Bayardo Jr., R.J.: Efficiently mining long patterns from databases. ACM Sigmod Rec. **27**(2), 85–93 (1998)

Research on Reasoning Model of Firefighting and Rescue Cases

Qingchun Kang[1,2(✉)], Lihong Lu[1,2], Libiao Pan[2], Yu Li[1,2], Jian Gao[2], and Danjun Lian[1,2]

[1] The Ministry of Public Security Key Laboratory of Firefighting and Rescuing Technology, Langfang, Hebei, China
kangqingchun@sina.com, {llh_qc, liandanjun, liyul0_01}@163.com
[2] The Chinese People's Armed Police Force Academy, Langfang, Hebei, China
{libiao429, 30491662}@qq.com

Abstract. Based on creating the digital scenario matrix of firefighting cases, this paper proposes four matrix-based drive reasoning models including "type - process" drive, "type - problem" drive, "type - countermeasure" drive and "type - lesson" drive, as well as hybrid reasoning model based on multi-base coordination, which will improve the rate and efficiency of matching cases, promote the utilization ratio of cases and accuracy of decision making, so as to lay a solid foundation for scientific decision-making on the firefighting and rescue scene.

Keywords: Firefighting and rescue · Case · Digitalization · Scenario matrix · Reasoning model

1 Introduction

Research on case-based reasoning (CBR) based on "scenario-response" has become a hotspot in present academic circle. The main idea of the method is to split a case into several scenarios [1–3] to obtain the solution to the current problem (targeted scenario) by consulting the solution of similar scenarios (source scenario) in the data base. It well solves the problems such as low threshold of matching similar cases in the case-based reasoning technology [4–6], and undesirability of case using. At present, the expression of scenario is more limited to text description while the cases are important documents of actual records and objective summaries of firefighting and rescue which contains a lot of information and data on fire and firefighting [7]. Compared with the digital language recognizable for computer, using text to describe case scenarios will make much more complicated calculation of scenario matching, more difficulty in quickly extracting the key elements of case matching, and waste of resources. In order to make the fullest use of these valuable resources, we apply the case-based reasoning technology in the firefighting case study. Taking oil tank fire as example, this paper puts forward case expression method by digital scenario matrix and case-based reasoning mode on the matrix, which lays a foundation for quick matching of firefighting cases in fire ground and building a computing-aided decision system of firefighting scenario based on digital cases.

© Springer International Publishing Switzerland 2014
Y. Chen et al. (Eds.): WAIM 2014, LNCS 8597, pp. 56–64, 2014.
DOI: 10.1007/978-3-319-11538-2_6

2 Construction of Digital Scenario Matrix of Firefighting Cases

Reasoning model of auxiliary decision system is closely related to the knowledge type, knowledge presentation and knowledge base establishment. Only the appropriate knowledge presentation will initiate correct reasoning model. Knowledge applied in the firefighting and rescue auxiliary decision system can be roughly classified as: white knowledge based on data and algorithm; Grey knowledge based on experience; scenario knowledge based on cases. The first two expressions have gone mature and will not be further discussed here. This section mainly discusses the expression method of cases and scenarios based on digital matrix, which lays a foundation for the four drive reasoning models based on digital scenario matrix in Sect. 3 in this paper.

Connected with the regular pattern of occurrence, development and evolution of an accident, a firefighting or rescue case can be split into a handful of scenarios according to the key points in the process such as the key nodes of basic information changes, or the key nodes of adjustment of making decision known to lower levels. Scenario is the key site condition that influences decision. Featured elements can best reflect nature of these states. After splitting a case into certain scenarios, some featured elements can be extracted from a scenario. Given value assignment to each type of featured elements, then its digital reconstruction and vector-expression can be realized. Therefore, the scenario can be expressed as a digital matrix constructed by a number of featured vectors.

Take oil tank fire as example here. 10 types of featured elements can be extracted from a certain scenario (symbolized by T), as shown in Table 1. The 10 element types are: object type (T_O), disaster type (T_D), environment (T_E), weather (T_W), type of water source (T_A), fire protection facilities (T_F), firefighting and rescue forces (T_R), measures (T_M), response strategy (T_S) and lessons learned (T_L). Each type can be further described by 10 factors (the featured element listed as the first one). For example, the object type can be described by the following factors: oil type, tank type and its material, structure form, tank volume, storage capacity, the number of the tank on fire, the number of exposure tanks, pipe pressure and pipe diameter. If the physical quantity of each elements type is assigned, the featured elements type will become a featured vector, so the scenario can be expressed as a 10×10 digital matrix shown below.

$$S = \begin{pmatrix} T_O \\ T_D \\ T_E \\ T_W \\ T_A \\ T_F \\ T_R \\ T_M \\ T_S \\ T_L \end{pmatrix} = \begin{pmatrix} O_1 & O_2 & O_3 & O_4 & O_5 & O_6 & O_7 & O_8 & O_9 & O_{10} \\ D_1 & D_2 & D_3 & D_4 & D_5 & D_6 & D_7 & D_8 & D_9 & D_{10} \\ E_1 & E_2 & E_3 & E_4 & E_5 & E_6 & E_7 & E_8 & E_9 & E_{10} \\ W_1 & W_2 & W_3 & W_4 & W_5 & W_6 & W_7 & W_8 & W_9 & W_{10} \\ A_1 & A_2 & A_3 & A_4 & A_5 & A_6 & A_7 & A_8 & A_9 & A_{10} \\ F_1 & F_2 & F_3 & F_4 & F_5 & F_6 & F_7 & F_8 & F_9 & F_{10} \\ R_1 & R_2 & R_3 & R_4 & R_5 & R_6 & R_7 & R_8 & R_9 & R_{10} \\ M_1 & M_2 & M_3 & M_4 & M_5 & M_6 & M_7 & M_8 & M_9 & M_{10} \\ S_1 & S_2 & S_3 & S_4 & S_5 & S_6 & S_7 & S_8 & S_9 & S_{10} \\ L_1 & L_2 & L_3 & L_4 & L_5 & L_6 & L_7 & L_8 & L_9 & L_{10} \end{pmatrix}$$

Table 1. Description of tank fire scenario matrix vector

Line number	Vector 1	Vector 2	Vector3	Vector4	Vector5	Vector6	Vector7	Vector8	Vector9	Vector10
Line 1(O)	Object type	Oil type	Tank type and material	Structure	Tank volume	Oil reserve	Number of tank on fire	Number of neighboring tank	Pipe pressure	Pipe diameter
Line 2(D)	Disaster type	Burning are	Position	Leakage quantity	–	–	–	–	–	–
Line 3(E)	Environment	Dike	Obstacle	Fire lane	Operation field	Adjacent place	–	–	–	–
Line 4(W)	Weather	Temperature	Wind power	Wind direction	Rain	Snow	Fog	Humidity	Thunder	–
Line 5(A)	Type of water source	Water flow of municipal pipe	Volume of fire pool	Water supply rate of fire pool	Natural water resource	–	–	–	–	–
Line 6(F)	Fire protection facilities	Number of fire hydrant	Fire hydrant functions well	Sprinkler functions well/not	Foam system functions well/not	–	–	–	–	–
Line 7(R)	Firefighting strength	Water load	Foam load	Number of fire fighter	Number of fire engine	Number of foam nozzle	Number of foam monitor	Number of water monitor	Number of fire nozzle	–
Line 8(M)	Measures	–	–	–	–	–	–	–	–	–
Line 9(S)	Response strategy	–	–	–	–	–	–	–	–	–
Line 10(L)	Lessons learned	–	–	–	–	–	–	–	–	–

3 Reasoning Model Based on Digital Scenario Matrix

For an auxiliary decision system, inference engine is the core components. Whether the reasoning model is appropriate or not directly influences reliability of application efficiency of knowledge and inference results of system. In the field of artificial intelligence (especially expert system), the study of reasoning mechanism is quite mature. It can be classified as inductive reasoning and deductive reasoning in theory, forward reasoning and backward reasoning in practice, precise reasoning and inexact reasoning in conclusion reliability. But for the practice of reasoning based on case scenario, there is no mature method yet. From the previous section, it can be known that a case can be split into several scenarios and each scenario can be expressed as the digital matrix recognizable for computer language, so that the case can be expressed as a scenario matrix set to accurately describe a complex process of firefighting and rescue and form into a digital case. A group of such cases will constitute a digital case base. As long as a matrix is retrieved, a key piece of the firefighting and rescue can be read. The core issue lies in how to realize the rapid and exact digital scene matrix matching based on digital scenario matrix reasoning. Combining with the characteristics of firefighting and rescue cases, the paper explores the four specific models of "type-process" drive, "type-problems" drive, "type-countermeasures" drive and "type-lessons" drive.

3.1 "Type-Process" Drive

The idea of "type-process" drive follows the forward reasoning thought and reasons out the specific command and tactical methods step by step according to the developing process of firefighting and rescue cases. Through matching the featured digital vectors described in the current basic situation and the corresponding vectors in the digital case scenario (history scenario), it can help commanders in decision-making by the response scheme (firefighting and rescue forces assembling, disposal methods and strategy) in history scenario. The drive includes the following two categories:

Firstly, for a certain disaster type happened in an object, vectors in the first two rows in Table 1 can be used as the main matching conditions and rapidly compared with historical situation. In this way, the historical scenario similar to the on-the-spot scenario can be obtained, so that corresponding disposal measures can be drawn, which corresponds to the 7th, 8th and 9th row vectors in historical scenario. The driving process is shown in path 1 in Fig. 1.

Secondly, the commander can quickly match corresponding eigenvectors in historical scenario database by taking different terrain, weather, water, fire protection facilities matching conditions as the main objects (i.e., 3rd–6th row vectors in Table 1). In this way, the historical scenario similar to the on-the-spot scenario can be obtained, so that corresponding disposal measures can be drawn, which corresponds to the 7th, 8th and 9th row vectors in historical scenario. The driving process is shown in path 2 in Fig. 1.

Single path drive may lead to multiple similar historical scenarios, which can puzzle the commander in making decision and affect commanding efficiency. Therefore, simultaneous multipath drive can be taken into consideration to improve the

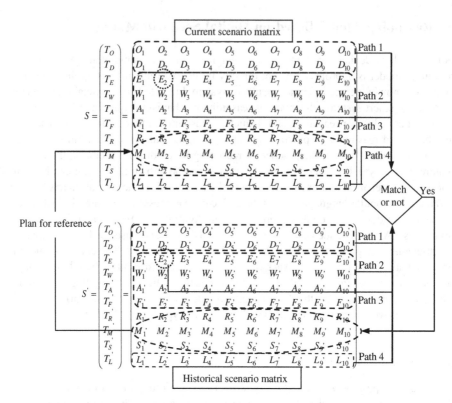

Fig. 1. Flow chart of reasoning model based on digital matrix

efficiency. For example, path 1 and path 2 in Fig. 1 can be merged two-way at the same time to produce better effect.

3.2 "Type-Problem" Drive

"Type-problem" drive follows the backward inference thought. By taking intractable problems in current situations as the highlighted matching conditions and by matching with historical scenarios, it can help commanders making decisions according to the historical response scheme. Given the dike in a certain tank area is bombed, then the specific location referred is E_2 in the matrix and the output plan corresponds to the measures in the 7th, 8th and 9th row vectors. The driving process is shown in path 3 in Fig. 1.

3.3 "Type-Countermeasures" Drive

"Type-countermeasures" and "type-problem" drives are two inverse drive processes. "Type-countermeasures" drive takes the measures expected by the commander (the featured vectors) as the main matching condition. It will help commander to evaluate whether the current decision is correct or not by matching similar historical scenarios with his situation and obtain the countermeasure adaptive to environment in the historical scenario such as information of solved problems and accomplished effect. Its drive process is a reverse application of "type - problem" drive.

3.4 "Type-Lessons" Drive

The "type-lessons" drive takes the worst results foreseen by the commander as the main matching condition. By matching them with similar lessons in historical scenarios, the commander can refer the scenario conditions and countermeasures, and compare with the available measures, then strives to minimize the inevitable loss in response after weighing the pros and cons. The essence of "type- lessons" drive mode is to choose the better from two evils, which belongs to the stratagem of a losing war philosophy. The driving process is shown in path 4 in Fig. 1.

4 Hybrid Reasoning Model Based on Multi-Base Coordination

Any reasoning type has its limitations, so it is best to use various methods complementary in reasoning mechanism. We might as well simulate a commander's way of thinking in a disaster scenario to better design the reasoning model.

4.1 Constant Drive Reasoning

Some basic decisions have been made by the commander on the disaster scene. For example, for a 5000 m^3 oil tank undergoing an open burning, the first thing that should occur to the commander is to consider what forces should be mobilized, how much force is needed. With the basic knowledge stored in his mind, he knows that 7 tons of foam, 8 foam nozzles and 4 fire companies are needed. And by his personal preference, the commander can decide the safety margin. All the data are already rooted in his brain without any calculation. It means to add a constant base in the auxiliary decision-making system, which can be directly called in the face of a similar situation.

4.2 Algorithm Drive Reasoning

On a real disaster scene, not all of the problems are as simple as above. Some of them are very complex and require a large amount of calculation, which needs support from a powerful algorithms base. For example, in the 7.16 Dalian gas pipeline incident, the explosion caused widespread flowing fire and ignited the oil tank as large as 100000 m^3. In such situation a large amount of calculation is required and it is difficult to be resolved merely by the commander's experience and memory. But all these belong to white knowledge and there is a complete calculation method in the textbook. As long as an algorithms base is established and with the help of a good drive model, the problem can be properly handled.

4.3   Rule-Driven

It would be better if the two methods above can solve all the problems. However, it is a pity that firefighting command and tactics is a complicated decision-making process, which is hard to conduct without commander's extraordinary experience and outstanding wisdom. The related knowledge is often gray that is difficult to express in scientific language, while they are the most valuable knowledge. Engineers research on the knowledge and formulate it with the language understood by computer. This is just the rule. For example, for 2 tanks in a 4-tank set undergoing an open burning, the experienced commander will be able to tell immediately which tank to cool first in the absence of sufficient recourse. This is the experience, the key to win. It can be further stated by the following form of rule:

IF (1 set of gasoline tank on fire, including one with open-type burning) condition 1 (Unburned gasoline tanks downwind, distance within 1.5 times the diameter).

THEN (cool the tank first).

By establishing the rule base, decisions can be made with the help of forward reasoning, and the answers to the problem can be obtained by backward inference reasoning.

4.4   Hybrid Reasoning Model

The hybrid reasoning thought is to simulate the commander's thinking process. For the most simple question, call data driven directly to get the answer (scheme); If a large amount of calculation needed, call algorithms base; If in need of support from experience, invoke rule base; If all issues need to refer to previous case scenarios, activate the four-drive-type mode based on digital scenario matrix reasoning. Another function of multi-base hybrid reasoning is for the strategy of conflict resolution. If there is a different data between the selected historical scenarios got from the above four drive models and the to-be-matched scenario, i.e. the element value in present scenario is different from that in historical scenario, then the different data can be processed by multi-base model, together with other reasoning method provided by knowledge base, so that to finally achieve the optimum solution. If there is no different data, the plan in the historical scenario can be output directly. Rule base, constant base (data base) and algorithms base in the knowledge base can be used to realize reasoning model based on rules, constants (data) and algorithms respectively. Multi-base coordination can integrate these models and form a hybrid reasoning model. Multi-base coordinated hybrid reasoning model can effectively improve the accuracy of the decision. The operational process is shown in Fig. 2.

As is shown in the figure, through the scenario matching, there are four different matrix element values between historical scenario matrix and the current scenario matrix. Among them, the difference of element 1 can be rewritten by reasoning based on algorithm provided by algorithms base; the difference of element 2 and 3 can be rewritten by reasoning based on rule provided by rule base; the difference of element 4 can be rewritten by reasoning based on data provided by data base. Through multi-base coordination model and integrating the reasoning results above, the plan in the historical scenarios are modified, so that the ideal solution is achieved.

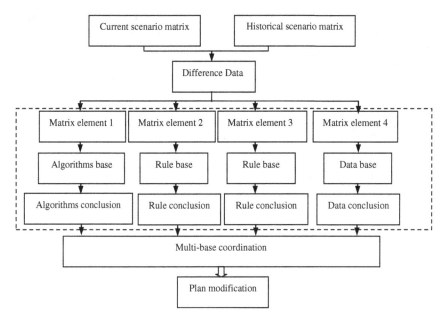

Fig. 2. Flow chart of hybrid reasoning model of multi-base coordination

5 Conclusion

The construction of digital case base is the foundation of scenario inference auxiliary decision system of firefighting and rescue. Reasoning model is an important means to solve digital case matching, improve emergency decision-making accuracy and enhance the decision-making efficiency. This paper proposes the four-types-drive reasoning model based on digital scenario matrix and the hybrid reasoning model based on multi-base coordination. If they can be applied to auxiliary decision system in firefighting and rescue, problems such as low matching rate, poor repeated utilization ratio and unscientific auxiliary decision-making will be appropriately handled, which will further increase the efficiency of the auxiliary decision-making system of firefighting and rescue.

Acknowledgement. The research is founded by major program of the national natural science foundation of china under Grant No.91024032.

References

1. Yuan, X.F., Li, H.X., Tian, S.C.: Research on key technology of emergency decision-making for unconventional emergency based on scenario analysis and CBR. Xi'an Technol. Univ. **6**, 101–137 (2011). (in Chinese)
2. Zhang, H., Liu, Y.: Scientific issues and integration platform based on the "scenario- response" national emergency system. Syst. Eng.-Theory Pract. **32**(5), 947–953 (2012). (in Chinese)

3. Wang, L., Tian, S.C., Yuan, X.F., Ma, X.M.: Research on gas explosion IDSS in coalmines based on CBR. In: International Conference on Advanced Computer Control, pp. 652–656 (2009) (in Chinese)
4. Wu, G.M., Zhao, W.C., Jiang, Y.P.: Research on a policy-making model of the recreation and development prediction of serious urban accidents. J. Southeast Univ. (Philosophy and Social Science) **13**(1), 18–23 (2011). (in Chinese)
5. Fang, W.J., Li, S.W., Yuan, Y., Wang, W.W.: The research and application of case-based reasoning technology. Agric. Netw. Inf. **1**, 13–17 (2005). (in Chinese)
6. Liu, T.M.: Study on scenes' construction of emergency planning. J. Saf. Sci. Technol. **8**(4), 5–12 (2012). (in Chinese)
7. Leng, L., Wei, H.D., He, N., Guo, Y.H., Chai, K.H.: Statistical analysis and countermeasure study of hundred conflagrations. Fire Sci. Technol. **26**(4), 439–443 (2007). (in Chinese)

Empirical Study of Online Public Opinion Index Prediction on Real Accidents Data

Xiao Long Deng[✉] and Yu Xiao Li

Key Laboratory of Trustworthy Distributed Computing and Service,
Beijing University of Posts and Telecommunications, 10 Xitucheng Road,
Beijing 100876, China
{shannondeng,liyuxiao}@bupt.edu.cn

Abstract. With the increased online public opinion data of real accidents in the last ten years. It has been important for us to attain and analysis the online public opinion data of real accidents. In this paper, an empirical study of online public opinion index prediction over more than 30 real accidents happened in China and other places in the world has been made and experiments results has proved that the online public opinion index prediction method is useful and we got some interesting results. And we have implemented GM (1, 1) model, GM (2, 1) model and chaotic prediction model based prediction method in online opinion for the first time.

Keywords: Grey prediction · Empirical study · Real accident · Public opinion · Index prediction · Data ming

1 Introduction

In the last decade, there have been much more public crisis events in China such as WenChuan earthquake in 2008, Mass protesting of Southern Weekend Newspaper event in 2013 and so on. It has been proved that China has come into the time of public crisis events fast growing.

At the same time, with the fast increase of Internet and data scale in Internet, Internet is playing a more important role in information spreading. And in nowadays, the online public opinion index is becoming into an important reason which is crucial to the country's stability and development of people's livelihood. The online public opinion index in China is featured with both a certainty of annual increment and an uncertainty of random variation. Thus, it can be seen as a typical grey system and shall be suitable for grey prediction modeling.

However, with the increasing complexity of online public opinion, the traditional grey prediction method gradually cannot meet the requirement of public opinion index prediction, and needs to be enriched and improved.

This article studies on various models and methods of grey prediction and their applications in some special area such as medium term and long term power load prediction [1], stock prices [2]. Furthermore, this article give a detailed introduction of time series prediction [3]. Grey system based prediction is always the branch of time series prediction which composed of AR (Auto Regressive), MA (Moving Average),

© Springer International Publishing Switzerland 2014
Y. Chen et al. (Eds.): WAIM 2014, LNCS 8597, pp. 65–76, 2014.
DOI: 10.1007/978-3-319-11538-2_7

ARMA (Auto Regressive Moving Average), ARIMA (Auto Regressive Integrated Moving Average) and so on.

In this paper, we proposed a empirical study on real accidents data based online public opinion index prediction of GM (1, 1) model, promoted GM (2, 1) model and chaotic prediction model [9]. It can be used to predict online public opinion index with more accuracy. In detail, our contributions are as follows:

1. We proposed a detailed introduction of time series prediction;
2. We systematically collected more than 30 real important accidents happened in China and every accident's online public index data such as the number of original news and number of tweeted news.
3. We systematically analyzed GM (1, 1) model, GM (2, 1) model and chaotic prediction model, proposed a new GM (2, 1) model to predict the online public opinion index of the 30 real important accidents and found some good and interesting results.

The outline of the paper is as follows: Sect. 2 surveyed the related work of time series prediction and grey prediction method. In Sect. 3, we introduced our method. In Sect. 4, we proposed the effective GM (1, 1), GM (2, 1) model and chaotic prediction model based prediction method. Section 5 showed more solid experimental results and Sect. 6 gives a conclusion.

2   Related Work

2.1   Time Series Prediction

Grey system based prediction is always the branch of time series prediction. A time series is referring a collection of data points which are generally sampled equally in time intervals. Time series prediction refers to the process by which the future values of a system is forecasted based on the information obtained from the past and current data points. Generally, a pre-defined mathematical model is used to make accurate predictions. Time series prediction models are always used in many areas, such as predicting stock price indexes in financial area, predicting future passenger's number in airport and so on. The ability to do prediction with a reasonable accuracy can make the economic policy of large companies and governments more reasonable and also make a more accurate analysis of the developing trends of online public opinion from data in BBS or Micro Blog.

Techniques for time series prediction is always divided into two main types of statistical and artificial intelligence based approaches in the literature. While AR (Auto Regressive), MA (Moving Average), ARMA (Auto Regressive Moving Average), ARIMA (Auto Regressive Integrated Moving Average) and Box–Jenkins models [4] can be mentioned as statistical models, neural network (NN) based models [5–7] are widely used as an artificial intelligence-based approach, back propagation being the most widely used technique for updating the parameters of the model. However, not only are the statistical models not as accurate as the neural network-based approaches for nonlinear problems, they may be too complex to be used in predicting future values of a time series. One major disadvantage about the NN model is that it demands lot of deal of training data and relatively long training period for robust generalization [8].

Other intelligent approaches seen in the literature for the analysis of time series includes Linear regression, Kalman filtering [9], fuzzy systems [10], hidden markov models and the support vector machines [11]. Some hybrid models are also seen in the literature: such as a combination method of genetic algorithms and neural network by Versace [12], and in support vector regression (SVR) a self-organizing feature map (SOFM) technique have been hybridized to reduce the cost of training time and to improve prediction accuracies by Huang and Tsai [13].

2.2   Grey System Theory

Grey system theory is an interdisciplinary scientific area that was firstly introduced in early 1980s by Deng [14]. Since then, the grey system theory has become quite popular with its ability to deal with the systems that have partially unknown parameters. As superiority to conventional statistical models, grey models require only a limited amount of data to estimate the behavior of unknown systems [15].

During the last two decades, the grey system theory has been developed rapidly and caught the attention of many researchers. It has been widely and successfully applied to various systems such as social, economic, financial, scientific and technological, agricultural, industrial, transportation, mechanical, meteorological, ecological, hydro-logical, geological, medical, military, etc., systems.

Some research studies are as follows: In one study [16], Wang introduced a support vector regression grey model (SVRGM) which combines support vector regression (SVR) learning algorithm and grey system theory to obtain a better approach to time series prediction. Another study [17], Peng researched an improved gray prediction model on some GDP data in a city of east China. Kong [18] used dynamic grey prediction and support vector machine to make middle-term and long-term power load forecasting. In these studies and the others, it is seen that grey system theory-based approaches can achieve good performance characteristics when applied to real-time systems, since grey predictors adapt their parameters to new conditions as new outputs become available. Because of this reason, grey predictors are more robust with respect to noise and lack of modeling information when compared to conventional methods.

In grey systems theory, a system can be defined with a color that represents the amount of clear information about that system. For instance, a system can be called as a black box if its internal characteristics or mathematical equations that describe its dynamics are completely unknown. On the other hand if the description of the system is, completely known, it can be named as a white system. Similarly, a system that has both known and unknown information is defined as a grey system. In real life, every system can be considered as a grey system because there are always some uncertainties. Due to noise from both inside and outside of the system of our concern, the information we can reach about that system is always uncertain and limited in scope [19, 20].

There are many situations in which the difficulty of incomplete or insufficient information is faced. Even a simple motor control system always contains some grey characteristics due to the time-varying parameters of the system and the measurement difficulties. Similarly, it is difficult to forecast the electricity consumption of a region accurately because of the various kinds of social and economic factors. These factors are generally random and make it difficult to obtain an accurate model.

3 Grey System Model

3.1 Grey System Based Prediction

Grey system theory is one of the major methods for studying and solving uncertain problems of the present world, and modeling technologies of grey prediction is an important component of grey system theory. In the last three decade, it has obtained many encouraging research achievements. However, as an emerging cross-discipline, its theory system is still expected to be further enriched and perfected, based on this, the paper in-depth studied grey prediction modeling technologies from modeling thought innovation, modeling object expansion, modeling method improvement, modeling sequences optimization, etc.

Instead of forming a knowledge base, the gray model constructs some differential equations to characterize the controlled system behavior. By using a few past output data and solving the differential equations, the gray model can predict the future behavior of the system accurately. And there are two kinds of gray algorithm model: single variable and first-order linear dynamic gray model, named as GM (1, 1); single variable and second-order, named as GM (2, 1).

3.2 GM(1, 1) Model

Consider a time sequence $X^{(0)}$ that denotes the public opinion index of a accident. Let $X^{(0)}$ be the original data series, that is the original numbers of our time series. For example, it may be the original index data of some online public opinion accident.

$$X^{(0)} = \left\{ x^{(0)}(1), x^{(0)}(2), \cdots, x^{(0)}(n) \right\} \tag{1}$$

Where $x^{(0)}(k) \geq 0, k = 1, 2, \cdots, n$.

where $X^{(0)}$ is a non-negative sequence and n is the sample size of the data. When this sequence is subjected to the Accumulating Generation Operation (AGO), the following sequence $X^{(1)}$ is obtained. It is obvious that $X^{(1)}$ is monotonically increasing. Let $X^{(1)}$ be the generated data series for the destination function:

$$X^{(1)} = \left\{ x^{(1)}(1), x^{(1)}(2), x^{(1)}(3), \ldots, x^{(1)}(n) \right\} \tag{2}$$

Where $x^{(1)}(k) = \sum_{i=1}^{k} x^{(0)}(i), k = 1, 2, \cdots, n$.

Let $Z^{(1)}$ be the generated mean sequence of $X^{(1)}$.

$$Z^{(1)} = \left\{ z^{(1)}(1), z^{(1)}(2), \cdots, z^{(1)}(n) \right\} \tag{3}$$

where $z^{(1)}(k)$ is the mean value of adjacent data, i.e.

$$z^{(1)}(k) = 0.5x^{(1)}(k) + 0.5x^{(1)}(k-1), k = 2, 3, \ldots, n \tag{4}$$

The least square estimate sequence of the grey difference equation of GM(1, 1) is defined as follows:

$$x^{(0)}(k) + aZ^{(1)}(k) = b \tag{5}$$

The whitening equation is therefore, as follows:

$$\frac{dx^{(1)}(t)}{dt} + ax^{(1)}(t) = b \tag{6}$$

Here, a, b are the parameters unknown, need to be resolve and $[a, b]^T$ is a sequence of parameter that can be found as:

$$[a, b]^T = (B^T B)^{-1} B^T Y \tag{7}$$

Where

$$Y = \begin{bmatrix} x^{(0)}(2) \\ x^{(0)}(3) \\ \vdots \\ x^{(0)}(n) \end{bmatrix} \tag{8}$$

$$B = \begin{bmatrix} -z^{(1)}(2) & 1 \\ -z^{(1)}(3) & 1 \\ -z^{(1)}(4) & 1 \\ -z^{(1)}(5) & \end{bmatrix} \tag{9}$$

According to Eq. (6), the solution of $x^{(1)}(t)$ at time k is:

$$x_p^{(1)}(k+1) = \left(x^{(1)}(0) - \frac{b}{a} \right) e^{-ak} + \frac{b}{a} \tag{10}$$

To obtain the predicted value of the primitive data at time (k + 1), the IAGO is used to establish the following grey model.

$$x_p^{(0)}(k+1) = \left(x^{(1)}(0) - \frac{b}{a} \right) e^{-ak}(1 - e^a) \tag{11}$$

and the predicted value of the primitive data at time $(k + H)$:

$$x_p^{(0)}(k + H) = \left(x^{(1)}(0) - \frac{b}{a}\right)e^{-a(k+H-1)}(1 - e^a) \tag{12}$$

3.3   GM(2, 1) Model

Second order single variable differential equation of GM (2, 1) model is also called the Verhulst model was first introduced by a German biologist Pierre Franois Verhulst. The main purpose of Velhulst model is to limit the whole development for a real system and it is effective in describing some increasing processes. GM(2, 1) model can be defined as:

$$\frac{dx^{(1)}(t)}{dt} + ax^{(1)}(t) = b(x^{(1)}(t))^2 \tag{13}$$

Grey difference equation of Eq. (13) is:

$$x^{(0)}(k) + aZ^{(1)}(k) = b(Z^{(1)}(k))^2 \tag{14}$$

Similar to GM(1, 1) model

$$[a, b]^T = (B^T B)^{-1} B^T Y \tag{15}$$

Where

$$Y = \begin{bmatrix} x^{(0)}(2) \\ x^{(0)}(3) \\ \vdots \\ x^{(0)}(n) \end{bmatrix} \tag{16}$$

$$B = \begin{bmatrix} -z^{(1)}(2) & (z^{(1)}(2))^2 \\ -z^{(1)}(3) & (z^{(1)}(3))^2 \\ -z^{(1)}(4) & (z^{(1)}(4))^2 \\ -z^{(1)}(5) & (z^{(1)}(5))^2 \end{bmatrix} \tag{17}$$

The solution of $x^{(1)}(t)$ at time k is:

$$x_p^{(0)}(k + 1) = \frac{ax^{(0)}(1)}{bx^{(0)}(1) + (a - bx^{(0)}(1))e^{ak}} \tag{18}$$

$$x_p^{(0)}(k) = \frac{ax^{(0)}(1)(a - bx^{(0)}(1))}{(bx^{(0)}(1) + (a - bx^{(0)}(1))e^{a(k-1)}} * \frac{(1 - e^a)e^{a(k-2)}}{(bx^{(0)}(1) + (a - bx^{(0)}(1))e^{a(k-2)}} \tag{19}$$

In Eq. (19), if $a < 0$, then $\lim_{k \to \infty} x_p^{(1)}(k+1) \to \frac{a}{b}$. In the GM(2, 1) model used in this article to predict opinion index, the GM(2, 1) model has been promoted to fit index variable distribution function.

4 Empirical Route Map

4.1 Empirical Framework

We designed and implemented crawler which can attain BBS news and other online public opinion news. The empirical system framework can be found in Fig. 1 which composed of Classifier Module, Prediction Module, PredictWarner Module, Decision Module and Evaluate Module. In The Prediction Module and PredictWarner Module, GM(1, 1) model, GM(2, 1) model and chaotic prediction model are implemented.

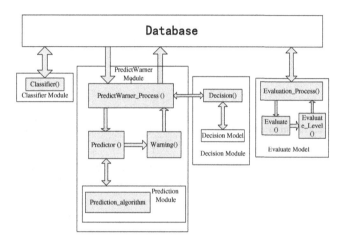

Fig. 1. Experiment system framework

4.2 Real Data of Some Accidents

There are some real data collected and refined by us. Here we defined three opinion index such as hot index, sensitive index and hot-sensitive index. Hot-sensitive index is a weighted index composed by hot index and sensitive index (Tables 1, 2, and 3).

5 Experiment Result

There is prediction result of one Accident from "EU Lift the embargo on China" and we can found the experiments results as follows. Furthermore, we can also found the result in Figs. 2, 3, and 4.

Table 1. Opinion data of accident "EU (European Union) Lift the embargo on China"

Accident: EU Lift the embargo on China			
Date	Hot Index	Sensitive Index	Hot-Sensitive Index
2011/1/10	17	10	11
2011/1/11	45	15	25
2011/1/12	40	15	23
2011/1/13	35	10	17
2011/1/14	41	30	33
2011/1/15	26	10	14
2011/1/16	26	20	21
2011/1/17	17	15	15
2011/1/18	26	25	24
2011/1/19	46	35	38
2011/1/20	41	20	26
2011/1/21	35	20	24
2011/1/22	21	10	13
2011/1/23	22	10	13
2011/1/24	41	20	26
2011/1/25	22	10	13

Table 2. Opinion data detail of accident "EU Lift the embargo on China"

Title	Author	Time	Link
欧盟再提解除对华军售禁令 2次解禁 滋味不同 症结在哪	News	2011/1/28	http://news.wuxi.cn/detail.asp?id=57&serial=795229
李克强会见西班牙首相萨帕特罗		2011/1/6	http://www.qjrb.cn/2011-01/06/content_212493.htm
2011年1月6日今日新闻午间版	马若冰	2011/1/6	http://news.ifeng.com/gundong/detail_2011_01/06/4037713_0.shtml
首页		2011/1/6	http://315online.com/news/qita/100556.html
李克强会见萨帕特罗		2011/1/6	http://www.jfdaily.com/a/1854120.htm
中国和西班牙将签署75亿美元合同!		2011/1/6	http://www.chnqiang.com/article/2011/0106/mil_35879.shtml
西班牙感谢中国支持 称尽快推动军售解禁		2011/1/6	http://bbs.tiexue.net/post_4786271_1.html
西班牙感谢中国支持 称尽快推动军售解禁		2011/1/6	http://bbs.tiexue.net/post2_4786271_1.html

Table 3. Opinion data of some important accident list from over 30 accidents

Accident Name	Time Scope
Riot in XinTang of ZengCheng in GuangZhou (广州增城新塘暴动)	2011.6.10-2011.7.1
Police help urban management in forced demolition (警察协助城管强拆)	2011.7.17-2011.8.1
Two years later after "2009.7.5" in XinJiang (新疆"7.5"两周年)	2011.6.5-2011.7.10
Beating abduct and trafficking people by MicroBlog (微博打拐)	2011.3.20-2011.5.20
Hujintao visit USA (胡锦涛访美)	2011.1.10-2011.4.24
EU Lift the embargo on China (欧盟解除对华武器禁运)	2011.1.10-2011.1.25
Jasmine flower revolution (茉莉花革命)	2011.2.1-2011.3.25

Table 4. Hot Index experiment result of accident "EU Lift the embargo on China"

Index	Data Type	Training data (2011.11-11.17) Evaluation data (2011.18-11.27)										Evaluation				
		18	19	20	21	22	23	24	25	26	27	u	σ	CC	TC	AC
Hot Index	Real Data	26	17	26	46	41	35	21	22	41	22	29.7	9.65			
	GM(1,1)	27	24	19	21	71	59	42	20	17	56	35.6	18.83	0.12	0.29	**High**
	GM(2,1)	27	24	19	45	100	75	42	1	42	56	43.1	27.21	0.55	0.33	Normal
	Chaotic prediction	26	21	21	29	42	42	38	20	24	40	30.3	8.75	0.24	0.18	**High**

• CC = Correlation Coefficient to Real data; TC = Tailor Coefficient to Real data; AC = Accuracy

Table 5. Sensitive Index experiment result of accident "EU Lift the embargo on China"

Index	Data Type	Training data (2011.11-11.17) Evaluation data (2011.18-11.27)										Evaluation				
		18	19	20	21	22	23	24	25	26	27	u	σ	CC	TC	AC
Sensitive Index	Real Data	20	15	25	35	20	20	10	10	20	10	18.5	7.43			
	GM(1,1)	3	35	19	26	43	22	12	7	6	27	20.0	12.50	0.16	0.31	Normal
	GM(2,1)	3	35	19	24	42	22	12	7	6	27	19.7	12.23	0.13	0.31	Normal
	Chaotic prediction	10	15	15	21	25	16	18	10	10	20	16.0	4.86	0.21	0.23	**High**

Table 6. Hot-Sensitive Index experiment result of accident "EU Lift the embargo on China"

Index	Data Type	Training data (2011.11-11.17) Evaluation data (2011.18-11.27)										Evaluation				
		18	19	20	21	22	23	24	25	26	27	u	σ	CC	TC	AC
Hot-Sensitive Index	Real Data	21	15	24	38	26	24	13	13	26	13	21.3	7.64			
	GM(1,1)	6	30	17	25	45	31	25	11	8	36	23.4	12.11	0.01	0.29	**High**
	GM(2,1)	6	30	17	21	90	44	37	11	8	36	30.0	23.56	0.05	0.43	Normal
	Chaotic prediction	14	24	15	23	27	24	24	13	12	25	20.1	5.52	0.01	0.22	**High**

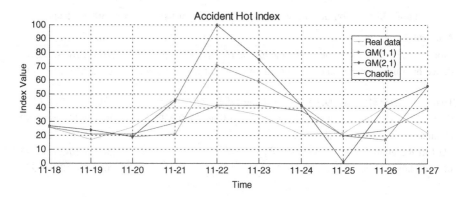

Fig. 2. Hot Index predict result by GM(1, 1), GM(2, 1), Chaotic

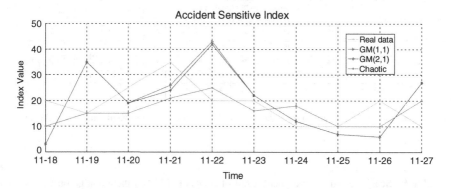

Fig. 3. Sensitive Index predict result by GM(1, 1), GM(2, 1), Chaotic

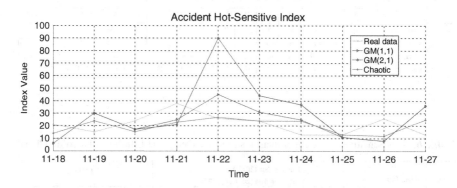

Fig. 4. Hot-Sensitive Index predict result by GM(1, 1), GM(2, 1), Chaotic

By analyzing Tables 4, 5, and 6 and Figs. 2, 3, and 4, we can find that the three mentioned predict algorithm GM (1, 1),GM(2, 1) and chaotic prediction model can predict the good result of public opinion index. We can summary that in Figs. 4, 5, and 6:

(1) Mean value u: the mean value u generated GM(1, 1),GM(2, 1) and chaotic prediction model are close to the mean value of real data while the mean value u of GM(1, 1) and chaotic prediction model is more closer.

(2) Standard deviation σ: from Tables 4, 5, and 6, the standard deviation σ of chaotic prediction model is the smaller than GM(1, 1) and GM(2, 1) while σ of GM(1, 1) is smaller than GM(2, 1). And the σ values of GM(1, 1) and GM(2, 1) are bigger than σ of real data. On this hand, chaotic prediction model does the best.

(3) Tailor Coefficient between predict data and Real data: Tailor Coefficient is a Coefficient among 0 and 1. When value of Tailor Coefficient equals 0 which means there is more predict accuracy than Tailor Coefficient value of 1. From the experimental results in Tables 4, 5, and 6, it can be found that the Tailor Coefficient values is in (0.18, 0.43) which means that the three method does not reached high accuracy but reach a acceptable accuracy Tailor Coefficient. The Tailor Coefficient value scope of chaotic prediction model is (0.18, 0.23) while (0.29, 0.31) for GM(1, 1) and (0.31, 0.43) for GM(2, 1). And the chaotic prediction model is also the best.

(4) Chaotic prediction model has some better performance on data of accident "Lift the embargo on China", but GM(1, 1) and GM(2, 1) has better performance on other accidents of attention continuous growing.

6 Summary

In this paper, we proposed a empirical study on real accidents data based online public opinion index prediction of GM (1, 1) model, GM (2, 1) model and chaotic prediction model. It can be used to predict online public opinion index with more accuracy. And we systematically collected more than 30 real important accidents happened in China and every accident's online public index data and has draw a new light to apply grey prediction method to online public opinion index prediction.

Acknowledgement. Thanks to the support of National Natural Science Foundation of China (NNSF) (Grants No. 90924029), National Culture Support Foundation Project of China (2013BAH43F01), and National 973 Program Foundation Project of China (2013CB329600), (2013CB329606).

References

1. da peng, W.: Research on grey prediction models and their applications in medium-and long-term power load forecasting. Doctor thesis, Huazhong University of Science and Technology (2013)
2. Wang, Y.F.: Predicting stock price using fuzzy grey prediction system. Expert Syst. Appl. **22**, 33–39 (2002)
3. Kayacan, E., Ulutas, B., Kaynak, O.: Grey system theory-based models in time series prediction. Expert Syst. Appl. **37**(2), 1784–1789 (2010)

4. Box, G.E.P., Jenkins, G.M.: Time Series Analysis: Forecasting and Control. Holden Day, San Francisco (1976)
5. Quah, T.S., Srinivasan, B.: Improving returns on stock investment through neural network selection. Expert Syst. Appl. **17**, 295–301 (1999)
6. Rabiner, L.R.: A tutorial on hidden markov models and selected applications in speech recognition. Proc. IEEE **77**, 257–286 (1989)
7. Roman, J., Jameel, A.: Back propagation and recurrent neural networks in financial analysis of multiple stock market returns. In: Proceedings of IEEE System Sciences Proceedings of the 29th Hawaii International Conference Hawaii, USA, vol. 2, pp. 454–460 (1996)
8. Jo, T.C.: The effect of virtual term generation on the neural based approaches to time series prediction. In: Proceedings of the IEEE Fourth Conference on Control and Automation, Montreal, Canada, vol. 3, pp. 516–520 (2003)
9. Ma, J., Teng, J.F.: Predict chaotic time-series using unscented Kalman filter. In: Proceedings of the Third International Conference on Machine Learning and Cybernetics, Shanghai, China, vol. 1, pp. 867–890 (2004)
10. Kandel, A.: Fuzzy Expert Systems. CRC Press, Florida (1991)
11. Cao, L.: Support vector machines experts for time series forecasting. Neuro Comput. **51**, 321–339 (2003)
12. Versace, M., Bhatt, R., Hinds, O., Shiffer, M.: Predicting the exchange traded fund DIA with a combination of genetic algorithms and neural networks. Expert Syst. Appl. **27**, 417–425 (2004)
13. Huang, C.L., Tsai, C.Y.: A hybrid SOFM–SVR with a filter-based feature selection for stock market forecasting. Expert Syst. Appl. **36**, 1529–1539 (2009)
14. Deng, J.: Control problems of grey system. Syst. Control Lett. **1**, 288–294 (1982)
15. Deng, J.L.: Introduction to grey system theory. J. Grey Syst. **1**, 1–24 (1989)
16. Chang, B.R., Tsai, H.F.: Forecast approach using neural network adaptation to support vector regression grey model and generalized auto-regressive conditional heteroscedasticity. Expert Syst. Appl. **34**, 925–934 (2008)
17. Peng, Y.: An Improved gray prediction model. Comput. Digit. Eng. **40**(1), 40–42 (2012)
18. Kong, F., Song, G.: Middle-long power load forecasting based on dynamic grey prediction and support vector machine. Int. J. Adv. Comput. Technol. **4**(5), 148–156 (2012)
19. Lin, Y., Liu, S.: A historical introduction to grey systems theory. In: Proceedings of IEEE International Conference on Systems, Man and Cybernetics, The Netherlands, vol. 1, pp. 2403–2408 (2004)
20. Liu, S.F., Lin, Y.: An Introduction to Grey Systems. IIGSS Academic Publisher, Grove City (1998)
21. Wu, A., Tian, Y., Song, Y., et al.: Application of the grey system theory for predicting the amount of mine gas emission in coal mine. J. China Coal Soc. **30**, 589–592 (2005)

Public Opinion Analysis and Crisis Response in Mass Incidents: A Case Study of a Flight Delay Event in China

Yi Liu[⊠], Yefeng Ma, Qing Deng, Yi Liu, and Hui Zhang

Institute of Public Safety Research, Beijing 100084, China
liuyi789@tsinghua.edu.cn

Abstract. Mass incidents caused by transportation malfunction generally refers to the course of conflicts which caused by abnormal weather or meteorological reasons in travel services. Because of the large scale and concentration of the group, and also the consistency of group's interest, such event always vulnerable to transformed into conflict or arising to a crisis, impact on social stability greatly. This article will take the mass incidents of passengers in Kunming Airport at 2013.1.3–2013.1.5 as an example. The social public feelings in the incidents will be analyzed, the reasons of the incidents will be given, and the decision making methods will be provided to help making decisions.

Keywords: Mass incident · Public opinion · Decision-making · Case study · Transportation malfunction

1 Introduction

Recent years, with the great increase of the amount of passengers travel, the mass incidents which caused by the delay of the transportation malfunction increased rapidly. The famous events, such as the riot of stranded passengers resulting from the freezing rain and snow weather in the Moscow Airport in December 2010, the large area stranded passengers caused by the American "Katrina" hurricane in August 2011, and, the group abusive and conflict incident in the Chengdu Shuangliu Airport in china in April 2012 are all the typical mass incidents caused by passengers stranded. The inevitability of this kind of event the weather changes and transportation malfunction leads to the occasional happening of the passenger conflict, and also put forward requirements to address such events.

But how to resolve such incidents scenically? the attention of experts and scholars from various fields, such as political science, sociology, law, management, psychology has given different conclusions. Xu Hong analyzed the incentives and status quo of our flight delays disputes [1]; Thomas and Todd studied the legal compensation of the passenger conflicts from the perspective of social legislation [2, 3]; Yang Lei discussed the disposal strategy of the flight delay incidents [4]; Zhao Bin analyzed the causes and countermeasures of the flight delay incidents from the perspective of social combustion theory [5].

Different fields have different research perspectives and concerns. But overall, they are generally refer the mass incidents caused by passenger's conflict as the group social

© Springer International Publishing Switzerland 2014
Y. Chen et al. (Eds.): WAIM 2014, LNCS 8597, pp. 77–86, 2014.
DOI: 10.1007/978-3-319-11538-2_8

incidents and considers the problem with the characteristic of group social conflict. These proposal method has analyzes the social causes of the passenger's conflict from the social level, But pays little attention on the micro level and ignores the decision making of disposal of the traffic. So, in this paper, we will try to solve the mass incident caused by transportation malfunction from the micro level of specific events to explore the practical measures for these mass incidents. And a new method big data- public opinion will be used and the form of case studies will be taken to explain how to make decision in mass incidents caused by transportation malfunction. The case will study we select the event which taken place on Jan. 1st–5th, 2013, in Kunming Changshui international airport as an example.

2 Case Introduction and Analysis

2.1 Case Introduction

On Jan 3rd, 2013, the last day of the New Year holiday, the Kunming Airport which had been put into use for just six months ushered the return of a large number of holiday travelers. And at this time, a thick fog enveloped the new airport and due to the visibility of the blind guide system, Passengers were stranded at the airport. Until 17:30 on Jan 3rd, flights had delayed accumulated 130 sorties. And at 8:30 pm, there were more than 7,500 passengers stranded in the airport.

The fog dissipated on the early morning of Jan 4th and the airport began to evacuate the stranded passengers on Jan 4th. After adding flights on Jan 4th, the flight plan was up to 940 sorties throughout the day while the capability was only 700 sorties.

Because of the transportation malfunction, during the waiting time, in the airport, the information display screen could not display correctly and hot water and food could not be supplied effectively. The passengers were so anxious and furious that they attacked staffs, demonstrated and occupied counters. Long waits and anxiously feeling led to conflicts.

The incident finished at 18:06 on Jan 5th with the last adding flight leaving the airport.

2.2 The Analysis of the Event Process and Decision-Making Methods

From the analysis of this case, the event in Kunming Airport is a typical mass incident caused by passengers' delay. Generally, in mass incident, the outbreak of the conflict derives from the inconsistent of interests is often the key node in the process of problem aggravated. So, in this case, to find the factors affecting the contradictory which like the effort of the technology, the passengers' demands and the modes of the intervention are all very important to need to be taken into consideration to resolve the conflict.

(1) The Analysis of Passengers' Public Opinion. The outbreak of mass incidents often results from the intensification of the social psychology conflicts. The analysis of the passengers' psychological is a key step to resolve the conflicts. The social network

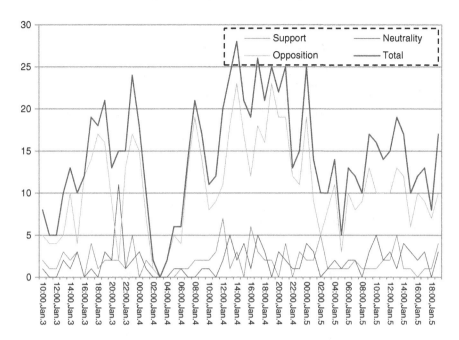

Fig. 1. The statistics chart of the public opinion from in the Kunming Airport incident.

under the background of big data provides a platform for the group expressing and also an access to obtaining the group demands.

In this paper, We choose the Sina weibo- an online social media under big data for the target website to collect the microblogs related the Kunming Airport incident from Jan 3rd to 5th, 2013 (a total of 36,500). The analysis of passengers 'public opinion shows in Fig. 1.

In this case, the negative emotion (opposition) has an absolute proportion and different small peaks appear with the development of the conflict and related intervention from emergency departments. And from analysis of content of public opinion, The keywords of "flights", "paralysis", "stranding", "delay", "flying off", "service", "cold", "eating and drinking" are the proportion of 19.4 %, 19.2 %, 18.2 %, 13.8 %, 5.1 %, 4.4 %, 3.3 % and 2.1 % in all microblogs, respectively. (Table 1 shows the frequency of several keywords).

If the content of public opinion which is the expression of weibo can be think of the group psychology, we can know that the demands of public in this event mainly focus on the airport operation capacity and the service capacity during waiting for the flights. (the "flights", "paralysis", "stranding", "delay", "flying off", "service", "cold", "eating and drinking "are all demands for the capacity of the airport) And of course, because these demands are not satisfied, the passengers' satisfaction reduces, and Conflict going broke.as shown in Fig. 2.

(2) How to Making Decisions? In the mass incident, how to find the factors which related to the outbreak of conflict is of great importance to solve the problem. Taking both the analysis results above and the case study into account, in this event, the airport

Table 1. The word frequency of some keywords

Keywords	Word frequency
Flights	7069
Paralysis	6993
Stranding	6626
Details	5688
Fog	5386
Delay	5047
Flying off	1874

Fig. 2. The relationship between the airport service and the passengers' satisfaction

bad service capacity, the passengers' anxiety and the resonance effects of large-scale groups, are all the important reasons to intensify the friction and escalate the conflicts. For convenience of description, we use a complex system which composite of the factors that influence the process of conflict in the airport to research. The complex system as this:

In the complex operation system of the Kunming Airport, there is an interaction between the airport disposal capacity (the airport capacity, service capability, emergency response capability et al.) and the passengers' demands (flight demands, psychological and emotional satisfaction et al.). The relative balance existing between the group needs and the airport operation ability. In normal state, the airport can operate normally; the balance is broken when in non-normal state. And in this process how to make decisions to maintain or restore the balance is a must for the relevant emergency departments, including airports, airlines, Civil Aviation Authority, and other government emergency departments, to calm the event. So in this paper, we will analyze the different decision-making methods in different process like occurrence and evolution process of this Kunming Airport conflict incident from complex systems perspective.

① The Incident Eruption Stage. In this event, the inclement weather occurring on Jan 3rd and the fog outside the airport ILS system level (the fog visibility 300 m, and the e visibility range of the blind guide system is 800 m), so the operation ability of the Kunming Airport cannot meet the flight needs of passengers and result to the gathering of the people. As shown in Fig. 3. The aircraft landing rate is mainly affected by airport hardware technical level, the airport management and the external environment. The objective factors of the fog make the aircraft landing rate reduction, which reduces the airport operation capacity. And in the other hand, there is little change in the passengers entering rate. Then the airport capacity cannot effectively match the passengers' demands for flying, so the number of passengers stranded is more and more.

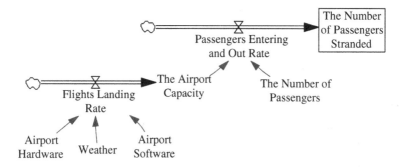

Fig. 3. The airport operation capacity chart in the Kunming Airport mass incident

In this stage, early warning, improving airport capacity and reducing the number of passengers are all effective measures to control the accumulation of passengers. The airport can improve the technology and management level to raise the operation efficiency from the long-term. But in the emergency situations, the limited speed of enhancing the hardware, software and management level determines the great difficulty to recover airport operation instantly. So in the case of not reducing airport profits and based on the complete contingency plan, there are some relatively reasonable measures to evacuate the stranded passengers as soon as possible, for instance, the real-time weather warning and the various needs preparation for long flight delays before the emergency occurring, and accurate estimating of the number of passengers stranded, offering real-time information to the passengers to switch other transportation as early as possible.

② The Conflicts Escalation Stage. On Jan 3rd and 4th, the operation capacity of the airport cannot match the passengers' demands for flying, and this is the direct cause of the outbreak of the conflicts. As shown in Fig. 4, when the passengers arriving the airport, they have not estimate sufficiently for their waiting time, and also in their long waiting time, Life necessities such as hot water, lunches are short of supply, the passengers' satisfaction reduces. When the angry passengers complain to protect their rights, their request is not met effectively, so they begin to attack and lead to the mass incident. And so, there even has a vicious circle in the system: Passengers attacking staffs result in the airport staff strike, which makes the whole airport service system more paralyzed, the airport more confusing and the contradiction more intensified.

In this stage, adjusting disposal strategy from the demands of the stranded passengers are the rational and effective methods to resolve the conflicts. Specifically, the airport should pay more attention on the needs of stranded passengers early when conflicts happen, then take the corresponding measures to ease the conflicts. From the long-term, enhancing the emergency service level and ready for all kinds of basic work, such as staff training and adequate infrastructure ensuring, are of great importance to improve the passengers' satisfaction and prevent such incidents.

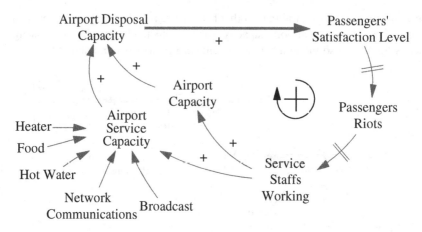

Fig. 4. The conflict intensified chart in the Kunming Airport mass incident

(3) The Whole Process of Event. In whole, this incident is a process from relative balance to the balance broken and then balance recovering. Such as when in Jan 3 the Kunming Airport is in the equilibrium state before the fog appearing, and the operation capacity can meet the basic needs of the passengers. When the fog appearing, the flights delaying caused large passengers gathering, and the operation & service capacity cannot satisfy the passengers' needs, the system broken into non-equilibrium state. And in this non-normal state, there is also an interaction between the service capacity and the passengers' satisfaction: the bad service makes passengers dissatisfactory, and the angry passengers makes the servicers dissatisfactory and service capabilities of the airport even worse, this vicious circle continues lead to the Intensification of the conflict. The balance recovers and the situation gets better when the intervention of the relevant departments and external injection after the fog dissipating improve the operation and service capability to satisfy the passengers' needs, In other words, conflict resolution until the system restore balance, as shown as Fig. 5.

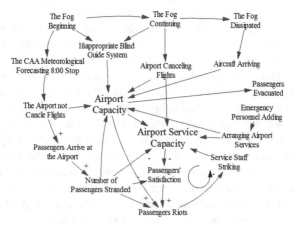

Fig. 5. The whole process analysis of the Kunming Airport mass incident

(4) The Incident Causes. The group conflict reflects the group interest contradictory directly. According to the case study, we conclude the factors affecting the incident as these five aspects: objective factors, airport facilities, response measures, group psychology and social environment, as shown in Fig. 6.

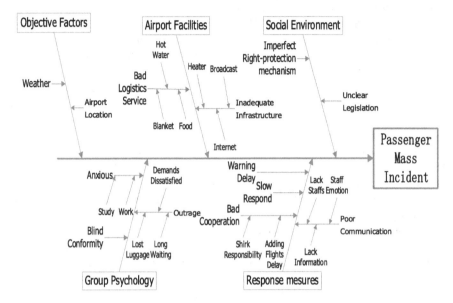

Fig. 6. The cause and effect diagram of the Kunming Airport incident

① Objective Factors. The objective factors include elements such as the weather, the airport location. Different from other modes of transportation, the air transport is more susceptible to objective factors, especially the weather. The bad weather is often the initial factor and the irresistible external factor causing the flight delay incidents. In the Kunming Airport incident, the objective factors are the initial causes of the group social conflicts.

② Airport Facilities. The airport facilities include the hardware, software and various logistics service of the airport. When large scale passengers gather, the service pressure and the lack of infrastructure intensify the contradiction between the airport and passengers and accelerate the deterioration of the incident. In this incident, about ten thousands passengers have been stranded for more than two days shorting of food, hot water, and heat preservation. And when the passengers request these services, they do not get them timely and so the service deterioration of the airport and airlines results in the passengers' dissatisfaction and the passengers protect their rights to cause the conflicts erupt.

③ The Response Measures. The condition of the Yunnan Provincial Meteorological Observatory releasing a fog warning is three or more regions appear visibility less than 500 m concurrently, so in the Kunming Airport incident, neither the airport nor the meteorological department makes weather warning decisions disposal. And even, after the fog appears to strand flights, the Kunming Airport does not make a full estimate of the time that the flight delays, neither communicate timely with the passengers according to the CAA meteorological judgment that the fog will dissipate soon. In the process of passengers waiting in the airport, the airport and airlines do not provide good arrangements for the anxious passengers gathered. After the flight landing rate reduces, the deficiency in early warning, emergency preparedness, timely emergency response intensifies the contradiction and leads to conflicts.

④ The Group Social Psychology. The social psychology, which is the intermediary between the social environment and social behavior, have a mental role in dominating and guiding the social behavior. The outbreak of the mass conflict incidents is affected by many factors, but in essence, the psychological imbalance is the most important motivation. In the conflicts caused by traffic, with the advancement of social democracy, passenger right-protection awareness has increased gradually and it is a normal phenomenon that the passengers protect their own interests through various channels, especially under the condition of flights delay. But in this incident, the emergency response of the airport and airlines cannot match the psychological needs of the group effectively. The passengers' request is not resolved reasonably and their rights to know and option are ignored to intensify the contradictions.

⑤ The Social Environment. The social environmental is mainly the legal environment for flight delay. There is no clear legislation on the time limits and compensation standards of flight delays, and the airlines and airports do not have a clear mechanism for rapid processing delays. Then coupling with the social environment is very easy to cause the expansion of passengers' dissatisfaction and the outbreak of group crisis.

Regarding the outbreak of the conflict as the result of the case, we include that the objective factors are predisposing factors to the incident, and airport facilities, emergency response and group psychology under the social environment are the key triggers of the contradictory. These factors are the causes of the incident, but also the disposal breakthrough point of passenger mass incident. From the decision management perspective, we can summarize some specific aspects to improve, such as the technical, service facilities, management, and social environment:

- The technical elements: The technical level of hardware facilities, such as the airports ILS systems, the broadcasting systems, etc.
- The service facilities: The effective delivery of the emergency supplies, such as food, water, heating, Internet and other items.
- The management elements: the early forecasting and warning, command decisions, emergency response, coordination of joint of relevant departments, the quality of airport staff, the completeness of airport contingency plan, etc.
- The social environment: The activist channels of impartiality and integrity, the social environment of fairness and justice, etc.

3   The Case Revelation

The mass incident caused by transportation malfunction often derived from the many factors. From the case study we know that, as a class of mass incident, the factors such as the accumulation of passenger size, the concentration of the positions and easy sharpening of the emotions are all the important causes to the conflict and breakout the mass incident in passengers. The case of the Kunming Airport is a typical traveler conflict and the analysis of the reason, process and disposal of this mass incident also have certain typical significance for our scientific decision in the similar incidents:

(1) Paying More Attention on Public Opinion. The outbreak of the conflict often results from the accumulation of passengers complaining and discontent. Flying delayed because of transportation malfunction, they often become anxious. And meanwhile, the large scale with complex composition are sentimental to perform the aggressive behavior and leads to conflict. So the psychology and emotion of the passengers are most important for the emergency response. Strengthening communication with the travelers to counsel their psychologies and public information can prevent the further expansion of the incident [6].

(2) Combination of Economic and Scientific. The outbreak of the group crisis is the result of a variety of subtle contradictions superimposed. Such as the abnormal weather often causes the transport delay to make passenger anxiety. The special features determine the mass incidents' impact on the social order within a short period. So the social prevention and preparedness must advance to the prevention, and focus on scientific decision-making in daily work to prevent any possibilities which can intensify the contradictions.

(3) Lean Management. The cumulative effect and hysteretic nature of the mass incident make the slow information release and emergency disposal. So the real-time evaluation is of great importance on accurately grasping the situation and resolve the contradiction at the beginning of the incident. The lean management requires forming the high-quality and lean disposal chain of whole process including the trigger process, evolution process and the emergency proposal of the incident. With the well preparedness of the prediction and warning from the root, real-time situation assessment and full forecasting the consequence, we can prevent the occurrence of the incident.

(4) The Organic Linkage of Various Departments. The mass incident often involves multiple participating subjects, so it needs the relevant departments collaborate and work together to evacuate the stranded passengers. Such as the subjects of the Kunming Airport incident, there are the CAA, airport, airlines, passengers, local emergency management departments, etc. when the fog appearing, the airport not counseling the flights according to the weather forecast from the CAA and when the passenger gathered, the different departments do not cooperate with each other and making the event very slowly to resolve. So the unified and coordinated linkage mechanism plays an important role in avoiding the improper disposal resulting from the lack information, poor communication and bad cooperation.

Acknowledgments. The authors appreciate the Project supported by National Natural Science Foundation of China (NO. 91024032; 91224008) and The National Science Foundation for Postdoctoral Scientists of China (NO.2013M540977)

References

1. Xu, H.: On Disputes Arising from Flight Delay: Present Situation, Causes and Corresponding Solutions. Law School of Hunan University, Changsha (2007)
2. Dickerson, T.A.: Flight Delays: The Airline Passenger's Rights & Remedies (2001). www.classactionlitigation.com (01 March 2001)
3. Curtis, T.: Compensation for Flight Delays and Overbookings (2002). www.airsafe.com (04 January 2002)
4. Yang, L.: Disposal Strategy for Mass Incidents Caused by Flights Delayed. International Relations and Public Affairs. Shanghai Jiaotong University, Shanghai (2010)
5. Zhao, B.: The Study of the Mass Incidents Caused by Flight Delays based on Theoretical Framework of Social Combustion Theory. Administration Institute of Jilin University, Jilin (2012)
6. Ma, B.: The principles of disclosure and true information in the emergency response. J. Law Econ. 284

An Effective Method to Detect People in Danger

Hongtai Zhang[1,2(✉)], Kuien Liu[1], Xiuli Wang[3], Limin Guo[1], and Zhiming Ding[1]

[1] Institute of Software, Chinese Academy of Sciences, Beijing 100190, China
{hongtai,limin}@nfs.iscas.ac.cn,
{kuien,zhiming}@iscas.ac.cn
[2] University of Chinese Academy of Sciences, Beijing 100049, China
[3] Central University of Finance and Economics, Beijing 100081, China
xlwang.cufe@gmail.com

Abstract. Instant disaster warning and accurate position information delivery are very helpful to reduce casualties and property losses in emergency management. Taking into account the fact that people and areas affected by disaster are always dynamically evolving, e.g., geo-areas suffered severe fire or infection are spreading and affected persons are increasing and moving. This paper proposes an effective method to support the location based query of PID (People in Danger), which can be used in guidance systems for emergency evacuation and surveillance. This method can deal with various kinds of moving objects like points and polygons that are widely used to express affected objects of disaster. The core of this method is to efficiently manage the dynamically changing status of all moving objects with the PIDA (People in Danger Algorithm) and accurately respond to various PID queries. The proposed algorithm consists of two major components: the TPR^H-tree, which indexes all moving objects with constant or changing shapes in a unique structure, and the optimized position judgment algorithm, which is used to rapidly determine the location relationship between people in danger. In one hand, taking advantage of effective data processing, this method can efficiently handle various PID queries in emergency cases. In the other hand, it supports mutual location query and evacuation route recommendation for moving objects. The experimental results demonstrate that the feasibility of proposed algorithms with a wide range of use.

Keywords: Emergency management · People in danger · Spatio-temporal data

1 Introduction

In the emergency management for public incidents, it is crucial to locate the affected crowd people and make evacuation announcement in disasters such as earthquakes, fires and harmful gas leak [1–3]. Disasters especially in the urban area, lead to congestion and personnel confusion, causing severe difficulties for disaster relief and evacuation [4]. We define such information service request as PID (People in Danger) query, which benefits a person by obtaining the accurate and instant positions of the disaster as well as the geometric relationships between the dangerous areas and this

© Springer International Publishing Switzerland 2014
Y. Chen et al. (Eds.): WAIM 2014, LNCS 8597, pp. 87–97, 2014.
DOI: 10.1007/978-3-319-11538-2_9

person's current positions or even future movement by predication. For example, when harmful gas leak happens and polluted gases are drifting, people would like to be notified with prompt positions and advices to escape [5]. The rapid development of mobile devices and sensor networks makes it possible to quickly locate and confirm the relationship between people and dynamic disaster areas.

We use the harmful gas leak case to explain in detail how to determine the location and make prompt warning in disaster situations effectively. At time t_0 in Fig. 1(a), the harmful gas cluster G_1 moves right and its shape is constantly changing with the wind. The person a is outside of the gas G_1, and consequently he is safe. At the same time, another harmful gas cluster G_2 is also moving with different direction. The two clusters keep changing shapes and, a moment later at t_1, mixed with each other. As shown in Fig. 1(b), the mixture may contain holes and irregular edges. While person b chooses the right moving direction, he easily gets rid of the harmful gases.

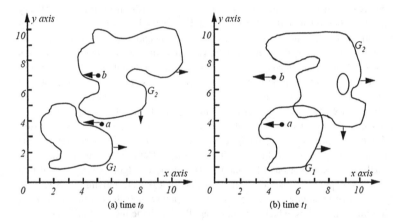

Fig. 1. An example of PID problem (harmful gas leak)

The main contributions of this paper are as follows. We proposed an algorithm PIDA (People in Danger Algorithm) which can detect the position relationship between people and emergency region dynamically. This method contains moving object index and position judgment algorithm. To handle the sudden change and frequently updated data in emergency scene, we propose a novel and efficient index TPR^H-tree (Hybrid Time Parameterized R-Tree) that is capable of indexing the moving point and polygon dynamically. The position judgment algorithm can determine the position relationship between people and dangerous regions quickly.

The rest of the paper is organized as follows. Section 2 discusses the related work. Section 3 gives the problem statement. The proposed algorithm is described in Sect. 4. The results of experiments and the analysis are presented in Sect. 5. Finally, conclusions and future work are summarized.

2   Related Work

2.1   Geometrical Location Problem

The similar situation of detecting static points and polygons' position is Geometrical Location problem [8] in GIS system and Computational Geometry. One of the famous problems is the Point-in-Polygon problem which is to find whether a given point in the plane lies inside, outside, or on the boundary of a polygon. This problem finds applications in areas of geometrical data processing, such as computer graphics, computer vision, geographical information systems (GIS), and CAD [9].

One definition of whether a point is inside of a region is the Jordan Curve Theorem. Essentially, it says that a point is inside of a polygon, if for any ray from this point, there exists an odd number of crossings points with the polygon's edges. The algorithms presented so far for this general problem have been O(n); the order of the problem is related to the number of edges. Preparata and Shamos (Preparata 1985) [10] presented a fascinating array of solutions which were theoretically faster. However, these algorithms have various limitations and tend to bog down when actually implemented due to expensive operations [11–13].

One method to speed up the algorithm is to classify the edges by their Y components and then test only those edges with intersecting the test point's $X+$ test ray [14]. The bounding box surrounding the polygon is split into a number of horizontal bins and the parts of the edges in a bin are kept in a list, sorted by the minimum X component.

2.2   Spatio-Temporal Index

Spatio-temporal index is an effective way to speed up the access to moving objects. R-tree is a well-known data structure used for spatial access methods [6]. The key idea of the data structure is to group nearby objects and represent them with their minimum bounding rectangle in the next higher level of the tree. It's also a balanced search tree, i.e. all leaf nodes are at the same height, which organizes the data in pages, and is designed for storage on disk (as used in databases). However, R-trees do not guarantee good worst-case performance and lack the support for moving objects. Satltenis et al. [7, 8] proposed the TPR-tree which is based on R-tree. TPR-tree has been widely used for the continuous moving objects. It is a kind of balanced tree derived from the R-tree which stores the MBRs as functions of time to index the continuous moving object and compute the future "MBR" as: $MBR(t) = MBR(t_0) + V(t)$.

Moreover, it's a great challenge to update the locations of moving object so TPR-tree uses a function of time $f(t)$ to store the instant location and update the database only when velocity changes [16].

3   Problem Statement

This section defines the moving object, dynamic dangerous region and spatial predicate (**INSIDE** and **WITHIN n**) in PID problem.

Definition 1 (Moving Object). The moving object MO is defined as:

$$MO = (id, t_i, long_i, lat_i) \quad (i \geq 1) \tag{1}$$

where moving object is organized as the trajectory of moving point at different time t_i, everyone can be identified by the unique id. The $long_i$ and lat_i denote the longitude and the latitude of the point.

Definition 2 (Dynamic Dangerous Region). The dynamic dangerous region R is defined as:

$$R = (id, t_i, seq_i) \quad (i \geq 1) \tag{2}$$

The region R can be any polygon (with or without interior rings) that is made up of sampling points sequence $seq_i = <long_1, lat_1>, <long_2, lat_2>, \ldots, <long_n, lat_n>$ (n is the number of sensors). Each R owns the unique id and sampling time t_i.

Definition 3 (INSIDE). The spatial predicate **INSIDE** is defined as:

$$INSIDE(MO, R, t_i) \in \{true, flase\} \quad (i \geq 1) \tag{3}$$

The predicate returns TRUE if a point is closed by the region. Notice that the point can be associated with one or more regions (i.e., two overlapping regions can contain the same point). A point can be **INSIDE** any regions with a timestamp less than the point's t where only the latest position of each region (up to that time) is considered.

Definition 4 (WITHIN n). The spatial predicate **WITHIN n** is defined as:

$$WITHIN(MO, R, t_i, n) \in \{true, flase\} \quad (i \geq 1) \tag{4}$$

The predicate returns TRUE if the point is at less than n units distance from the region. Notice that each point may be associated with one or more of the regions, as the same point can be within n units distance from several regions.

Definition 5 (PID). The PID query is defined as:

$$PID(MO, R, t_i) \in \{INSIDE, WITHIN\,n\}(i \geq 1) \tag{5}$$

The query gets the MO's position from $long_i$ and lat_i, then judges the spatial position relationship with the dangerous region R from the sampling data. PID return **INSIDE or WITHIN n** predicate to the front end application and they can define different kinds of services for person in emergency management.

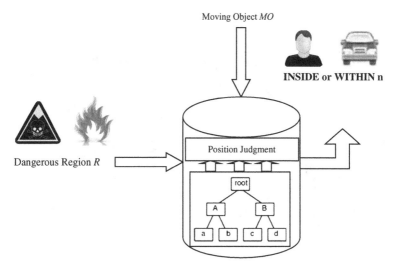

Fig. 2. Framework of PIDA

4 Proposed Algorithm

4.1 Overview of PIDA

Figure 2 shows the framework of the PIDA algorithm. It contains two main components: TPRH-tree, which indexes all moving objects with constant or changing shapes in a unique structure and the position judgment algorithm, which is used to rapidly determine the position relationship for people in danger.

We assume that the sampling data containing location and shape information are collected by the fixed sensor network.

4.2 TPRH-Tree Index Structure

The TPRH-tree improves the TPR-tree's index by optimizing the maintenance and query cost with some assistant structure like Filter-buffer and Grid-Table

(1) Filter-buffer. Every node except the leaf node in TPRH-tree has one filter buffer which is consist of some pointers to the updated moving polygons whose velocity and shape are always changing. The filter buffers are stored in the memory and are updated timely which can reduce the frequency of modifying the index tree.

(2) Grid-table. One effective way to determine the position relationship between static point and polygon is ray method. It preserves the moving function of the polygon's feature point with $G(t)$ and $H(t)$ (Fig. 3).

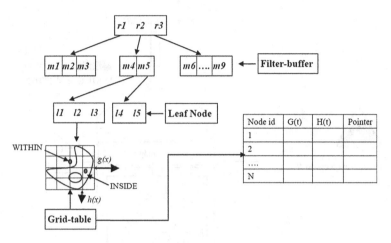

Fig. 3. TPRH-tree index structure

4.2.1   Insertion

In the TPRH-tree index structure, every interior node has one filter buffer. At first, the object in the root node's buffer calls the insert algorithm to push the moving object node downward the child node until the buffer overflows. The lower node will recursively filter the paired buffer if it is overflowed after pushing down to the leaf node. The algorithm of insertion is as follows:

Algorithm Insert（moving object m）

/* NE is the path to insert the node. PQ is the queue to choose the insertion node.

Get the root's buffer A;

For $(a_i \in A)$ do

  $R(n) =$ **group_choose**();

  For $(all\ R_i \in R(n))$ do

  $R_i.costDegradation \leftarrow SR(Area(R_i.boundingBox, NE.boundingBox), q_T)$

  $PQ. push(R_i.costDegradation)$ and Choose the $PQ's$ smallest intersect area NE;

  If ($NE's$ buffer is full) then

    filter_choose (R_i);

  Else

    Insert the R_i into $PQ.push's\ NE$ buffer with the smallest value;

4.2.2   Deletion

The TPRH-tree does not support the batch delete operation. The table records the id of every moving object and the page number in the disk. The search table combines the deleting node with the same page number into the same entry. Delete operation based lookup table, the algorithm firstly finds the target page, and then it removes the node whose page number matches directory table rectangular page information in the end of D-Table. One specific process in this algorithm is that the deleted items with the same page number would be saved into groups, and then the algorithm selects all groups of maximum number of entries.

Algorithm Delete (moving object *m*)

/*lookup Table in the memory stores the moving objects' page number.

/*D-Table stores the moving object will delete

If (Traverse (*lookup Table*) == *m*)

   D-Table ← *m*

If(*D-Table* is full) then

  Searchentry(m);

  If (*Targetpage* is a buffer page)

      return;

  else

         retract the bounding rectangle as the TPR-tree;

4.3   Position Judgment Algorithm

This section proposes a position judging algorithm which is optimized for detecting people in danger. It defined two circles for the danger region: incircle and excircle which respond to most PID queries efficiently. Then the rest of people would be tested by the method bin test which builds bins between two circles and conducts ray test in certain direction. For example, the people P_a's position can be ensured by the incircle immediately as the distance between the center of circle and P_a is less than r_i. It will take another test for the people P_b when it is moving toward the ring between two circles. The algorithm is following:

Algorithm Position Judgment (moving object *m*)

for each moving object *m*

  if *isIncircle(m)* == *true*

     pos = INSIDE; return;

  else if *isExcircle*(m) == *true* && *dist* > r_e

     pos = OUTSIDE; return;

  else if *rayBinTest*(m) == *true*

     pos = INSIDE; return;

  else *pos* = WITHIN; return;

When there are many concurrent requests, the algorithm can be optimized as the people's PID query locations always form a trajectory and the algorithm can skip intermediate ones. In Fig. 4(b), if we have confirmed the p_{a1} is outside of the dangerous region, the next detecting point should be p_{a3}. If it is still outside of the region, we can directly respond to p_{a2}'s query. Otherwise we need to further consider p_{a2}.

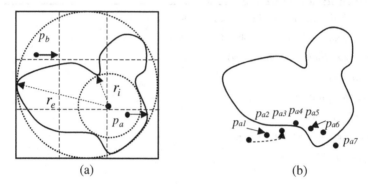

(a)   (b)

Fig. 4. Position judgment algorithm

5   Experiment and Analysis

This section focuses on experiments to analyze the performance of the method. In this paper, the experimental data are generated randomly using the method in [16]. The data in the 100 K mobile data object set are uniformly distributed in the space 10000 * 10000 unit distance (the unit distance value can be any size, such as 1 m, 1 km, etc.). In the other hand, the paper defines some features of the spatio-temporal query. (1) The query about the safe distance and the affected region window (QRlen) is changed from 100*100 to 1000*1000 which represents the safe distance of danger threshold for different degrees of disaster. (2) The position prediction range is expressed as $[t_{sub}, t_{sub} + t]$, the time of duration t (QTlen) is assumed from 10 to 1000(s). The efficiency is tested by different queries to access the node number of the index tree.

5.1   Efficiency

The first section compares the efficiency of moving object position query between PIDA and traditional algorithms, e.g. TPR-tree based and Grid-based Position Judgment algorithm. The node number accessed by the query can effectively reflect the response time and efficiency. The experimental results are shown in Fig. 5.

From the curves, we find PIDA can greatly reduce the response time. In Fig. 5, we used fixed time window and spatial query window respectively. In the subfigure (a) and (c), the experiment tests the accessed node numbers' change with the QRlen as the QTlen is set 200 and 1000 unit time. The PIDA's performance is better than TPR-tree at an average of 60 % and Grid at 40 %. The results of another group are similar. The

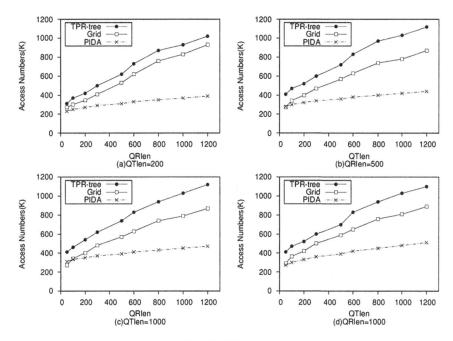

Fig. 5. Efficiency

PIDA performs more efficiently than the other indices combined with PIP as the latter ones index the leaf node with PIP every time to position query for people in danger.

5.2 Memory Cost

Based on the foregoing experimental comparison, we know PIDA is more efficient than TPR-tree and Grid in terms of response time. However, the PIDA's memory cost does not increase more than TPR-tree and less than Grid. In this part, we set the two group update operation parameters, in first group the QRlen = 200 and QTlen = 400, and for the other group the QRlen = 1000 and QTlen = 2000.

In Fig. 6, the PIDA's memory cost increases with low growth rate. Based on the auxiliary data structure in the TPRH-tree, the memory cost increases linearly with the Update numbers. As the buffer set in the middle node of TPRH-tree, it can dynamically adjust the cached sensor data to reduce the leaf node access frequency and memory cost. The Grid Table in the leaf node can also filter some duplicate data and merge many movement sampling data in the same direction.

5.3 The Amount of Concurrent Queries

This part of experiment is aimed to determine the max number of queries that different data volume supports at a time unit. In emergency management, the PID method should not only support the bursts of the sampling data, but also the bursts of queries.

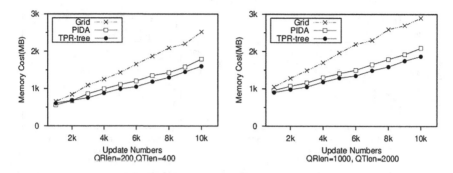

Fig. 6. Memory cost

The available emergency management system must respond to PID queries as many as possible. In this part, we define the *fill rate* to measure the degree supported for concurrent queries if the method can respond to the PID query within less than 1 s.

This part sets the different kinds of entity to simulate the concurrent update operation. We change the QRlen from 200 to 1000 and the QTlen from 500 to 1000. The result in Fig. 7 shows the PIDA can deal with the operation at the high level with more than 99 % fill rate. It is effective in disasters, like fire or pollution gas leak.

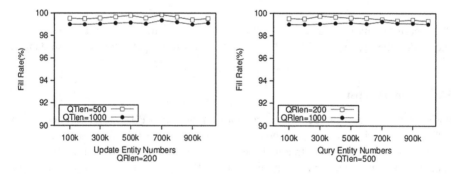

Fig. 7. Concurrent queries

6   Conclusion

Based on the analysis in emergency management, the paper proposes an algorithm PIDA to answer the location based PID query effectively. The algorithm contains two relevant parts, the moving object index TPRH-tree and the position judgment algorithm. The TPRH-tree can update and maintain the affected cases effectively, and manages to perform better than TPR-tree and Grid in terms of node splitting and IO throughout. In addition, the position judgment algorithm can better deal with large-scale disaster zone data than other methods. At last, the experimental analysis demonstrates that the proposed method perform better than existing methods and could be used in many cases.

Acknowledgement. We would like to thank Dr. Jia Zhu and Mr. Yaguang Li for providing valuable suggestions. This work was supported by the National Natural Science Foundation of China (Nos. 91124001 and 91324008), and the Strategic Priority Research Program of the Chinese Academy of Sciences (No. XDA06010600).

References

1. Tong, Z., Zhang, J., Luo, X., Bao, Y., Liu, X.: Study on grassland fires disaster emergency spatial decision support system based on case-base. In: Proceedings of the 4th International Conference on Risk Analysis and Crisis Response, Istanbul, Turkey, 27–29 August 2013, p. 397. CRC Press (2013)
2. Du, H., Gone, H., Jiang, Z., Zhang, J.: Key Technique Study of Emergency Spatial Data Management and Service System, p. 024 (2012)
3. Ertug, G.A., Kovel, J.P.: Using GIS in emergency management operations. J. Urban Plan. Dev. **126**(3), 136–149 (2000)
4. Murray, A.T., Tong, D., Grubesic, T.H.: Spatial optimization: expanding emergency services to address regional growth and development. Studies in Applied Geography and Spatial Analysis: Addressing Real World Issues, 109 (2012)
5. Wu, J., Yang, H., Gao, X., Ma, J., Liu, K.: Where to find help when you are in an emergency? In: Proceeding of the 2nd International Workshop on Emergency Management in Big Data Age (BIGEM 2014) (2014)
6. Hadjieleftheriou, M., et al.: R-trees–a dynamic index structure for spatial searching. In: Encyclopedia of GIS, pp. 993–1002. Springer US (2008)
7. Šaltenis, S., Jensen, C.S., Leutenegger, S.T., Lopez, M.A.: Indexing the positions of continuously moving objects. ACM Sigmod Rec. **29**(2), 331–342 (2000)
8. Saltenis, S., Jensen, C.S.: Indexing of moving objects for location-based services. In: Data Engineering 2002, pp. 463–472. IEEE (2002)
9. Haines, E.: Point in polygon strategies. In: Graphics Gems IV, vol. 994, pp. 24–26 (1994)
10. Žalik, B.: Point-in-polygon tests for geometric buffers. Comput. Geosci. **31**(10), 1201–1212 (2005)
11. Preparata, F.P., Ian Shamos, M.: Geometric Searching (1985)
12. Jimenez, J.J., Feito, F.R., Segura, R.J.: Robust and optimized algorithms for the point-in-polygon inclusion test without pre-processing. Comput. Graph. Forum **28**(8), 2264–2274 (2009)
13. Jiménez, J.J., Feito, F.R., Segura, R.J.: A new hierarchical triangle-based point-in-polygon data structure. Comput. Geosci. **35**(9), 1843–1853 (2009)
14. Hu, Y., Ravada, S., Anderson, R.: Geodetic point-in-polygon query processing in oracle spatial. In: Pfoser, D., Tao, Y., Mouratidis, K., Nascimento, M.A., Mokbel, M., Shekhar, S., Huang, Y. (eds.) SSTD 2011. LNCS, vol. 6849, pp. 297–312. Springer, Heidelberg (2011)
15. Pedersen, R.N.: Point-in-polygon target location. U.S. Patent 8,416,122 (2013)
16. Sinha, S., Nanetti, L.: Inaccessibility-Inside Theorem for Point in Polygon. *arXiv preprint arXiv:1010.0552* (2010)

Structured and Quantitative Research on Scenario of Firefighting and Rescue Cases

Lihong Lu[(✉)], Qingchun Kang, Shaofang Yang, Yu Li, Danjun Lian, and Sicheng Li

The Ministry of Public Security Key Laboratory of Firefighting and Rescuing Technology, The Chinese People's Armed Police Force Academy, Langfang, Hebei, China
{llh_qc,yshf2008,liyul0_01,liandanjun}@163.com,
kangqingchun@sina.com,
lsich@mail.ustc.edu.cn

Abstract. Structurization and quantization is the key to establish the "scenario-response" reasoning system based on digital scenario matrix of firefighting and rescue cases. Hierarchical index structure based on scenario expressing was studied firstly in this paper. On this basis, the methods of decomposing the case scenario, extracting the characteristic elements and assigning value to them were researched respectively. The author puts forward a universal scenario model based on 10 characteristic elements which will be represented by vectors and meanwhile digitally reconstructed. The work lays the foundations for building a scenario database based on the expression of digital vectors.

Keywords: Scenario · Firefighting and rescue case · Structurization · Quantization

1 Introduction

Firefighting and rescue cases generally contain a large number of data on fire and firefighting and rescue combat process [1], which is significant for dealing with similar disasters and accidents. However, the low threshold of matching similar cases in traditional case-based reasoning technology applied by the auxiliary decision system resulted in low reutilization rate of cases and waste of case resources. Scenario analysis is a flexible and dynamic strategic idea [2] which can deal with the uncertainties in the process of conventional or unconventional emergencies. Therefore, the construction of "scenario-response" model has been widely recognized in the emergency academic field home and abroad [3–6]. Its application in firefighting and rescue case deducting or reasoning will greatly improve the matching efficiency. One of the keys in "scenario-response" reasoning model is the expression method of firefighting and rescue cases, which is still in the stage of exploration. Based on structurization of cases, the paper illustrates the structuring and quantitative methods for firefighting and rescue case scenarios in order to lay foundations for computer auxiliary decision system of digital scenario reasoning.

Y. Chen et al. (Eds.): WAIM 2014, LNCS 8597, pp. 98–105, 2014.
DOI: 10.1007/978-3-319-11538-2_10

2   Hierarchical Index Structure Based on Scenario Expression

Hierarchical Classification is the basis of firefighting and rescue scenario structurization. The article ten in "The State Council's Opinions on Further Strengthening the Fire Protection Work" ([2006] No. 15), issued by the State Council on May 10, 2006, for the first time made clear the disaster types that fire departments should take an active part in response or disposal with other agencies [7], as shown in Table 1.

Table 1. Disaster types that fire departments involved in

Role of fire departments	Disaster types
Main rescue forces	Fire, hazardous chemicals spill, traffic accident, earthquake and its secondary disasters, building collapse, major production accidents, air crash, explosion and terrorist incidents and mass incident
Supporting rescue forces	Flood and drought disasters, meteorological disasters, geological disasters, forest, grassland fires and other natural disasters, mine, water accidents, serious environmental pollution, nuclear and radiation accidents, emergent events of public health

The "Law of the PRC on the emergency response" was implemented in November 1, 2007. Article 3 in the law defines emergency events as including sudden natural disasters, accidents, public health and social security incidents. According to degree of social harm and the scope of influence, for the first time the law put the emergency into 4 levels [8] – extraordinarily big accidents, major accidents, accidents and minor accidents, and into 4 categories as shown in Table 2.

Table 2. Emergency classification

Disasters and accidents types	Contents
Natural disaster	Meteorological disasters, marine disasters, floods, geological disasters, earthquakes, forest disaster, crop disaster etc.
Accident disaster	Accident in production safety, public facilities and equipment accident, hazardous chemicals accident, traffic accidents, etc.
Public health event	Epidemic of infectious diseases, mass unidentified diseases, animal epidemic, other serious public health events
Social security event	Economic security events, diplomatic incidents, terrorist attacks

According to Tables 1 and 2, with reference to literature [1], the paper established the firefighting and rescue case hierarchical index structure based on scenario expression, as shown in Fig. 1. As can be seen from the graph, a case can be divided into a number of scenarios. So the firefighting and rescue expert system based on cases reasoning can be transferred into a "scenario-response" expert system based on scenario reasoning.

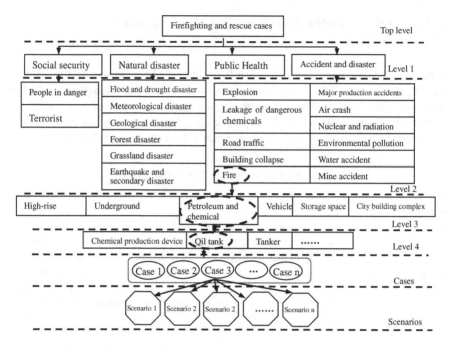

Fig. 1. Hierarchical index structure of firefighting and rescue cases

3 Structured and Quantitative Methods Based on the Scenario in Firefighting and Rescue Cases

Structured analysis and quantification of scenarios is the basis of firefighting and rescue cases structurization and quantification, as well as the key to the scenarios-response reasoning technology. A good scenario expressing method will greatly reduce the calculation processing and improve the matching speed and efficiency.

3.1 Scenario Decomposing Method

Scenario decomposition is the premise of cases structurization and quantification. The occurrence, development and evolution of any accident are continuous function changing with time. However, not all corresponding scenario fragments are key elements. Only those playing a vital part in the development of the whole accident can be extracted to reappear the development and evolution process of the accident. Therefore, case scenario decomposing method is going by dividing the key time points in the axle of time. Then the key scenarios can be extracted around the core points. Take the oil tank fire as an example, division of key time points and selection of key scenarios are shown in Fig. 2, which mainly includes the following three aspects:

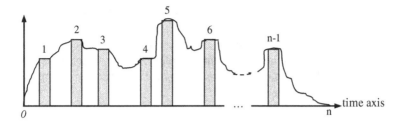

Fig. 2. Division of scenario in oil tank fire along the time axis (0- occurrence; 1-timely first response; 2-sudden boiling over; 3-reinforcements; 4-wind shift; 5-explosion of nearby tanks; 6-readjustment of disposition; n-1-general offensive; n-termination)

(1) The points of abrupt change that disaster begins to expand or the disaster bearing bodies get into sudden changes. The former includes points such as boilover occurs during the stable combustion in tank fire, or the area of flowing fire expands instantaneously. The latter includes point such as the radiation effect from burning oil tank results in explosion of the exposure tank.

(2) The time points when the rescue is seriously influenced by the changes in environment or firefighting facilities including water, fire control facilities, meteorological conditions etc. For example, in the response to hazardous chemicals spill, the sudden change in wind direction will bring great impact on the disposal operations. In oil tank fire fighting, the points of destruction of fixed foam extinguishing system, exhaustion of water supply, and so on, will all affect the rescue.

(3) The key time point when decision is issued or adjusted. For example, the key points of adjusting the recourse deployment on the scene including the arrival and turn-out of the first responders, the arrival of reinforcement force, launch of offensive fighting, scene clearing up and transferring.

3.2   Extraction and Assignment of Characteristic Elements

Characteristic elements (featured elements) are key factors which can best reflect the scenario. The combining of the specific characteristic elements extracted from a scenario can achieve very good effects in case scenario representing. By extracting, duplicate removing, clustering the characteristic elements from a large number of case scenarios, groups of general characteristic elements that are able to describe different firefighting case scenarios at the same time can be summarized. Hence the general characteristic factor model will be established. Through large amount of case studies, our research group extracts 10 types of general featured elements including object (T_O), damage (T_D), environment (T_E), weather (T_W), water (T_A), fire protection facilities (T_F), firefighting strength (T_R), measures (T_M), response strategy (T_S) and lessons (T_L). This general scenario model (S) based on the 10 types of characteristic elements is shown as follows:

$$S = (T_O, T_D, T_E, T_W, T_A, T_F, T_R, T_M, T_S, T_L) \qquad (1)$$

Each element has different significance and nature in different case scenarios, so different values should be assigned to them according to the practical situation. Take the element types of object, disaster, and meteorological conditions as examples, values assigned to them are as shown in Table 3.

Table 3. Example of value assignment for characteristic elements

Case scenario	Types of characteristic elements	Assignment
High rise building fire	Object (T_O)	High rise building
	Disaster (T_D)	Fire
	Weather (T_W)	Wind
Oil tank explosion accident	Object (T_O)	Oil tank
	Disaster (T_D)	Explosion
	Weather (T_W)	Thunder

3.3   Vector Expression of Characteristic Elements

Only with the assignment to the above 10 types of characteristic elements as shown in Table 3, the type get to possess practical significance and can be further described with more detailed elements. For example, after the object "T_O = oil tank" is determined, it can be describe by featured elements such as oil type, tank type and its material, structure form, tank volume, the number of the tanks on fire, the number of exposure tanks, storage capacity, pipe pressure and pipe diameter. Taking the object type as vector element 1, taking its describing elements as vector element 2, element 3......, the vector expression of this element type can be fulfilled.

For example, vector expression of characteristic elements for tank fire object (T_O) is:

$$T_O = (\text{oil type, tank type and its material, structure form, tank volume,}$$
$$\text{the number of the tank on fire, the number of exposure tanks,} \qquad (2)$$
$$\text{storage capacity, pipe pressure, pipe diameter})$$

Vector expression of characteristic elements also can be expressed in letters. For example, vector expression for weather is:

$$T_W = (w_1, w_2, r, s, f, t, l, h, c, m) \qquad (3)$$

Where: "w_1" means wind power, "w_2" means wind direction, "r" means rain, "s" means snow, "f" means fog, "t" means thunder, "l" means lightning, "h" means hail, "c" means temperature, "m" means humidity.

3.4 Digital Reconstruction of Characteristic Elements Vector

With the vector description of 10 kinds of characteristic elements in case scenarios, these elements can be featured as knowledge units and further knowledge database on case scenario can be built. However, compared with digital language easily recognized by computers, expressing characteristic elements vectors in text words is more complicated and results in much more calculation. Therefore, in order to simplify the expression, enhance the computing speed and case matching efficiency, we put forward new thought of digital reconstruction to the featured vectors.

Digital reconstruction means assigning values to the non numeric vector elements among the featured vectors so as to get all vector elements can be indicated by numbers. The vector elements which are originally presented by numbers can be expressed with their actual value.

Still take oil tank fire as an example, the digital reconstruction of its object characteristic vector (formula (2)) and meteorological characteristic vectors (formula (3)) were shown in Tables 4 and 5 respectively. If a vector element has no specific significance in the current scenario, it can be set to 0. If the related information source is incomplete or inaccurate, the element can also be set to 0.

Table 4. Assignment of object type characteristic vector

Vector element position	Quantitative index	Contents and assignment
1	Oil tank (Object type)	1 (other type such as high rise building can be set as 2......)
2	Oil type	1-Crude oil; 2-Gasoline; 3-Diesel; 4-Residue; 5-Asphalt oil; 6-Kerosene; 7-Heavy oil; 8-Residue
3	Tank type and its material	1-Overground and metal; 2-Over ground and non-metallic; 3-Semi-underground and metal; 4-Semi-underground and non-metallic; 5-Underground and metal; 6-Underground and non-metallic
4	Structure form	1-Arch-roof tank; 2-Internal floating roof tank; 3-Floating roof tank; 4-Horizontal tank; 5-Oil basin
5	Tank volume (m^3)	True value
6	Storage capacity	1-Empty tank; 2-Half –filled tank; 3-Full tank
7	Number of burning tank	True value
8	Number of exposure tank	True value
9	Pipeline pressure (kp)	True value
10	Pipe diameter (m)	True value

Table 5. Assignment of weather type characteristic vector

Vector element position	Quantitative index	Contents and assignment
1	Wind power	Actual wind series (0 no wind)
2	Wind direction	0-no wind 1 East wind; 2-South wind; 3-West wind; 4-North wind; 5-Southeast wind; 6- Northeast wind; 7-Southwest wind; 8-Northwest wind
3	Rain	0-Clear; 1-Light rain; 2-Rain; 3-Heavy rain; 4-Storm
4	Snow	0-Clear; 1-Slight snow; 2-Snow; 3-Heavy snow; 4-Snowstorm; 5-Rain and snow
5	Fog	0-Clear; 1-Light fog; 2-Fog; 3-Heavy fog; 4-Smog
6	Thunder	0 Clear 1Thunder
7	Lightning	0 Clear 1Lightning
8	Hail	0 No hail 1Hail
9	Temperature (°C)	True value
10	Humidity (%)	True value

For example: 2 over-ground metal external-floating-roof oil tank burst into fires, with the following characteristics elements: volume 50000 m^3, full tank, 6 nearby tanks, unknown pipeline pressure and diameter, sunny weather, three-grade east wind, temperature of 27 °C, humidity of 30 %. Then according to Table 4, the object type vector of characteristics elements (T_O) can be expressed as(i.e. the formula (2)):

$$T_O = (1 \quad 1 \quad 1 \quad 3 \quad 50000 \quad 3 \quad 2 \quad 6 \quad 0 \quad 0)$$

The formula (3) of weather type vector expression of characteristics elements (T_W) can be expressed as:

$$T_W = (3 \quad 1 \quad 1 \quad 0 \quad 0 \quad 0 \quad 0 \quad 0 \quad 27 \quad 30).$$

4   Conclusions

With the help of structuring and quantitative analysis presented in this paper, the firefighting case can be divided into several scenarios, each of which is described by 10 types of characteristic elements. With the vector description and digital reconstruction of characteristic elements, the case scenario can be featured as 10 digital vectors thus they can be composed of a 10 × 10 digital matrix, and accordingly the case scenario knowledge database can be established based on digital matrix. In this way, large amount of processing calculation in "scenario-response" reasoning system is simplified while the computing speed and efficiency is greatly enhanced, and also the case utilization rate is improved. The work lays the foundations for building a firefighting and rescue case "scenario-response" decision-making system based on digital scenario matrix reasoning.

Acknowledgement. The research is founded by major program of the national natural science foundation of China under Grant No. 91024032.

References

1. Leng, L., Wei, H.D., He, N., Guo, Y.H., Chai, K.H.: Statistical analysis and countermeasures study of hundred conflagrations. J. Fire Sci. Technol. **26**(4), 439–443 (2007) (in Chinese)
2. Yuan, X.F.: Research on Key Technologies of Emergency Decision-making for Unconventional Emergency Based on Scenario Analysis and CBR. D. Xi'an University of Science and Technology 6, 10 (2011) (in Chinese)
3. Li, J.W.: Research on Unconventional Emergency Scenario Model Based on Knowledge Element. Master Degree thesis of Dalian University of Technology 61 (2012) (in Chinese)
4. Zhang, H., Liu, Y.: Scientific issues and integration platform based on the "Scenario-response" national emergency system. J. Syst. Engineering-theory Pract. **32**(5), 947–953 (2012) (in Chinese)
5. Guo, Y.M.: Research on Unconventional Emergency Scenario Model Based on Knowledge Element. Master Degree thesis of Dalian University of Technology. 6, 1 (2012) (in Chinese)
6. Wang, Y.X.: Research on Context Reconstruction Model of Unconventional Emergency. Doctoral Dissertation of Harbin Institute of technology. 7, 1–2 (2011) (in Chinese)
7. Yunnan Daily. Officials in the Ministry of Public Security Answered Reporters' Questions in Terms of "Opinions of the State Council on Further Strengthening Fire Work" [EB/OL]. http://news.sina.com.cn/c/2006-06-04/
8. Manual for Chinese Firefighting, vol. 10. Shanghai Scientific and Technical Publishers, Shanghai (2006) (in Chinese)

A Cross-Reasoning Method for Scenario Awareness and Decision-Making Support in Earthquake Emergency Response

Jing Qian, Yi Liu[✉], Gangqiao Wang, Ni Yang, and Hui Zhang

Department of Engineering Physics, Institute of Public Safety Research,
Tsinghua University, Beijing, China
liuyi@tsinghua.edu.cn

Abstract. In this paper, the multi-dimensional scenario space method (MDSS) is applied in case study of earthquake, and a modified MDSS is developed. For complex disasters such as earthquake, one "space" is not enough to describe the disaster, and a multi-level multi-dimensional scenario space method (MLDSS) is developed. With MLDSS, a certain scenario may comprise several sub-spaces each representing an element object in the disaster. Thus, a "case-scenario-element" model with MLDSS is developed for case study taking earthquake as example. Ontology method is applied in developing the case model. Detail scenario data and information may be collected both from documents materials and the website searching. Scenarios deduction process with MLDSS is provided in this paper, and two types of scenarios evolution processes are discussed: one is single scenario process and the other is multi-scenarios process. Brief view of scenarios' evolution and the reasoning relationship are given as results, which may provide reference and recommendation to decision-making support in emergency responses.

Keywords: Case study · Multi-dimensional scenario space method (MDSS) · Multi-level multi-dimensional scenario space method (MLDSS) · Scenario evolution · Decision-making support

1 Introduction

Decision-making support is one of the challenges in emergency response and management. Case study is an efficient approach to learn from historical cases and to find possible solutions. However, in large-scale and complex disasters such as earthquake, the disaster is often too complicated to be described. As emergency response decisions are often based on and aimed to certain scenarios, developing suitable models that can describe scenarios exactly and precisely is one of the most fundamental works for decision-making in emergency response.

Roger Schank [1] first proposed a method of case-based reasoning (CBR) as an approach of knowledge-based problem solving and learning methods in 1982. In recent years, case-based reasoning method is often used in emergency decision-making research. Ke Jing [2] proposed emergency rescue decision making method on the basis of overall dominance. Ensley [3] found the pilot's ability to perceive scenarios is an

© Springer International Publishing Switzerland 2014
Y. Chen et al. (Eds.): WAIM 2014, LNCS 8597, pp. 106–118, 2014.
DOI: 10.1007/978-3-319-11538-2_11

important factor which determines the effectiveness of decision-making process in emergency situations. Jiang Hui [4] has pointed out that real-time decision is based on the scenarios of emergencies. Only on the basis of knowing current situation well, can decision makers obtain scientific solutions. Wu Guangmou [5] applied historical case analysis to rebuild scenarios in order to achieve scenario deduction by case study. Hristidis [6] introduced how to use present data mining methods to extract and analyze different types of data. In the area of data retrieving, ontology method [7] is one of the popular approaches. Ontology model provides an effective way for expanding the scope of information, and it has a wide range of applications [8–11] in the field of emergency management.

In this paper, a modified MDSS is developed as a multi-level multi-dimensional scenario space method (MLDSS) for complicated disasters such as earthquake. Ontology method is applied in developing the case model for data collecting and processing. As to scenarios deduction process, two types of scenarios evolution processes are discussed: Then, scenarios' evolution and the reasoning relationship are given, which may provide reference and recommendation to decision-making support.

2 Ontology Model

It has been found there are so much factors affect decision-making process and the effectiveness of solutions responding to scenarios by studying earthquake emergency response process. Ontology method has excellent advantages in expressing concepts and complex relationships, and is accepted in this paper. For the earthquake cases, scenarios are clarified into four categories: environment-related scenarios, disaster carrier-related scenarios, derivative disaster scenarios and rescue-related scenarios. These four categories can be further divided into many sub-categories which are shown in Fig. 1. Derivative disasters will probably take place after earthquake disaster; some environment-related scenarios can also lead to derivative disasters. These scenarios have great impact on how scenarios develop, thus risk analysis procedure needs to be added into emergency response system. The possibility of derivative disasters and environment-related scenarios can be estimated through the risk assessment process as well as influence caused by these secondary scenarios.

According to the classification approach and ontology method mentioned above, accepted concepts in earthquake can be expressed such as: living materials, relief supplies, meteorology, and so on. When new data is tapped into system, related data and scenarios can be sorted very quickly on the basis of well-established concepts; On the other hand, these accepted concepts and semantics can be used as key words to filter and extract useful information from data resources which derive from internet.

3 MLDSS Model

As there are always so much factors that affect decision-making process and the effectiveness of solutions responding to scenarios, related earthquake scenarios often form a complex scenario network. In order to solve the problem of complexity brought

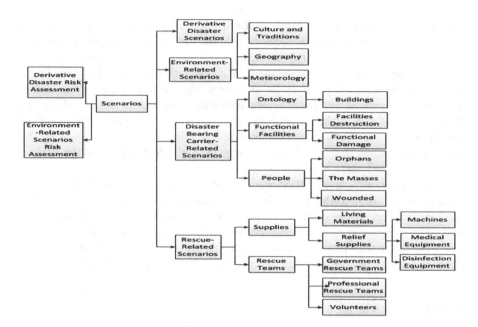

Fig. 1. Scenario ontology model

by scenarios and scenario networks, this paper modified previous multi-dimensional scenario space (MDSS) method [12] and developed the multi-level multi-dimensional scenario space method (MLDSS). It divides scenarios into some related elemental objects and each object has an own sub-space (MDSS). To establish a more comprehensive model, this model has a six-layer framework of "case-scenario-element", the MLDSS model is shown in Fig. 2.

Elements, the fundamental information of database, are located in the bottom of the case base. Objects in earthquake emergency response process are the bearers of disaster and response actions. An object may contain several different elements, namely dimensions in MLDSS. Different dimensions correspond to various statuses; these statuses may affect associated scenarios. Each object corresponds to a sub-space; there is a relationship among different dimensions of an object. Scenario space is composed of sub-spaces. Figure 2 shows sub-spaces are independent of each other with shared elements. Taking meteorological service interruption as an example, research has established a meteorological "sub-space-object" space which is shown in Fig. 3.

Decision-making process is based on the status of objects and scenarios, and relationship among objects and scenarios directly influence decision-making process and effectiveness. There exists relevance among different dimensions of one object; likewise, the different dimensions of different objects, the relationship is shown in Fig. 4. The degree of meteorology monitoring equipment damage associates with the quantity of them, and similarly, defects existing in meteorological station may cause related scenarios happen such as damage of monitoring equipment. Relationship of different objects and different dimensions of the same object can compose relation

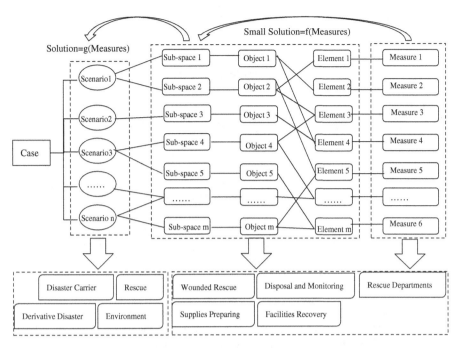

Fig. 2. Framework of MLDSS model

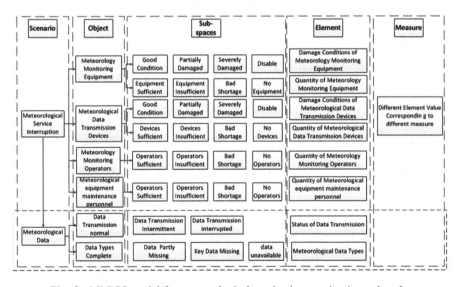

Fig. 3. MLDSS model for meteorological service interruption in earthquake

network, further form sub-space space. Efficient decision-making process relies on good understanding and analysis of scenarios and relationship networks which is also the basis of decision-making.

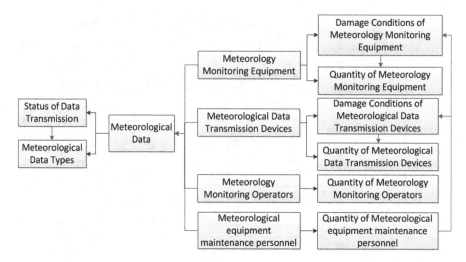

Fig. 4. Relationship network of sub-spaces

In order to keep the structure of "measure" and "scenario" consistent with each other, the framework of measure space is designed as: "elemental measure"–"measure"–"solution". Each element corresponds to an elemental measure, the implementation of elemental measures depends on rescue departments' action, "measure" corresponds to sub-space. "Measure" and "solution" are both composited of elemental measures, but redundancy sometimes even conflicts may occur among elemental measures corresponding to the same scenario, so solutions to scenarios are not simply putting "elemental measure" and "measure" together, adjustments of elemental measures need to be done in order to get solution.

4 Scenario Evolution Recovery with MLDSS Model

The relationships among scenarios are very complex which forms scenario networks, the association among scenarios influence decision-making process, so scenario expression method and relationships among scenarios are the basis of scenario evolution and emergency response. Considering information of elements is dynamic in scenario evolution process, scenario which consists of those elements is also dynamic. The scenario evolution process is discussed below from the perspective of elements. The scenarios will be assorted into two types: single scenario evolution and multi-scenario evolution. Scenario set S contains n scenarios and element set E includes m elements, the two sets S and E can be expressed as follows:

$$S = \{S_1, S_2, S_3, \ldots, S_n\} \tag{1}$$

$$E = \{E_1, E_2, E_3, \ldots, E_m\} \tag{2}$$

An element corresponds to a coordinate axis, assuming there are 1, 2, 3 ... j ... k points, and different value symbolizes different status of elements. If element E_1 involves j points, E_1 can be described as formula (3):

$$E_1 = \{E_{11}, E_{12}, E_{13}, \ldots, E_{1j}\} \tag{3}$$

Similar, element set E can be expressed as formula (4):

$$E = \{[E_{11}, E_{12}, \ldots, E_{1j}], [E_{21}, E_{22}, \ldots, E_{2k}], \ldots, \\ [E_{f1}, E_{f2}, \ldots E_{fy}], \ldots, [E_{m1}, E_{m2}, \ldots E_{mz}]\} \tag{4}$$

Development of elements causes the evolution of relevant scenarios, so the scenario set can be expressed as:

$$S = \{[S_{11}, S_{12}, \ldots, S_{1j}], [S_{21}, S_{22}, \ldots, S_{2k}], \ldots, \\ [S_{f1}, S_{f2}, \ldots S_{fy}], \ldots, [S_{n1}, S_{n2}, \ldots S_{nz}]\} \tag{5}$$

Scenario evolution includes development both in time and space, the single scenario evolution means the severity of the scenario itself becomes better or worse, the essence of scenarios does not change. For example, "communications intermittent" may appear after earthquake, and aftershocks can cause the scenario of "communications interrupted" happen. If the repairment is timely, communication can be completely restored. Scenarios varies in these kinds of status: "intermittent", "completely interrupted", "partially restored" and "full recovery", however, the connotation of the scenario does not change. Namely scenarios' status change among m classes, S_{i1}, S_{i2} ... S_{ij}, ... S_{im} are shown in Fig. 5.

Multi-scenario evolution indicates different scenarios appearing in different stages or different evolutionary relationships. It means the interaction among distinct scenarios, for example, landslides cause road damage, road damage links with traffic congestion, traffic jams and other scenarios can bring about troubles in transferring the injured people. This series of scenarios deal with the evolution of different nature of scenarios, these scenarios develop in n classes: S_1, S_2 ... S_k, ..., S_n, the network of multi-scenario evolution is presented in Fig. 6.

Scenario evolution in time and space is a combination of single scenario evolution and multi-scenario development, scenarios are a series of fragments as well as the "points" in multi-dimensional space, and the process of scenario evolution is shown in Fig. 7.

Fragments of scenarios make up the graph of scenario evolution; every fragment illustrates the development of itself. It should be noted the changes in values of elements lead to changes in associated scenario, any change of element may cause mutations in scenarios. In the background of earthquake, there are so many scenarios and elements, thus the relationships among scenarios are complicated and sensitive.

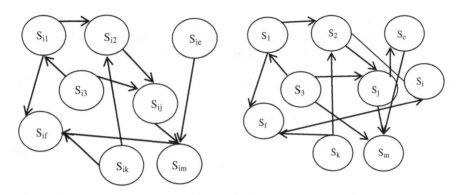

Fig. 5. Single scenario evolution   **Fig. 6.** Multi-scenario evolution

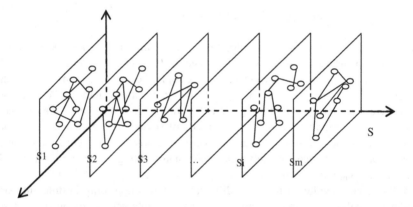

Fig. 7. Scenarios evolution

5   Decision-Making Based on MLDSS

If emergency response decisions are made for every scenario, the decision-making process will become very difficult, so researchers try to find some key scenarios or critical decision points to do right choices. In the view of event tree, these critical points and key scenarios usually locate in the upper layer of the tree; decisions made according to the top event can suppress the occurrence of subsequent disasters, in order to enhance the efficiency of emergency response. Scenario tree for decision-making is shown in Fig. 8.

The top layer scenario is the root scenario which exists in the disaster spot, the root scenario can cause the subsequent evolution of the scenario, so in the emergency response process. It's more important to inhibit the occurrence of the upper scenarios rather than the lower; moreover a certain scenario will reach its end and no more evolution, such as scenario S_{92}. The root scenario S_{11} may arouse the occurrence of the second layer scenarios: S_{31}, S_{24}, S_{61} and S_{23}, the number of inputs and outputs of S_{31}

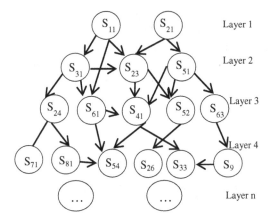

Fig. 8. Scenario tree for decision-making

can be represented as a coordinate point $(1, 3)$, in the same manner, the number of inputs and outputs of S_{23} can be expressed as $(3, 2)$. Comparing scenario S_{31} with S_{23} find input $3 > 1$, that means scenario S_{23} is more likely to occur than S_{31}, therefore the left coordinates characterize the initial sensitivity of scenarios, the greater the values of the left coordinate points are, the more sensitive the scenarios are, namely the occurrence of the scenarios are more difficult to be controlled. The right coordinates indicate the number of outputs, the more outputs there are, the easily the scenarios occur, therefore controlling the occurrence of such a scenario can improve the efficiency of decision-making process, and at the same time avoid scenarios develop. From the perspective of reducing the number of scenarios, the effect of controlling occurrence of Scenario S_{31} will be better than restraining scenario S_{23} under the same decision-making conditions. Next, this article will elaborate the evolution pattern of scenarios with an emergency response case.

Rescue mission varies in different stages of emergency rescue process; the order of decisions affects the efficiency of decision-making and scenario evolution. Considering timeline of earthquake emergency response procedure, emergency rescue disposal will be divided into four stages:

1. Stage one: From the occurrence of earthquake to activating contingency plans, stage one often lasts a few hours. Main tasks of this stage are starting contingency plans, collecting relative information and materials, organizing rescue and preparing for publishing notification.
2. Stage two: seventy two hours after activating contingency plans, this stage is called the golden rescue period. State Council takes charge of planning and arranging emergency rescue operations, relative ministries and units arrange into relief work. Emergency command center has to be established to coordinate the work of various agencies during the emergency response process.
3. Stage three: last to the 15th after earthquake happening, the rescue action involves many departments and people. The rescue mission includes critical infrastructure recovery in early period and the placement of people.

4. Stage four: This stage starts from the 15th day after the occurrence of earthquake. Tasks in the 4th stage are deployment of long-term victim's resettlement plans, disaster site cleaning and restoration of infrastructure systems. This paper focuses on scenario relationship in stage one and two as well as decision-making support. According to the timeline of rescue, the main tasks of rescue process are shown in Fig. 9:

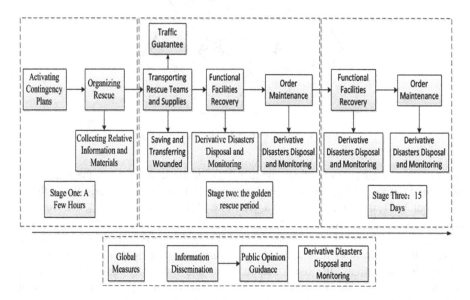

Fig. 9. Timeline of rescue

Figure 9 indicates the main tasks of stage one are activating contingency plans, organizing rescue. Collecting relative information and materials is one of the most important things in this stage. The complexity of information matters the situation grasping of disaster area and deployment of rescue forces. Thus, the significance of information related scenarios is very high. In the second stage of rescue, namely the golden rescue period, supplies and rescue teams need to get into the disaster area. In this period, the wounded and rescue forces need to be transferred, the transformation of the wounded and relief materials need transportation insurance. Transport facilities include: roads, bridges, railways and airport, therefore, in the initial period of the second phase of rescue, the importance transportation-related scenarios is obvious, at this time, the disposal of traffic-related scenarios should rank in the front of disposition order. On the basis of timeline, knowing the relationship among scenarios and determining importance of every scenario at some time point can help decision makers do the right and efficient choices. The figure shows the derivative disasters disposal and monitoring which affect all aspects of emergency response exist in whole secondary and third rescue stage, likewise, derivative disasters disposal and monitoring focus on different objects at different time points. For example, as to the functional facilities recovery, disposal and monitoring mission of derivative disasters focus on the ratio of the number of professionals and equipment related scenarios like lack of medical personnel and medical equipment. Considering the whole emergency rescue process, there are some global measures, such as information dissemination and public opinion guidance. In the early stage of rescue, information dissemination is more important compared to public opinion guidance, and with the development of scenarios over time, the focus should be shifted to public opinion guidance at home and abroad.

6   Example: Case Analysis of Yushu Earthquake

In the earthquake emergency procedure, the information about rescue publishing by networks and media is usually timely and comprehensive, so this article will take news as an important source of scenarios in Yushu earthquake. In order to solve problems, for instance, inconsistent format between news, redundancy, conflicts and other issues, the study grabs news and pictures from the website by using python programming. This study has crawled about 5000 news of Yushu earthquake, we have extracted three types of scenarios from the related news according to the classification method of disaster scenario, and a few scenarios are shown in Table 1.

Table 1. A case of scenario classification

Disaster related scenario	Element	Element coordinate	Element value	Measures
Lack of first hand data	Rescue data integrity	Completed Shortage	0, 1	Forays into disaster areas to meet demands
Lack of mapping data	Mapping data integrity	Completed Shortage	0, 1	Low-altitude UAV operation; accessing images; remote sense
Communication instability	Communication state	Stable Unstable Break	0, 1, 2	Set downtime; shortwave connection; satellite phone help
Emotional instability	Emotional state of wounded	Stable Unstable	0, 1	Medical help; subsidy policy making; care material; medical team prepared

Table 1 shows a portion of disaster related scenarios, elements, measures, element points and values of the coordinate points, and values of coordinate points lay foundation for case-based reasoning and deduction. The expression method of environment related scenarios and disaster bearing carriers scenarios is similar to the Table 1. A scenario network can be got by selecting a portion of the scenarios in earthquake to do research on scenario relationship and interaction, the network is shown in Fig. 10.

Figure 10 explains the relationship among scenarios succinctly, the complicated evolution patterns forms a scenario network. There are several key scenarios, such as "traffic jam", "road damage", "power failure" and so on, all the scenarios can lead to the occurrence of a few scenarios, and at the other hand, the evolution of these scenarios themselves can be controlled. However, the appearance of scenarios related to

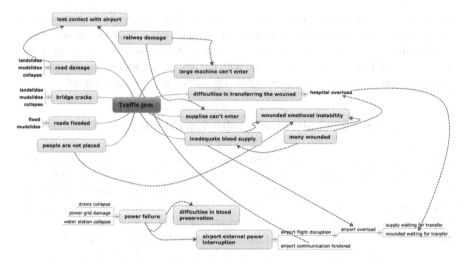

Fig. 10. Scenario network

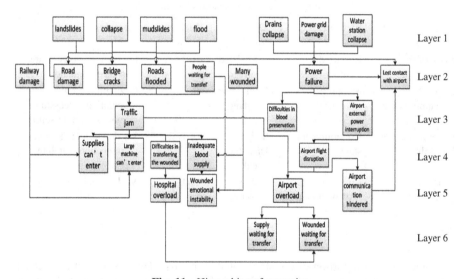

Fig. 11. Hierarchies of scenarios

natural disasters like "landslides", "mudslides" are hard to be prevented. Because Fig. 10 does not show the hierarchy between the scenarios, according to Fig. 10, we can map the hierarchy of scenarios on the grounds of scenario tree. Figure 11 is the hierarchy diagram which reflects the priority of decision-making.

Root scenarios are the natural disaster interrelated scenarios, such kind of scenarios are difficult to be controlled, the decision-making priorities of the second layer scenarios like "road damage" and "bridge cracks" are high, the consequences of disposing such scenarios determines the controllability of subsequent scenarios' evolution, if the disposition of the second layer scenarios is not appropriate, that can easily lead to occurrence of the third layer scenarios. "Traffic jam" is a sensitive scenario, easily triggered and can cause a series of scenarios. If the traffic jam has been timely and effectively treated, the associated problems "supplies can't enter", "difficulties in transferring the wounded" and other scenarios can be easily resolved. Proper disposal of key nodes will relief the severity of the entire disaster greatly. Similarly, if the key scenario is not got correct disposal, the negative impact will present by related scenarios, resulting in deterioration of the disaster. The hierarchy diagram shows the key scenarios and priorities of decision-making objects which provide a good reference for decision-makers to make emergency plans.

7 Conclusions

In this paper, taking earthquake as example, a modified MDSS is developed as a multi-level multi-dimensional scenario space method (MLDSS). With MLDSS, a certain scenarios may comprise several sub-spaces each representing an element object in the disaster. Thus, a "case-scenario-element" model with MLDSS is developed for case study. Ontology method is applied in developing the case model. Detail scenario data and information may be collected both from documents materials and the website searching. Scenarios deduction process with MLDSS is provided, and two types of scenarios evolution processes are discussed: one is single scenario process and the other is multi-scenarios process. Brief view of scenarios' evolution and the reasoning relationship are given as results, which may provide reference and recommendation to decision-making support in emergency responses.

An example case analysis with Yushu earthquake is performed with MLDSS. The study proved that emergency decisions made for crucial scenarios and upper nodes can effectively block the evolution of related scenarios. On the other hand, the scenario network diagram can achieve the goal of predicting the evolution of scenarios and valuing the likelihood of the occurrence of associated scenarios.

Acknowledgement. This work was partially supported by the National Natural Science Foundation of China (No. 91324022, No. 91224008, No. 91024032, No. 70601015, No. 70833003).

References

1. Schank, R.C.: Dynamic Memory: A Theory of Reminding and Learning in Computers and People. Cambridge University Press, Cambridge (1982)
2. Ke, J.: Emergency rescue & T. based on the overall dominance reasoning case decision. Syst. Eng. **26**(9), 15–20 (2008)
3. Endsley, M.R.: Situation awareness in aviation systems. In: Garland, D.J., Wise, J.A., Hopkin, V.D. (eds.) Handbook of Aviation Human Factors, pp. 257–276. Erlbaum, Mahwah (1999)

4. Hui, J.: Scenario and relationship expression in real-time emergency response. Univ. Electron. Sci. Technol. (Soc. Sci. Edn.) **01**, 48–52 (2012)
5. Wu, G., Chuan, Z.: Jiang million square decision model. City scene of major accidents and the trend deduced reproduction of Southeast University. Philos. Soc. Sci. **13**(1) (2011)
6. Hristidis, V., Chen, S.C., Li, T., Luis, S., Deng, Y.: Survey of data management and analysis in disaster situations. J. Syst. Softw. **83**(10), 1701–1714 (2010)
7. Gruber, T.R.: Toward principles for the design of ontologies used for knowledge sharing. Int. J. Human-Comput. Stud. **43**(5–6), 907–928 (1995)
8. Klien, E., Lutz, M., Kuhn, W.: Ontology-based discovery of geographic information services - an application in disaster management. Comput. Environ. Urban Syst. **30**(1), 102–123 (2006)
9. Halliday, Q.: Disaster Management, Ontology and the Semantic Web (2007). http://polaris.gseis.ucla.edu/qhallida/DisasterManagementFinal.html, UCLA
10. Klann, M.: Tactical navigation support for firefighters: the LifeNet ad-hoc sensor-network and wearable system. In: Löffler, J., Klann, M. (eds.) Mobile Response. LNCS, vol. 5424, pp. 41–56. Springer, Heidelberg (2009)
11. Xu, M., Duan, L.-Y., Xu, C.-S., Tian, Q.: A fusion scheme of visual and auditory modalities for event detection in sports video. In: Proceedings of the 2003 International Conference on Multimedia and Expo, pp. 333–336 (2003)
12. Liu, Y., Feng, Y., Zhang, H., Yang, R., Zheng, L.: Study on multi-dimensional scenario-space method for case-based reasoning. In: The 10th International Conference on Cybernetics and Information Technologies, Systems and Applications, Orlando, FL, USA, 9–12 July 2013

PM2.5 Air Quality Index Prediction Using
an Ensemble Learning Model

Wei Xu[1(⊠)], Cheng Cheng[1], Danhuai Guo[2], Xin Chen[2], Hui Yuan[1],
Rui Yang[3], and Yi Liu[3]

[1] School of Information, Renmin University of China, Beijing 100872, China
weixu@ruc.edu.cn, chengcheng_ruc@126.com,
yuanhui91414@163.com
[2] Computer Network Information Center, Chinese Academy of Sciences,
Beijing 100872, China
{guodanhuai, chx}@cnic.cn
[3] Department of Engineering Physics, Institute of Public Safety Research,
Tsinghua University, Beijing 100084, China
{ryang, liuyi}@tsinghua.edu.cn

Abstract. PM2.5 has a significant influence on human health. And with the modern society developing, PM2.5 has been becoming a severe problem for people. In this paper, an ensemble learning method for PM 2.5 prediction is proposed. The assumption is that the information inside the historical data of PM2.5 in the selected station and other stations orderly from the one can be beneficial for the prediction of PM2.5. The results show that the more information, the more accurate the predictions are. Moreover, there are a balance between the good performance and the costs of modelling.

Keywords: PM2.5 air quality index · Prediction · Ensemble learning · Emergency management

1 Introduction

Fine particulate matter is the suspended particulate with the diameter less than or equal to 2.5 in the air. The higher the concentration of PM2.5 is, the more severe the pollution is. Compared with the traditional thick particulate, PM2.5 has smaller partical size, bigger area and stronger activity and attaches the toxic and harmful substance more easily. What's more, the long time PM 2.5 stays and the distance it transfers influence the air quality and body health more.

Fine particulate matter has drawn the increasing attention in recent years. Throughout the world, fine particulate matter has been studied in different domains, such as environment itself and the human health. The emissions from cooking are related with the concentration of PM2.5 and are quantified according to He et al. in 2004 [1]. Kleeman et al. analyzed PM2.5 from several sources, such as wood burning, from the viewpoint of the distribution of size and constitution [2]. Pope evaluated the effects to human health caused by PM2.5 [3]. And then he explored the changes of life expectancy influenced by the PM2.5 in U.S from the 1980s to the 1990s [4]. Similarly,

© Springer International Publishing Switzerland 2014
Y. Chen et al. (Eds.): WAIM 2014, LNCS 8597, pp. 119–129, 2014.
DOI: 10.1007/978-3-319-11538-2_12

this kind of research is being studied in China. Guo et al. deliberated the relationship between PM2.5 and the cardiovascular diseases quantified through the visits of emergency room [5].

Recently, PM2.5 prediction is becoming a hotspot area. Researches about PM2.5 aforementioned are most related to the association between PM2.5 and different diseases. Forecasting PM2.5 can contribute to the researches of disease and the control of epidemiology. Cobourn summarized that there are two categories of models in air quality predicting [6]. One is deterministic model and the other one is empirical model. The deterministic model is to predict the data through simulating the process of air pollution while the empirical model is to evoke the statistical or numerical methods to excavate the association between the independent and dependent variables. Moreover, in the empirical model, plentiful methods are employed. One of the most widely used linear methods is linear regression and after introducing the parameter called land use, the land use regression (LUR) is used to forecast [7–10]. Meanwhile, the data mining methods are widely used, especially neutral networks (NN). Ordiere et al. employed NN with three topology structures, which are respectively Multi-layer Perceptron, Radial Basis Function and Square Multi-layer Perceptron to probe the prediction of PM2.5 and compared the performance of the three different models [11]. McKendrya evoked NN to predict the particulate both PM10 and PM2.5 with meteorological factors and so on [12]. Also, Markov model is also used in PM2.5 prediction. A hidden Markov was invoked to forecast PM2.5 of every day with three kinds of emission distribution [13]. Dong et al. proposed a novel method called hidden semi-Markov models to eliminate the limitation of the temporal structures in traditional Markov model [14].

In this paper, we propose an ensemble learning method for PM2.5 air quality index prediction. The bi-dimensional exploration using Support Vector Regression (SVR) is applied to expect the more accurate prediction. Meanwhile, not only the PM2.5 historical data of the station is used, but also the data from adjacent stations is incorporated in the model. What's more, the selected scope and time lag also influence the prediction.

The rest of paper is organized as follows. Section 2 proposes a novel PM2.5 air quality index prediction method using an ensemble learning model. Then the proposed method is verified in Sect. 3. Section 4 summarized the whole paper and provides the drawbacks and future work of this research.

2   The Proposed Method

In this section, a SVR based bi-dimensional exploration framework is proposed for PM2.5 prediction. More specifically, time lag of PM2.5 time series and geographical locations of the different monitor stations are explored to find out the best-suited time lag the geographical scope. In the proposed framework, the PM2.4 time series data is firstly crawled from pm25.in, a website which provides PM2.5 information covering 190 cities in China. Secondly, PM2.5 series of one station is chosen as the prediction target, the PM2.5 time series of other stations around the prediction station is provided as attributes. Thirdly, as it is nature to assume that value of PM2.5 of one station is affected by the ones of stations nearby and their historical records. Therefore, with increasingly adding higher time-lagged PM2.5 time series of peripheral monitoring stations and including PM2.5 time series from peripheral stations with larger scope, we trained different models so that to figure out the best settings of time lag and geographical scope. In particular, if the number of monitoring stations included is more than five, random subspace is introduced for exploring the capability of ensemble algorithms in increasing forecasting performance. At last, best-performed model is selected and analyzed.

The aforementioned framework is illustrated in Fig. 1 below.

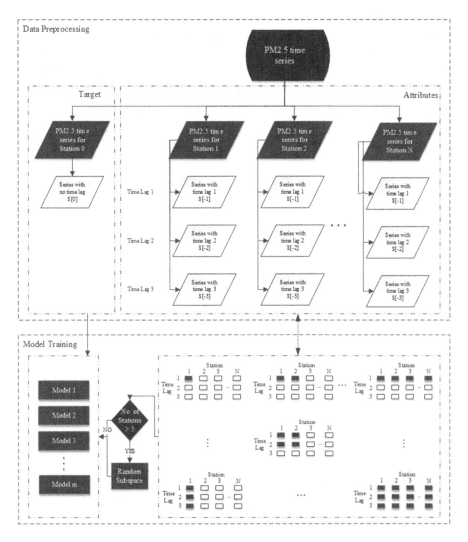

Fig. 1. The process of the proposed method for PM2.5 air quality index prediction

2.1 Data Retrieving

In this first stage, we designed and implement data crawling script with the API provided by PM25.in to crawl PM2.5 time series data. PM2.5 is a small project run by Best App Studio and distributed to the public free for public interests. PM25.in covers PM2.5 data of 190 cities with 945 monitoring stations. The data is hourly updated. However, since PM25.in does not provide dataset and API for looking for historical data, all PM2.5 time series data can only be collected hour by hour.

2.2 Data Preprocessing

At this second stage, data is filtered and reorganized for our research framework. Since we only focus on PM2.5 data in one city, data of rest cities are filtered out. As illustrated in research framework, PM2.4 time series of center station is selected as prediction target, which is calculated and figured out as center of all interested stations based on their geographical distances of each other. Data from other stations (attributes stations) are then ordered by the geographical distances between their position and that of target station. The nearest station is labeled as 1, and the second nearest station is labeled as 2, and so forth. Figure below illustrate the labeling process (Fig. 2).

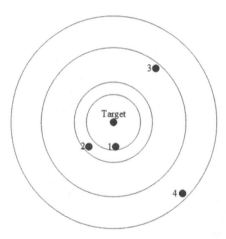

Fig. 2. Geographical scope exploration demonstration

Furthermore, time lagged series with maximum value of 3 is included in each station. Therefore, all data of attribute stations with different geographical distances and time lag are used as attributes for research.

2.3   Model Exploration

Once data are prepared, bi-dimensional exploration models are constructed to find out the best suited time lag of PM2.5 time series and geographical locations of the different monitor stations. As illustrated in figure below, on one side, in terms of time lag, series with larger time lag are added into attributes set for exploration; on the other side, in terms of geographical location, increasingly more data of attributes stations with larger distances are added as attributes for exploration. It is mentionable that, when attributes station included in larger than 5, random subspace is introduced for exploring the capability of ensemble algorithms in increasing forecasting performance with larger attributes set (Fig. 3).

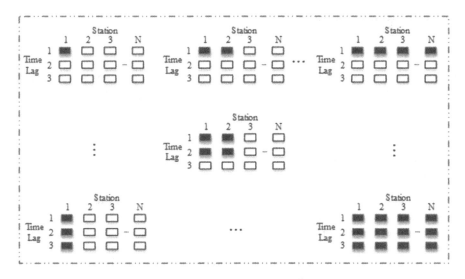

Fig. 3. Demonstration of exploring attributes in time lag and geographical scope dimension

2.4   Results Analysis

After all the experimental results are collected, RMSE (Root Mean Squared Error), RRSE (Root Relative Squared Error), and RAE (Relative Absolute Error) are employed for performance evaluation and analysis. Different evaluation criteria measure performance of models in different aspects. RMSE is most commonly used one for measuring numeric results, which punishes outliers (predicted value is far from real one) for larger weights by the square. Furthermore, RRSE can be interpreted as RMSE normalize errors by dividing the squared error of simply predicting average values rather than number of instance. In addition, RAE normalize errors by dividing summed error of simple prediction from summed error of evaluated predictor, where outliers are punished with the same weight of the other.

$$RMSE = \sqrt{\frac{(p_1 - a_1)^2 + \ldots + (p_n - a_n)^2}{n}} \qquad (1)$$

$$RRSE = \sqrt{\frac{(p_1 - a_1)^2 + \ldots + (p_n - a_n)^2}{(a_1 - \overline{a})^2 + \ldots + (a_n - \overline{a})^2}} \qquad (2)$$

$$RAE = \frac{|p_1 - a_1| + \ldots + |p_n - a_n|}{|a_1 - \overline{a}| + \ldots + |a_n - \overline{a}|} \qquad (3)$$

where a_i is the actual values and p_i stands for predicting values.

3 Empirical Analysis

3.1 Data Retrieving

As aforementioned, we first collected data from PM25.in. Our data collection began at December 20th 2013. The data we collected involves PM2.5 and related records (e.g. O3, SO2 and etc.) of 945 monitoring stations of China.

3.2 Data Preprocessing

The research interest of this paper is focus on PM2.5 prediction in Beijing. Therefore, data from all other cities are filtered out. In addition, records of O3 and other related ones are not of our research interests here either, therefore they are eliminated. Thus, 1944 records of PM2.5 time series of 12 monitoring stations in Beijing remains for our research. The date of data ranges from December 20th 2013 to March 11th 2014.

Olympic Sports Center Monitoring Station is calculated as the center of all stations, therefore data from Olympic Sports Center Monitoring station is selected as prediction target. The figure and table below demonstrate the distribution of attribute stations, whose data are used as attributes for our models (Fig. 4).

Time lag records then is calculated by the PM2.5 time series data we have. The maximum time lag we set is 3, therefore 1826 records remains for our research.

It is noticeable that, due to transmission error and other reasons, some values are missing in our data. Missing data is replaced with the average of nearest known value. For those attributes who start with or end with missing data, we simply drop the records (Table 1).

3.3 Model Exploration

Weka 3.6.9 is employed in our experiment. Weka (Waikato Environment for Knowledge Analysis) is a free data mining tool based on java, which is developed by University of Waikato. Weka includes a collection of machine learning and data mining algorithms which enable researchers employing data mining tools more easily.

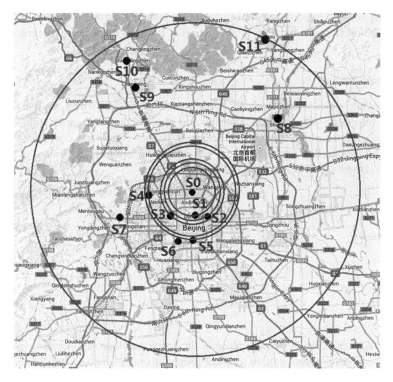

Fig. 4. Distribution of monitoring station in Beijing

Table 1. Name and stations and distances between them and target station

Station label	Station name	Distance (KM)
S0	Olympic Sports Center	0
S1	Dongsi	6.134933
S2	National Agriculture Exhibition Hall	7.402177
S3	Guanyuan	7.692285
S4	Wanliu in Haidian	9.388431
S5	Temple of Heaven	10.70871
S6	Wanshouxigong	12.18422
S7	Gucheng	19.66866
S8	New town in Shunyi	27.24251
S9	Changping Village	29.74192
S10	Dingling Mausoleum	37.61129
S11	Huairou Village	43.19275

As aforementioned, SVM is utilized in our model exploration process, which is implemented as SMOReg in Weka. Since our research focus on how the value of PM2.5 is influenced in dimension of time and space, parameters of SMOReg is set as default. Moreover, Random subspace is also implemented through Weka with default settings. 10-fold cross validation is also introduced in our empirical experiment.

3.4   Results Analysis

The following tables demonstrate the performance of models which includes PM2.5 series as attributes with different time lag and space scope, measured with different evaluation criteria, which are RMSE, RRSE, RAE and correlation coefficient respectively. The percentiles between values of evaluation criterion indicate the rate of change as larger scope or time lag (Table 2).

Table 2. Experiment result in terms of RMSE

Time lag	Scope				
	1	2	3	4	5
1	31.96	29.67	28.67	27.88	27.95
2	31.59	29.02	27.87	26.93	27.18
3	31.57	28.97	28.03	27.07	27.30

Generally speaking, RMSE decrease as the scope and time lag increase. However, there are some exceptions, e.g., in terms of scope dimension, as the scope increase to 5 (including historical PM2.5 information of 5 nearest five stations), the RMSEs increase. This indicate that the inclusion of PM2.5 information of fifth station introduces some extra interference rather than helpful information. In addition, in terms of time lag dimension, as larger time lag information included, RMSE decrease except for those whose scope is larger than 2 and time lag is larger than 2. This phenomenon may also indicate the information of farer station with larger time lag is interference. The two facts observed is easy to interpret. The larger time lag and the larger distance, the information is less relevant and the inclusion of such information will impair the performance of model. Moreover, another fact observed from experiment result enhances the interpretation above. As we can see, as the scope and the time lag increase, the performance improvement decrease, and performance is even impaired as scope and time lag reach certain point (Table 3).

Table 3. Experiment result in terms of RRSE

Time lag	Scope				
	1	2	3	4	5
1	30.85	28.64	27.67	26.91	26.98
2	30.49	28.01	26.90	26.00	26.23
3	30.47	27.96	27.05	26.13	26.35

The observation of the performance in terms of RRSE provides us the exact same result. (1) RRSE decrease as the scope and time lag increase, (2) as the scope increase to 5 (including historical PM2.5 information of 5 nearest five stations), the RRSEs increase, (3) as larger time lag information included, RMSE decrease except for those whose scope is larger than 2 and time lag is larger than 2, and (4) as the scope and the time lag increase, the performance improvement decrease, and performance is even impaired as scope and time lag reach certain point. It is not surprising the observations are the same. As indicate by the formulas of calculation of RMSE and RRSE, they both punishes outliers for larger weights by the square. The difference is that RRSE normalize errors by dividing the squared error of simply predicting average values rather than number of instance (Table 4).

Table 4. Experiment result in terms of MAE

Time lag	Scope				
	1	2	3	4	5
1	22.01	20.84	20.16	19.89	19.74
2	21.85	20.53	19.75	19.32	19.10
3	21.82	20.52	19.73	19.30	19.06

The performance result evaluated by RAEis generally same with those of RMSE and RRSE. However, the result is much more ideal. RAE decrease as the scope and time lag increase rigidly. In addition, as the scope and the time lag increase, the performance improvement decrease without exceptions. The inconsistences of the observation in terms of different criteria can be explained from their different focuses on error calculation. It is easier to interpret the differences from formula of RRSE and RAE. Their difference comes from squared error is added up in RRSE however absolute error is added up in RAE. Errors come from prediction ofoutliers is exaggerated in RRSE and RMSE. Therefore, regardless of the exceptions introduced by outliers, two facts can be assured, (1) prediction errors decrease as the scope and time lag increase, and (2) as the scope and the time lag increase, the performance improvement decrease (Table 5).

Table 5. Comparison of Performance with/without random subspace method

	RMSE	RRSE	RAE
Model with RS	24.73	23.87	17.71
Model without RS	24.75	23.89	17.65

Our last experiment is training models with all information of 11 attribute stations and all different time lags. In this case, there are 11*3 that is 33 attributes are introduced. For exploring the capability of performance enhancement of random subspace, models with and without random subspace, are trained for comparison. As we can see from the table above, model performance is slightly improve by the introduction of random subspace. This fact implies that the choice of whether including random subspace is depend on actual needs. As we know, ensemble algorithms are very time consumptive, and it is a question of balancing the performance improvement and time consumption in application. In addition, the phenomenon of prediction errors decrease as the scope and time lag increase, is reinforced again by the experiment result. As observed, performance of model with all information of 11 attribute stations and all different time lags is better than all of the models aforementioned with information of maximum 5 attribute stations.

4   Conclusions and Future Work

According to the empirical experiment result analysis above, we can draw conclusion that (1) prediction errors decrease as the scope and time lag increase, and (2) as the scope and the time lag increase, the performance improvement decrease. However, even though the more information, the better performance, it is noticeable that collection of information model training with massive information are expensive. The balance of them should be struggled in real application. In addition, whether including ensemble algorithms in model training is another balance of performance improvement and time consumption.

Moreover, this research proposes some questions for further exploration. (1) The prediction of PM2.5 is one hour forward only, which provides little implications in real life. Therefore, how to build prediction model which can predict one day or even one week forward is remained for research. (2) With hourly PM2.5 time series of more than 945 monitoring stations which covers 190 cities in China in hand, is that possible to construct propagation mode of PM2.5 in China? (3) This research only takes account historical PM2.5 time series data of peripheral monitoring stations as attributes. However, in practice, value of PM2.5 is affected by other elements such as SO2, how prediction model can be improved by introducing information of these elements is another question remains for further research.

Acknowledgment. This research work was partly supported by 973 Project (Grant No. 2012CB316205), National Natural Science Foundation of China (Grant No. 71001103, 91224008, 91324015), Beijing Social Science Fund (No. 13JGB035), Beijing Natural Science Foundation (No. 9122013), Beijing Nova Program (No. Z131101000413058), and Program for Excellent Talents in Beijing.

References

1. He, L.Y., Hu, M., Huang, X.F., Yu, B.D., Zhang, Y.H., Liu, D.Q.: Measurement of emissions of fine particulate organic matter from Chinese cooking. Atmos. Environ. **38**(38), 6557–6564 (2004)
2. Kleeman, M.J., Schauer, J.J., Cass, G.R.: Size and composition distribution of fine particulate matter emitted from wood burning, meat charbroiling, and cigarettes. Environ. Sci. Technol. **33**(20), 3516–3523 (1999)

3. Pope III, C.A.: Epidemiology of fine particulate air pollution and human health: biologic mechanisms and who's at risk? Environ. Health Perspect. **108**(Suppl. 4), 713 (2000)
4. Pope III, C.A., Ezzati, M., Dockery, D.W.: Fine-particulate air pollution and life expectancy in the United States. N. Engl. J. Med. **360**(4), 376–386 (2009)
5. Guo, Y., Jia, Y., Pan, X., Liu, L., Wichmann, H.: The association between fine particulate air pollution and hospital emergency room visits for cardiovascular diseases in Beijing, China. Sci. Total Environ. **407**(17), 4826–4830 (2009)
6. Cobourn, W.G.: An enhanced PM2.5 air quality forecast model based on nonlinear regression and back-trajectory concentrations. Atmos. Environ. **44**(25), 3015–3023 (2010)
7. Kloog, I., Koutrakis, P., Coull, B.A., Lee, H.J., Schwartz, J.: Assessing temporally and spatially resolved PM2.5 exposures for epidemiological studies using satellite aerosol optical depth measurements. Atmos. Environ. **45**(35), 6267–6275 (2011)
8. Ross, Z., Jerrett, M., Ito, K., Tempalski, B., Thurston, G.D.: A land use regression for predicting fine particulate matter concentrations in the New York City region. Atmos. Environ. **41**(11), 2255–2269 (2007)
9. Moore, D.K., Jerrett, M., Mack, W.J., Künzli, N.: A land use regression model for predicting ambient fine particulate matter across Los Angeles, CA. J. Env. Monit. **9**(3), 246–252 (2007)
10. Henderson, S.B., Beckerman, B., Jerrett, M., Brauer, M.: Application of land use regression to estimate long-term concentrations of traffic-related nitrogen oxides and fine particulate matter. Environ. Sci. Technol. **41**(7), 2422–2428 (2007)
11. Ordieres, J.B., Vergara, E.P., Capuz, R.S., Salazar, R.E.: Neural network prediction model for fine particulate matter (PM2.5) on the US–Mexico border in El Paso (Texas) and Ciudad Juárez (Chihuahua). Environ. Model Softw. **20**(5), 547–559 (2005)
12. McKendry, I.G.: Evaluation of artificial neural networks for fine particulate pollution (PM10 and PM2.5) forecasting. J. Air Waste Manag. Assoc. **52**(9), 1096–1101 (2002)
13. Sun, W., Zhang, H., Palazoglu, A., Singh, A., Zhang, W., Liu, S.: Prediction of 24-hour-average PM2.5 concentrations using a hidden Markov model with different emission distributions in Northern California. Sci. Total Environ. **443**, 93–103 (2013)
14. Dong, M., Yang, D., Kuang, Y., He, D., Erdal, S., Kenski, D.: PM2.5 concentration prediction using hidden semi-Markov model-based times series data mining. Expert Syst. Appl. **36**(5), 9046–9055 (2009)

Cloud Service-Oriented Modeling and Simulation of Regional Crowd Evacuation in Emergency

Weiqing Ling[✉], Jian Wang, and Xinquan Wei

CIMS Research Center, Tongji University, Shanghai, China
{lingweiqing, jwang}@tongji.edu.cn,
xinquan1103@126.com

Abstract. In the situation of unexpected incidents, how to quickly evacuate the people at the risk to the safe areas, is an important target for emergency preparedness and is also the key composition for emergency plans. A two stage crowd evacuation strategy—"from the accident area to refuge areas, and then to the safety areas", is put forward in this paper. Correspondingly, the refuge area evacuation model that aims at the first stage and the combination traffic evacuation model which focuses on the second one are established. Based on the speedy calibration of the locations of the refuge areas and safety areas, a cloud service-oriented visual prototype system for regional crowd evacuation is designed and developed. By conceiving a hypothetical incident scenario, the system functions of modeling, simulating and deriving are presented and thus the feasibility of the system is verified.

Keywords: Emergency · Crowd evacuation · Simulation and derivation · Cloud services

1 Introduction

Unexpected events cover a wide range, and involve in many factors. Once the treatment of emergency have been ineffective, the government will have a negative impact on the credibility and ultimately will lead to social instability, economic stagnation, and many other consequences that people do not want to see. In the period of transitions of social and economic structures, the research on management and tactics of regional crowd evacuation in emergency is not only an interdisciplinary foreland field, but also a very serious challenge to the governments at different levels [1].

A number of influence factors, strong randomness, and rich behavior features are involved in the study of regional crowd evacuation in emergency, thus how to quickly evacuate the people at the risk to the safe areas, is an important target for emergency preparedness and is also the key composition for emergency plans. Therefore, the research on regional crowd evacuation has become the focus and difficult problems of public protection fields. The current research may be divided into two aspects, one is on the characteristics of psychological behavior in the evacuation and the other on the evacuation modeling and simulation technology.

© Springer International Publishing Switzerland 2014
Y. Chen et al. (Eds.): WAIM 2014, LNCS 8597, pp. 130–140, 2014.
DOI: 10.1007/978-3-319-11538-2_13

Bryan studied behavioral response patterns in the fire by investigation of parties in the several major fire accidents of United States, the "return" behavior of family members was confirmed and the "herding phenomenon" was found [2]. Sime J.D. studied personnel behavior of evacuation in all aspects from a psychological level, such as spatial cognitive memory, sense of fire signal, audible alarm systems, exchange of information under the state of emergency, emergency assessment in fire, escape behavior logic, fire psychological interviews and post-disaster psychological assistance and so on [3]. After the 9.11 terrorist attack on the United States, the escape behavior of the staff of the New York World Trade Center was studied by the scholars from the U.S., Canada and UK [4, 5].

In terms of microscopic evacuation simulation, Helbing presented social force model for pedestrian dynamic, and studied relationship between disorder phenomenon of a high-density population and human disaster [6, 7]; Hughes proposed continuous model based on high density characteristics of the flow of human crowds [8]; Still G.K. used the mobile cellular automaton model to develop the software for simulating the crowd movement [9]; Schelhorn et al. studied and developed an "Agent model" that based on individual behavior model of crowd evacuation [10]. Zhao et al. aimed at the integral behavior, avoidance behavior and guide behavior of microscopic evacuation, built a microscopic evacuation model based on swarm intelligence theory [11]; Nuria and Norman researched on personnel escape behavior when emergency event occurred, mainly for personnel how to quickly find safe evacuation path in the situation of communicating with personnel and having self-awareness [12]. Gregor L. et al. mainly studied the impact of free stream, bottleneck area and space constraints on the evacuation in large-scale microscopic evacuation simulation [13]. With the optimization goal of minimizing the distance of pedestrian and export, considering a number of static constraints and factors such as the exit location, exit size, use-rate of exit and so on, Izquierdo J. et al. forecasted pedestrian evacuation times by using swarm intelligence theory [14]. Mohammad S. et al. established a multi-objective evacuation planning using image information system to specify the escape plan, demarcate the safe area, and finally determine the optimal route to each safety zone for individuals [15].

Through the analysis of the current research situation of regional crowd evacuation in emergency, it is obvious that there does not exist a set of theories and methods of public protection to deal with emergencies. Especially in the situation of unexpected incidents, the research of tactics of regional crowd evacuation, regional evacuation model and combination traffic evacuation model is considerably rare. Aimed at the issues discussed above, a two-stage crowd evacuation strategy—"from the accident area to the refuge area first, and then to the safety area", is put forward. Correspondingly, the regional crowd evacuation model from the accident area to refuge area that based on swarm intelligence theory and combination traffic evacuation model from the refuge areas to safe areas are established. By modeling the accident scene, calibrating speedily the location and amount of the refuge area and safe area, a cloud service-oriented visual prototype system for regional crowd evacuation is designed and developed. The efforts this paper contributed are to provide a novel evacuation scheme for the public protection in emergency and to supply the theoretical basis and method guidance for evacuating the public at risk to the safe area. In addition, the service-oriented idea is adopted in the system which is given accent by cloud services.

2   The Evacuation Model from Accident Area to Refuge Area

The crowd evacuation movement from the accident area to refuge area mainly depends on walking. By considering the evacuation individual and small groups as the unit intelligent particle and intelligent composite particles correspondingly, then establishing the mapping from the behavior characteristics of individual and group (degree of authority, family, force, etc.) to the properties of the particle (speed, position, size, etc.), the classical Particle Swarm Optimization (PSO) model is improved. Thus the evacuation model from the accident area to refuge area is built, which targets the shortest evacuation time and the most evacuation amounts under the dual effects of the small crowd evacuation behavior and information guidance.

According to the family behavior, herd behavior, authoritative compliance behavior, guiding behavior, exclusion behavior and obstacle-avoidance behavior, four kinds of behavior that increase the particle size and two kinds of behavior decrease the particle one are analyzed and summarized as follows.

The behavior that increase the particle size includes:

(1) Family behavior, indicates that the effects scope between individuals will cover the entire evacuation group. Once the individual distance is small enough, they immediately merged with a particle, and the vulnerable members of the small group (such as elderly, children, etc.) decide the speed of new particles.
(2) Herd behavior, it is the phenomenon that the people will come close to the larger crowd blindly in the process of evacuation. In this case the particles with small size will merge into the larger size one under the condition of the distance between them is close enough.
(3) Authoritative compliance behavior, namely to the people often obey the person of authority (such as police, place manager, etc.) and comply with the customary rules. Correspondingly when a particle approaches enough the one with higher authority, the former will be combined and provided with the speed as same as the latter.
(4) Guide behavior can also cause the increase of particle size. When the particle speed is more approximate and the distance is closer, if the same guide information is received at the same time by multiple particles, they will be merged with a unit.

In succession, the behavior that cause the decrease of the particle size are listed as:

(1) Exclusion behavior, means that the exclusion between individuals and between individual and boundary (such as walls and obstacles). The particles tend to become a number of new small particles with different speed.
(2) Obstacle-avoidance behavior, this indicates that a particle with greater size will often be separated into several smaller ones to avoid the obstacle, and then they meet again and restore to the original one after respectively flowing around on both sides of the obstacle.

According to the analysis, the formula of the particle's updating velocity and position are followed as:

$$v_{id+1} = \omega \times v_{id} + c_1 \times rand() \times (p_{id} - x_{id}) + c_2 \times rand() \times (g_{id} - x_{id}) \qquad (1)$$

$$x_{id+1} = x_{id} + v_{id+1} \qquad (2)$$

where v_{id+1} and x_{id+1} represent the particle speed and position at the next moment; is the inertia weight coefficient of the particle; c_1 and c_2 respectively denote the weight coefficient of complied with the most optimal individual position and the best global position, they can be expressed as:

$$c_1 = c_{1s} + \sigma_r \times \frac{(c_{1\varepsilon} - c_{1s}) \times t}{t_{max}} \qquad (3)$$

$$c_2 = c_{2s} + (1 - \sigma_r) \times \frac{(c_{2\varepsilon} - c_{2s}) \times t}{t_{max}} \qquad (4)$$

3   The Evacuation Model from Refuge Area to Safety Area

The evacuation model from refuge area to safety area concerns how to evacuate the people in the refuge area to the safety area orderly and effectively by using the combination mode of transportation. Considering the shortest evacuation time, the largest evacuation amount and the fewest casualties as the optimization goal, utilizing the combination transport (including bus, rail transportation, ship and other vehicles), the model for the second stage evacuation is established which aims to provide the types and amounts of combination vehicles.

Assumed that the refuge area amount is m and the evacuation numbers in each refuge area are $N_1^R, N_2^R, N_3^R, N_4^R, N_5^R$, the unit capacity of each transportation vehicle respectively are: Bus c_1, Light rail c_2, Tube c_3, Train c_4 and Ship c_5. The preparation time of personnel going up and down the transportation is $T_1^\gamma, T_2^\gamma, T_3^\gamma, T_4^\gamma, T_5^\gamma$.

Notated the total evacuation time as T, the total number of evacuation personnel as N, and the total casualty probability as P, they are expressed as:

$$T = \sum_{i=1}^{5}\sum_{j=1}^{m} t_j^{ai} p_j^{ci} \lambda_i^j + \sum_{i=1}^{5}\sum_{j=1}^{m} t_j^{\beta i} f(p_j^{ci}, \varphi_i) \lambda_i^j$$
$$+ \sum_{i=1}^{5}\sum_{j=1}^{m} t_j^{\gamma i} p_j^{ci} g(\omega_i) \lambda_i^j, \qquad j = 1, 2, \cdots, m \qquad (5)$$

$$N = \sum_{i=1}^{m}\sum_{j=1}^{5} c_i p_j^{ci} \lambda_i^j, \quad i = 1, 2, \cdots, m \qquad (6)$$

$$P = \sum_{i=1}^{m} \frac{1}{\sqrt{2\pi}} \int_{-\infty}^{Y-5} e^{-\frac{u^2}{2}} du, \quad i = 1, 2, \cdots, m \qquad (7)$$

The optimization goal can be written as Min T, Max N, Min P. The constraints are given as $\sum_{j=1}^{5} c_j p_i^{cj} \lambda_j^i \geq N_i^R, i = 1, 2, \cdots, m$ and $\sum_{i=1}^{m} p_i^{cj} \leq N_{cj}, j = 1, 2, 3, 4, 5$, in which $\lambda_i^j = \begin{cases} 0 \\ 1 \end{cases}, i = 1, 2, \cdots, m, j = 1, 2, 3, 4, 5$.

4 Speedy Calibration of the Refuge Area and Safe Area

In order to evacuate safely and effectively, it is vitally important to calibrate quickly the location and amount of the refuge area and safety area. The siting principles for refuge areas include: close to the accident point as far as possible in safe conditions, convenient for the evacuation of crowd in the building-concentrated area, has a certain holding capacity and convenient transportation. The siting principles for safety areas include: meet safety and stability principle, near the light rail train, transport terminals and other transport stations, as well as possessing a certain carrying capacity and so on.

Based the calibration principles described above, a multi-objective optimization model is built, which may be solved by genetic algorithm or complex networks method. The optimization goals are: the shortest transportation distance, the best utilization rate of the refuge areas and safety areas and the fewer number of the refuge areas and safety areas. The optimization goals can be formulated as:

$$\begin{cases} Min \sum_{i=1}^{n} \sum_{j=1}^{n} a_i d_{ij} x_{ij} \\ MinMax \sum_{j=1}^{n} \frac{a_i x_{ij}}{c_j} \\ Min \sum_{j=1}^{n} y_j \end{cases} \tag{8}$$

where a_i represents for the ith node weight coefficient of the population, d_{ij} is the distance between the ith and jth node, c_j acts as the capacity coefficient of the jth node. The value of x_{ij} is decided by the conditions that whether the node is allocated to the refuge area or not, while the value of y_j is ascertained by the situation that whether the node is selected as the refuge area or not.

The constraints are followed as:

$$\begin{cases} \sum_{j=1}^{n} x_{ij} \geq 1 \\ x_{ij} \leq y_j \\ \sum_{i=1}^{n} a_i x_{ij} \leq c_j \\ \sum_{j=1}^{n} y_j x_{ij} \leq d_{max} \end{cases} \tag{9}$$

where:

$$x_{ij} = \begin{cases} 0, & \text{otherwise} \\ 1, & \text{the } i\text{th node allocated to the } j\text{th refuge area} \end{cases}$$

$$y_j = \begin{cases} 0, & \text{otherwise} \\ 1, & \text{the } i\text{th node selected as the refuge area} \end{cases}$$

5   Design and Implementation of the System

5.1   System Design

As shown in Fig. 1, the function of the system is divided into two parts.

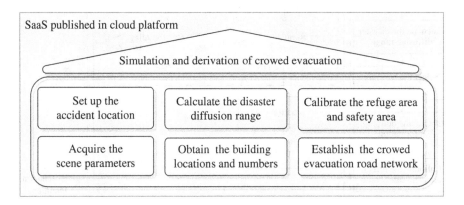

Fig. 1. The function of the system

The first part concerns the evacuation modeling, which includes the setting of accident point, acquirement of environmental parameters, calculation of disaster boundary, extraction the buildings in the disaster scope and the number of people in the buildings, calibration of the refuge area and safety area, and modeling the road network.

The second one which is based on the crowd evacuation model built above will realize the finally evacuation plan by simulating and deriving the process of evacuation. The final evacuation plan is comprised of the required evacuation time, the total number of people, the number of people in each refuge area and safety area, the transport capacity of transportation vehicles, and so on.

In addition the whole system is designed for the cloud computing environment to provide the application level service. It means that the system can be published towards the cloud platform (Google App Engine, Amazon EC2, Microsoft Azure, etc.) as a SaaS application service.

Figure 2 illustrates the system operation process which consists of three steps. In the cloud services environment, each step can be encapsulated as a service and published to the cloud platform in which provides the cloud services for other application. They are evacuation information service, simulation and derivation service and evacuation planning service.

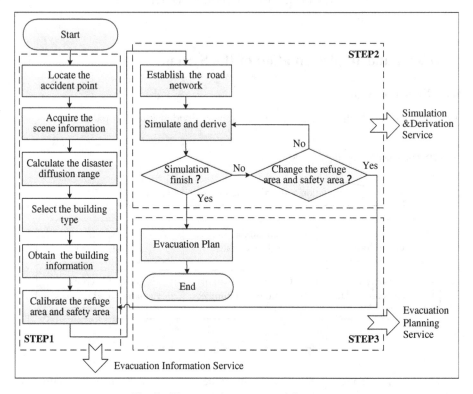

Fig. 2. The operation process of the system

The three procedures are described below:

Step1: firstly set up the accident location and scene parameters (including the accident time, temperature, relative humidity, atmospheric stability, wind speed, wind direction, surface roughness, gas type, gas leakage intensity and rate, etc.); then based on the model of gas leakage, such as Gauss puff diffusion model, calculate the disaster diffusion range, which will be divided into the death area, injured area and flesh-wound area; according to the specific scenario, select the building types (community, school, office building, etc.) within the scope of disaster area and obtain the corresponding building location and number information from web services or other third-party applications; finally, the locations and numbers of refuge area and safety area are solved by using the speedy calibration algorithm.

Step2: according to the location information of the buildings and refuge areas, establishing the area crowed evacuation network by the web-based GIS system; thereafter simulate and derive the two stage evacuation models; in the derivation process, dynamic adjustment of the quantity and location of the refuge areas and safety areas is supported in case of the evacuation don't have the expected effects.

Step3: when the simulation and derivation process is end, the detailed evacuation plan is implemented which consists of the evacuation information, the simulation and derivation information, as well as the final evacuation schema.

5.2   System Implementation

The Browser/Server (B/S) architecture is employed in the system which divides into the view layer, business logic layer and data layer. The view layer, in charge of interaction with the user interface, is developed by the web technologies (HTML, CSS and DIV, etc.); the business logic layer that interacts with the Web GIS system (such as Google Map), the database system, as well as the response for user request is realized by JavaScript, JQuery and Ajax technologies; the data layer storages and manages the application data (accident scene, simulation and deduction information, evacuation plans, etc.) and the necessary system data (user information, authority and authentication, etc.)

Assuming a scenario that a chlorine leakage incident occurred in a chemical industry park, Figs. 3, 4 are the system screenshots that illustrate the evacuation from the accident area to refuge areas, where the accident position, disaster range, the location and number of refuge areas, the evacuation path network are exhibited in

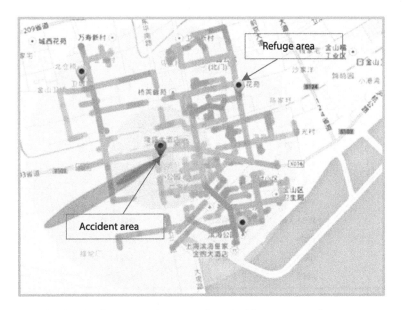

Fig. 3. The first evacuation stage: from accident area to refuge areas

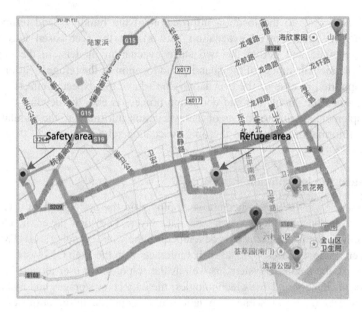

Fig. 4. The second evacuation stage: from refuge areas to safety areas

Fig. 3, while the evacuation road networks from the refuge areas to safety areas are shown in Fig. 4. The simulation and derivation information of the crowed evacuation as well as the final evacuation plan are shown respectively in Figs. 5, 6.

事故发生时间	08:00		
推演开始时间	08:00	推演结束时间	08:30
预计疏散结束时间	10:51		
避难区个数	3	安全区个数	2
	名称	经纬度	
避难区1	金山区枣泉街	经度: 121.34467 纬度: 30.72365	
避难区2	金山区卫生局	经度: 121.34997 纬度: 30.71182	
避难区3	金山区卫城村	经度: 121.32226 纬度: 30.72539	
安全区1	金山区新卫公路	经度: 121.29547 纬度: 30.73121	
安全区2	金山区卫清东路	经度: 121.36302 纬度: 30.73686	
疏散总人数	38634		
	需进入人数	当前已进入人数	待进入人数
避难区1	20102 人	13202 人	6900 人
避难区2	15932 人	8862 人	7070 人
避难区3	2600 人	1760 人	840 人

Fig. 5. The simulation and derivation information

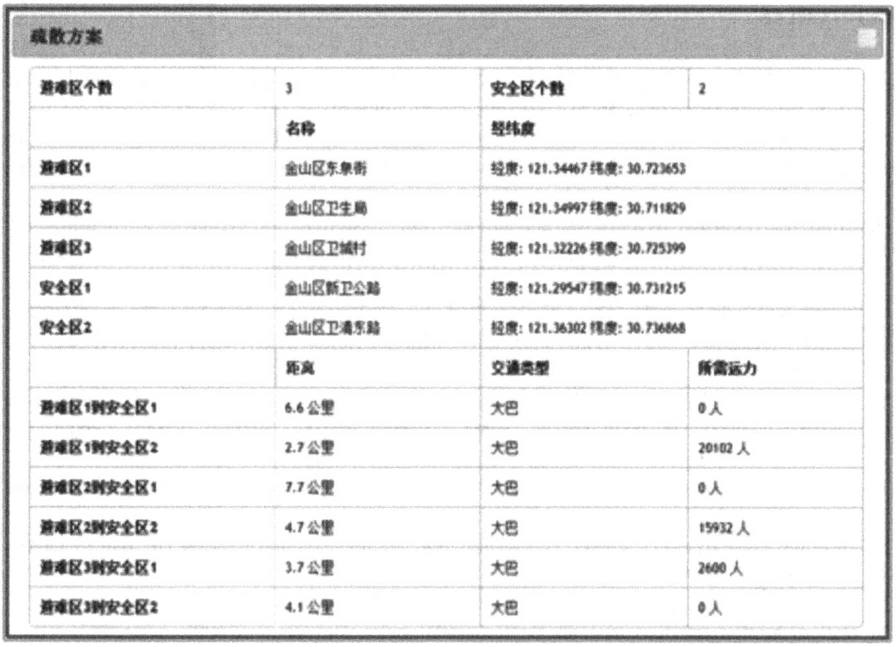

疏散方案

避难区个数	3	安全区个数	2
	名称	经纬度	
避难区1	金山区东泉街	经度: 121.34467 纬度: 30.723653	
避难区2	金山区卫生局	经度: 121.34997 纬度: 30.711829	
避难区3	金山区卫城村	经度: 121.32226 纬度: 30.725399	
安全区1	金山区新卫公路	经度: 121.29547 纬度: 30.731215	
安全区2	金山区卫清东路	经度: 121.36302 纬度: 30.736868	
	距离	交通类型	所需运力
避难区1到安全区1	6.6 公里	大巴	0 人
避难区1到安全区2	2.7 公里	大巴	20102 人
避难区2到安全区1	7.7 公里	大巴	0 人
避难区2到安全区2	4.7 公里	大巴	15932 人
避难区3到安全区1	3.7 公里	大巴	2600 人
避难区3到安全区2	4.1 公里	大巴	0 人

Fig. 6. The final evacuation plan

6 Conclusions

This paper focuses on how to quickly evacuate and transfer the people at the risk to the safe areas in the unexpected incidents. Through the analysis of the current research situation of regional crowd evacuation in emergency, a two stage evacuation strategy — "from the accident area to the refuge area, and then to the safety area", is proposed. The evacuation models correspond to two stage evacuation are set up. By the aid of the speedy calibration of the refuge areas and safety areas, a cloud service-oriented visual prototype system for regional crowd evacuation is designed and developed. The feasibility of the system is verified by a hypothetical incident scenario.

At present, the system has been published in the Google cloud platform as a SaaS application. The next work we are preparing is to encapsulate the system partial function to produce the cloud services, for instance, evacuation information service, simulation and derivation service and evacuation planning service etc.

Acknowledgements. This work was financially supported by the National Natural Science Foundation of China (No.71273188).

References

1. Kou, G., Li, S.M., Wang, S.Y., Yang, L.X.: Emergency management for incidents and crisis. Syst. Eng. Theory Pract. **35**(5), 1–4 (2012). (in Chinese)
2. Bryan, J.L.: A review of the examination and dynamics of human behavior in the fire at the MGM Grand Hotel, Clark County, Nevada, as determined from a selected questionnaire population. Fire Saf. **5**, 233–240 (1983)
3. Sime, J.D.: Crowd psychology and engineering: designing for people or blabbering. In: Proceeding of Engineering for Crowd Safety, Elsevier Science Publishers, pp. 119-133. Elsevier Science Publishers Press, New York (1993)
4. Gershon, R., Hogan, E., Qureshi, K., Doll, L.: Preliminary results from the world trade center evacuation study—New York City. Morb. Mortal. Wkly. Rep. (MMWR) **53**(35), 815–817 (2004)
5. Galea, E.R., Lawrence, P., Blake, S., Dixon, A.J.P., Westeng, H.: The 2001 world trade centre evacuation. In: Nathalie, W., Peter, G., Hermann, K., Michael, S. (eds.) Pedestrian and Evacuation Dynamics 2005, pp. 225–238. Springer, Heidelberg (2005)
6. Helbing, D., Molnar, P.: Social force model for pedestrian dynamics. Phys. Rev. **E51**, 4282–4286 (1995)
7. Helbing, D., Farkas, I., Vicsek, T.: Simulating dynamical features of escape panic. Nature **407**(6803), 487–490 (2000)
8. Hughes, R.L.: The Flow of Human Crowds. Ann. Rev. Fluid Mech. **35**, 169–182 (2003)
9. Still, G.K.: Crowd dynamics. Ph.D. Thesis, Warwick University (2000)
10. Schelhorn, T., O'Sullivan, D., Haklay, M, Thurstain-Goodwin, M.: STREETS: An Agent-Based Pedestrian Model, Working Paper 9, Centre for Advanced Spatial Analysis, University College London (1999)
11. Zhao, R.Y., Wang, J., Ling, W.Q., Ren, Z.H.: Model for high-density pedestrian evacuation in public building based swarm intelligence. Comput. Sci. **37**(10A), 85–88 (2010). (in Chinese)
12. Nuria, P., Norman, I.: Modeling crowd and trained leader behavior during building evacuation. IEEE Comput. Graph. Appl. **26**(6), 80–86 (2006)
13. Gregor, L., Dominik, G., Kai, N.: The representation and implementation of time-dependent inundation in large-sale microscopic evacuation simulations. Transp. Res. Part C Emer. Technol. **18**(1), 84–98 (2010)
14. Izquierdo, J., Montalvo, I., Pérez, R., Fuertes, V.S.: Forecasting pedestrian evacuation times by using swarm intelligence. Phys. A Stat. Mech. Appl. **388**(7), 1213–1220 (2009)
15. Mohammad, S., Ali, M., Mohammad, T.: Evacuation planning using multi-objective evolutionary optimization approach. Eur. J. Oper. Res. **198**(1), 305–314 (2009)

Strategy Study on Mass Evacuation with LBS Information

Yi-zhou Chen, Rui Yang$^{(\boxtimes)}$, and Yi Liu

Institute of Public Safety Research, Department of Engineering Physics,
Tsinghua University, Beijing, China
ryang@mail.tsinghua.edu.cn

Abstract. The primary goal of mass evacuation in one region is to evacuate as many evacuees as possible to safe areas in the shortest time when emergency events occur. This paper chooses one scenic spot in Beijing as the research object, and uses a study method which combines by an application of real-time heat map from Baidu map in the big data era and computer simulation of the BuildingEXODUS software to put forward the emergency evacuation plan or guided strategy for evacuees. We present the orderly index to improve the efficiency of mass evacuations, and bring up a formula that demonstrates when the value of evacuation speed of individual gets closer to that of average evacuation speed of group, orderly index becomes higher. And we build a model and obtain the results of the simulation, that is, evacuation time is from 589.25 s to 443.70 s in the four case, and 443.70 s are saved by orderly guided and effective management strategy for mass evacuation. Besides, we can get the values of optimal performance statistic respectively are 0.309, 0.211, 0.128 and 0.088 according to the four cases. So it is necessary to make a reasonable guidance in the every evacuation exits to improve the efficiency of evacuation.

Keywords: Mass evacuation · Heat map · Strategy · LBS

1 Introduction

Location Based Services (LBS) is a value-added service that provides users with the corresponding service to obtain the position information of mobile terminal through the networks with the support of electronic map platform [1]. LBS are used in a variety of contexts, such as health, indoor object search [2], entertainment [3], work, personal life, etc. [4]. LBS include services to identify a location of a person or object, such as discovering the nearest banking cash machine (a.k.a. ATM) or the whereabouts of a friend or employee. LBS include parcel tracking and vehicle tracking services. LBS can include mobile commerce when taking the form of coupons or advertising directed at customers based on their current location. They include personalized weather services and even location-based games. They are an example of telecommunication convergence [5].

The heat map is a product, which real-timely describes distribution, density, and change trends of the crowd with overlaid blocks of different colors in the Baidu map, and also gives a convenient travel service based on LBS in big data era. As a form of data visualized, heat map indicates the different crowded degree of pedestrians in one site with different colors and brightness, and can comprehensive provide the backtracking

© Springer International Publishing Switzerland 2014
Y. Chen et al. (Eds.): WAIM 2014, LNCS 8597, pp. 141–150, 2014.
DOI: 10.1007/978-3-319-11538-2_14

heat value of 7 days before the day and the predicted population flow of the time before 24 h that day. Besides, the data of heat map are updated per 10 min and we can master the situation of one region. At present, the change of pedestrians density is fully displayed in the hundreds of cities, thousands of scenic spots and business circles through mining and analysis of the LBS big data of Baidu in China, and shows the national audience the value of trend judgment of the whole society with Baidu big data [6, 7].

Recently, computer simulation technology developed overseas is commonly applied in working out safety management scheme, and researchers domestic and overseas have conducted a great deal of researches in the evacuation of buildings, subways, stadiums and traffic networks [8–10]. In the process of planning, simulation model provides the assessment network clearance time or detects bottleneck problems for emergency evacuation [11, 12]. When researchers pass through the streets safely by underground passages, studies show that pedestrians subjective conditions worse than objective conditions [13]. Jorge A. Capote, et al. present a real-time model integrated in a Decision Support System (DSS) for emergency management in road tunnel, the main difference is that the proposed model can provide results faster than real-time (less than 5 s) while the run time of the other models is really higher [14]. A novel approach to represent space, which called the 'Hybrid Spatial Discretisation' (HSD), in which all three spatial representations can be utilized to represent the physical space of the geometry within a single integrated software [15]. A loudspeaker, which provided people with an alarm signal and a pre-recorded voice message, was found to perform particular well in terms of attracting people to the exit, independent of which side of the tunnel the participants were following [16]. To test the predictive capabilities of different evacuation modeling approaches to simulate tunnel fire evacuations, the study is based on the a priori modeling vs. a posteriori modeling of a set of tunnel evacuation experiments performed in a tunnel in Stockholm, Sweden [17]. Without crowding within the aisle, it would reduce the evacuation time by optimal design for the width of the entrance, opening exits and layout of escaping passages [18, 19].

However, the emergency evacuation study based on big data is scant, of which the relevant theoretical basis is still relatively weak, especially in the application of heat map based on big data for mass evacuation which has not been reported yet. In order to master the real-time dynamic changes of crowded density in one zone, making plans or providing strategies for pedestrians evacuation in one region have become key subjects when an emergency event happened. So it can make a far-reaching implication on the urban safety and social stability.

2 Impact Factors for Mass Evacuation

2.1 Evacuation Order

Mass evacuation is different between emergency and general state. Because of the difference of personal emotions, and the subjective judgment when emergencies happen, crowding and trampling are prone to occur. Therefore, this article puts forward the index of evacuation order under state of emergency evacuation, namely, orderly index. Orderly index refers to an indicator of the staff in evacuation zone maintaining an orderly evacuation, after receiving the evacuation instruction or judging by themselves to avoid local congestion, stampede and casualty incident. In the process of population evacuation, the number of personnel evacuation is nearly equal per unit time, and gets higher when the speed of staff going out of the exit is stable.

The higher the evacuation ordering, the better consistency of evacuation speeds among individuals in the evacuation group. Given the correlation between orderly index and individual evacuation speed is robust, this article chooses the evacuation speed as the index of evacuation ordering.

It's assumed $V = (V1, V2, V3, \ldots, Vn)$, Vi is evacuation speed of individual i, the average speed of evacuation group $\overline{V}$ can be expressed in terms of the following formula:

$$\overline{V} = \frac{\sum_{i=1}^{n} V_i}{n} \tag{1}$$

The deviation value of individual evacuation speed relatively to the average evacuation speed $\overline{S}$ can be expressed as follows:

$$\overline{S} = \frac{\sum_{i=1}^{n} |V_i - \overline{V}|}{n} \tag{2}$$

The orderly index r can be expressed as follows:

$$r = 1 - \frac{\overline{S}}{\overline{V}} \tag{3}$$

From above, it can be obtained from the following formula:

$$r = 1 - \frac{\sum_{i=1}^{n} |V_i - \overline{V}|}{\sum_{i=1}^{n} V_i} \tag{4}$$

Formula (4) shows when the value of Vi gets closer to that of $\overline{V}$, orderly index becomes higher, and vice versa. So, the higher the evacuation order, the better the orderly index.

2.2   Potential and Exit Attractiveness

The exit potential for an external exit seeds the overall potential map, it reflects the catchment area of the exit on the potential map. The seeding of the potential map will have no impact upon the use of the target exit system or the use of the local familiarity system. For internal exits, the exit potential can be used to control the potential map within the compartment concerned. The higher the exit potential, the less attractive the exit becomes, as occupants head for positions of lower potential. The default value for the exit potential of all exits is 100.0, it mean all exits are equally attractive. To make an exit more attractive, its exit potential must be decreased relative to the other exits. The exit attractiveness attribute refers to the probability that an occupant will be aware of the existence of this exit. The exit attractiveness figure is interrogated by each occupant individually to determine whether they are aware of a particular exit [20].

2.3   Relevant Parameter

Exit efficiency in the evacuation is evaluated by two parameters which respectively are Optimal Performance Statistic (OPS), and Mean Non-flow Statistic (MNS) [20].

The OPS is to evaluate whether the utilizing efficiency of every exit is optimum. OPS can be expressed as follows:

$$OPS = \frac{\sum_{i=1}^{n}(TET - EET_i)}{(n-1)TET} \qquad (5)$$

Where n is the total number of exits for personnel evacuation; EET_i is the time cost by the last person, s; TET is the total evacuation time, that is, the maximal of EET_i, s.

MNS is to measure the time nobody go through the passage during evacuation so as to evaluate the efficiency of the evacuation of whole pedestrian. For every exit, MNS is obtained from the expressions:

$$MNS_i = \frac{TNF_i}{TFT_i} \qquad (6)$$

Where MNS_i is the calculated time nobody pass the exit i; TNF_i is the time exit i has nobody to go through, s; TFT_i is the time people use exit i, and is equal to subtract EFT_i from EET_i, s; EET_i is the time cost by the last person, s; TFT_i is the time cost by the first person, s.

3   Evacuation Model

3.1   Model Building

In the case, the mass evacuation is studied based on one square (see Fig. 1(a)) with the length of 250 m and width of 90 m in Beijing, and a real-time heat map (see Fig. 1(b)) for the crowd density is used in this paper. There are ten exits and Fig. 1(a) shows the layout of the square of which width of exit 1 and exit 2 is 6 m, exit 3 and exit 10 is 8 m, exit 4, exit 5, exit 6 and exit 7 is 4 m, and exit 6 and exit 7 is 5 m.

3.2   Simulation

Then, we use the Building EXODUS software to simulate the place with 20000 people that was estimated by the real-time heat map from Baidu, and Fig. 2(a) shows the initial

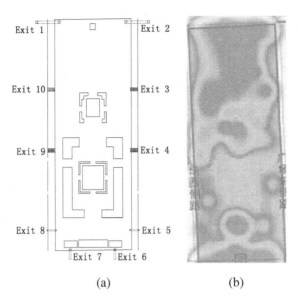

(a)   (b)

Fig. 1. A square layout and heat map

density and distribution of pedestrians in the square. According to the results of domestic and international research relevant, it shows that due to the differences of pedestrians age, physiology (height, weight, etc.) and investigating locations, the pedestrians speed is usually from 0.9 m/s to 1.5 m/s. We adopt pedestrians speed is 0.9 m/s in general condition and 1.5 m/s in emergency situation. In addition, the other related parameters will be assumed by Building EXODUS software with default Values. So the changes of simulating scenarios in general condition can be seen in Fig. 2(b–j) at different time.

Some related values according to simulation results in cases are shown in Tables 1, 2, 3 and 4 based on above studies.

3.3 Results Analysis

The simulation results show the evacuation time of ten exits is not identical. Therefore, the efficiency of evacuation can be enhanced by increasing the attraction (attract people to the exits evacuate in advance), improving the orderly index, and increasing the guiding measures. It is not hard to figure out that the longest time is 589.25 s, and the shortest is 443.70 s in the four evacuating cases, that is, 145.55 s is shortened through optimal adjustment to save the precious time for emergency rescue work.

Based on the different cases above, according to OPS formula, we can get the values of OPS 0.309 in case 1, 0.211 in case 2, 0.128 in case 3 and 0.088 in case 4. Calculation results show the efficiency of the first three is lower, and in case 4, it is increased by adding necessary guiding measures to make the exits used more evenly. But Mean Non-flow Statistic of each exit is less than 0.1, which illustrates traffic efficiency of some exit is low while all the exits have been used, so the traffic effect is

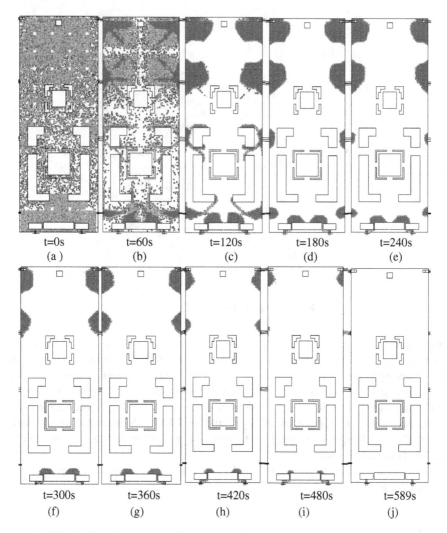

t=0s   t=60s   t=120s   t=180s   t=240s
(a)   (b)   (c)   (d)   (e)

t=300s   t=360s   t=420s   t=480s   t=589s
(f)   (g)   (h)   (i)   (j)

Fig. 2. Simulation evacuation scenarios in general condition in one square

Table 1. The related value in case 1

Exit No.	Evacuees	Initial through time(s)	Last through time(s)	The whole time(s)	MNS
1	3345	5.12	589.25	584.13	0.009
2	3381	5.12	578.11	572.99	0.009
3	3478	4.63	482.20	477.57	0.010
4	1067	7.52	248.44	240.92	0.031
5	1137	4.99	261.42	256.43	0.019
6	908	8.23	559.05	550.82	0.015
7	920	6.25	549.76	543.50	0.011
8	1120	3.78	248.42	244.64	0.015
9	1071	5.10	248.71	243.61	0.021
10	3573	5.24	489.57	484.33	0.011

Table 2. The related value in case 2

Exit No.	Evacuees	Initial through time(s)	Last through time(s)	The whole time(s)	MNS
1	2044	5.16	384.92	379.76	0.014
2	2063	6.23	364.15	357.92	0.017
3	3186	4.22	393.50	389.28	0.011
4	1692	5.06	375.89	370.83	0.014
5	1969	5.58	436.09	430.50	0.013
6	1035	4.59	521.67	517.08	0.015
7	1065	2.52	513.69	511.17	0.009
8	2048	2.73	452.09	449.36	0.006
9	1793	3.50	399.92	396.42	0.009
10	3105	3.74	384.68	380.94	0.010

Table 3. The related value in case 3

Exit No.	Evacuees	Initial through time(s)	Last through time(s)	The whole time(s)	MNS
1	2241	4.51	422.09	417.57	0.011
2	2229	3.10	400.97	397.87	0.008
3	2982	3.19	377.42	374.23	0.009
4	2087	2.36	472.00	469.64	0.005
5	1750	4.54	397.75	393.21	0.011
6	981	7.55	462.57	455.02	0.015
7	899	3.79	418.86	415.06	0.017
8	1748	3.63	382.59	378.96	0.015
9	2091	3.89	470.06	466.17	0.041
10	2992	3.04	372.22	369.19	0.008

Table 4. The related value in case 4

Exit No.	Evacuees	Initial through time(s)	Last through time(s)	The whole time(s)	MNS
1	1864	5.16	347.75	342.59	0.015
2	1904	6.47	335.42	328.95	0.020
3	3576	4.22	443.70	439.47	0.010
4	1814	5.06	405.26	400.20	0.013
5	1757	5.58	396.25	390.67	0.014
6	889	4.59	443.48	438.89	0.010
7	929	2.52	441.09	438.57	0.006
8	1842	2.73	407.25	404.52	0.007
9	1945	3.50	434.97	431.47	0.008
10	3480	3.74	372.22	430.58	0.009

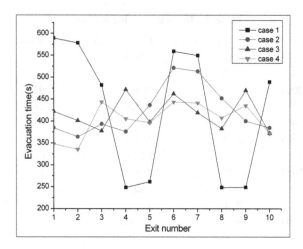

Fig. 3. Relation of exit number and evacuation time

poorer. Therefore, guidance should be increased to induce the staff to choose appropriate exits to evacuate.

Figure 3 shows the evacuation time of every exit in case 4 is more balanced and the total evacuation time is shortened by 145.55 s than that of the general case to achieve the best result when guiding measures are taken.

4 Conclusion

In this paper we have presented a new study method and considered many factors to make a optimum plan for pedestrians evacuation in one region.

(1) Many factors should be considered for the design of evacuation exits, in order to improve their comprehensive efficiency and reliability.
(2) The demand for quantity of evacuees has a prodigious impact on the evacuation efficiency. So effective analysis of the relationship between quantity of evacuees and evacuation speed is conducive to provide a reference basis for emergency evacuation in a place where has high-density public.
(3) Pedestrians factors such as age, gender and physical condition have a great influence on the evacuation speed in the emergency event. The evacuation speed or evacuation speed range should be comprehensively evaluated by experiments data when those factors have a big span.
(4) We put forward the orderly index for indicating the evacuation state in one region evacuees being guided or judging by themselves to avoid local congestion, stampede and casualty incident in this paper. In reality, we can use the guided strategy to explain or match this parameter. Optimally orderly index is 1.0, in fact, the index is less than 1.0. So, the evacuation order is better when the value tends to be 1.0.

(5) In fact, the quantity of people in one region can be easy got by data acquisition methods, while it is difficult to estimate population density and distribution real-timely. Amazingly, baidu made the breakthrough with the heat map to solve the problem. Heat map data, however, can only make the preliminary estimate, and the proportion of people who use baidu map is changing with time. We present a preliminary study on mass evacuation with LBS information in this paper, and should be further studied in the future.

Acknowledgement. The authors appreciate the Project supported by National Natural Science Foundation of China (NO. 91024032;91224008) and The National Science Foundation for Post-doctoral Scientists of China (NO.2013M540977).

References

1. Location Based Services. http://hi.baidu.com/wangdong8611/item/9458849e1708b7dc1e 4271fd3
2. Guo, B., Satake, S., Imai, M.: Home-explorer: ontology-based physical artifact search and hidden object detection system. Mob. Inf. Syst. **4**(2), 81–103 (2008). (IOS Press)
3. Guo, B., Fujimura, R., Zhang, D., Imai, M.: Design-in-play: improving the variability of indoor pervasive games. Multimedia Tools and Applications (2011). doi:10.1007/s11042-010-0711-z
4. Deuker, A.: Del 11.2: Mobility and LBS. FIDIS Deliverables, **11** (2) (2008)
5. Location-based service. http://en.wikipedia.org/wiki/Location-based_service
6. Heat map and big data. http://www.donews.com/it/201404/2743379.shtm
7. Heat map in Baidu. http://baike.baidu.com/view/3668215.htm?fr=aladdin
8. Jiang, C.S., Yuan, F., Chow, W.K.: Effect of varying two key parameters in simulating evacuation for subway stations in China. Saf. Sci. **48**, 445–451 (2010)
9. Zhong, M.H., Shi, C.L., Tu, X.W., et al.: Study of the human evacuation simulation of metro fire safety analysis in China. J. Loss Prev. Process Ind. **21**, 287–298 (2008)
10. Chen, Y.Z., Cai, S.J., Deng, Y.F.: Simulation study on main affect factors to evacuation corridor. Appl. Mech. Mater. **170–173**, 3533–3538 (2012)
11. Jiang, C.S., Zheng, S.Z., et al.: Experimental assessment on the moving capabilities of mobility-impaired disabled. Saf. Sci. **50**, 974–985 (2012)
12. Shi, C., Zhong, M., et al.: Modeling and safety strategy of passenger in a metro station in China. Saf. Sci. **50**, 1319–1332 (2012)
13. Mahdavinejada, M., Hosseinib, A., Alavibelmana, M.: Enhancement HSE factors in pedestrian underpass regarding to chemical hazards, Mashhad, Iran. Soc. Behav. Sci. **51**, 10–14 (2012)
14. Capote, J.A., Alvear, D., Abreu, O., Cuesta, A., Alonso, V.: A real-time stochastic evacuation model for road tunnels. Saf. Sci. **52**, 73–80 (2012)
15. Chooramun, N., Lawrence, P.J., Galea, E.R.: An agent based evacuation model utilising hybrid space discretisation. Saf. Sci. **50**, 1685–1694 (2012)
16. Fridolf, K., Ronchi, E., Nilsson, D., Frantzich, H.: Movement speed and exit choice in smoke-filled rail tunnels. Fire Saf. J. **59**, 8–21 (2013)
17. Ronchi, E.: Testing the predictive capabilities of evacuation models for tunnel fire safety analysis. Saf. Sci. **59**, 141–153 (2013)

18. Jingwei, J., Zhen, W., Ke, W., et al.: An investigation of effects on evacuation of fire-protection evacuation walk caused by internal and external door width. J. Shenyang Jianzhu Univ.(Nat. Sci.) **29**, 698–702 (2013)
19. Jeon, G.-Y., Kim, J.-Y., Hong, W.-H., Augenbroe, G.: Evacuation per formance of individuals in different visibility conditions. Build. Environ. **46**, 1094–1103 (2011)
20. Fire Safety Engineering Group University of Greenwich, buildingEXODUS V4.0 (2004)

Online Social Network as a Powerful Tool to Identify Experts for Emergency Management

Wei Du[1], Wei Xu[2(✉)], Jianshan Sun[1,3], and Jian Ma[1]

[1] Department of Information Systems, City University of Hong Kong,
Tat Chee Avenue, Kowloon, Hong Kong
duwei_0127@163.com, isjian@cityu.edu.hk
[2] School of Information, Renmin University of China, Beijing, China
weixu@ruc.edu.cn
[3] School of Management, University of Science and Technology of China,
Hefei, Anhui, China
sunjs9413@gmail.com

Abstract. Widespread use of social network has changed people's daily life as well as the way of emergency management.As a useful tool for information dissemination, communication and collaboration, social network plays an important role in the process of emergency management: mitigation, preparedness, communications, response and recovery. Emergency problem solving nowadays often need experts from various domain areas such as medicine, nuclear, chemistry, information technology and so on. It's difficult and costly for local emergency management databases to be well prepared since emergency disasters are small probability events. This research captures the advantage of social network to tackle such issues. Expertise on online social network is available and trustworthy with supervision of crowds. Therefore, we propose a method though the integration of social position analysis and expertise level analysis to profile online individuals. Social position analysis captures the importance or prominence of an individual in social network, and expertise level analysis measures one's expertise relevance and expertise level. Expert finding on online social can also capture one's interest in the specific emergency disaster. We also give an empirical analysis in a selected small group. Outperformed individuals can be identified. More work is needed to validate the method on the specified social network in the future.

Keywords: Expert finding · Online social network · Emergency management

1 Introduction

Widespread use of web 2.0 has changed people's daily life in communication and collaboration. Online social network consisting of social events, involved individuals and social relations alters the traditional working way as well as the research strategy. Facebook, LinkedIn, Sina Weibo, We Chat and other online social networks are acting beyond their original function in communicating with friends, but being well used in various areas such as marketing, economics analysis and finance prediction. As a

© Springer International Publishing Switzerland 2014
Y. Chen et al. (Eds.): WAIM 2014, LNCS 8597, pp. 151–162, 2014.
DOI: 10.1007/978-3-319-11538-2_15

powerful tool to organize local individuals and local communities in a unified and dynamic system, online social network also reveals its utility in emergency management in every stage [1]. The most intuitive use of online social network is to regard it as an online information repository [1, 2]. Information exchange, dissemination and propagation can be fast in online social network by involvement of billions of users. Response in online social network is immediate with huge amount of online users, this advantage makes online social network succeed most professional emergency management information systems [3]. In April 2013, Sina Weibo plays an important role in alerting dangers, asking help and seeking for survivors during Ya'an Earthquake in China. There are 64 million micro-blogs supporting the donation to Ya'an and raising 10 million Yuan in just 90 min. Besides, online social network functions as a crowdsourcing platform for disaster relief [4]. Online individuals contribute their wisdom to the correction of fraud information and problem solving, which can be extremely useful for emergency manager. Also, online social network facilitates communication and collaboration in emergency management. Social network makes it possible for local individuals and local communities from everywhere to communicate conveniently and collaborate without obstacles, through which the emergency management efficiency can be largely improved.

Emergency Management Information Systems (EMIS) are built by emergency management agencies to response disasters with prepared materials and human resources. Domain experts in EMIS are responsible to provide professional solutions for problems followed by disasters. Emergency disasters often need professionals from various fields such as medical area, computer science, biology, etc. in both industry and academia [3], which are often lack in local EMIS database. Especially for certain disasters requiring multi-area knowledge, it's difficult for local EMIS database to be well prepared [5] with experts from all necessary knowledge domains. Therefore, many emergency agencies tend to take the advantage of online social network to integrate the scattered experts from various knowledge domains [1, 3–5]. There are several reasons to do so. First, it facilitates the communication and collaboration of experts as well as agencies from different geographical locations. Besides, expertise of experts and wisdom of crowdsin online social network are available to public. It's more time and cost efficient to seek for experts in online social network than in local database. Maintaining experts in local database costs high because they are only utilized when emergency happens. Last, the specific feature of online social network makes identifying important actors easily. When emergency happens, experts and institutions with high expertise and reputation, and certain organizational media as main information channel are to be the focus in online social network. These important actors can be identified through different network analysis methods.

Sina Weibo with 500 million registered users, as the most popular online social network in China, has played an important role in emergency relief and response. In reality, it not only functions as information channel or information repository, but also can be an online expertise network. Emergency disasters in China would incur a widespread dissemination and discussion on Sina Weibo, which also involves lots of relevant professional experts and institutions. The missing Malaysia Airlines Flight 370 from Malaysia to Beijing on 8 March 2014, as an emergency disaster, led to a huge challenge to the emergency management system of Chinese and Malaysia government as well as other countries. The critical problem people care most is tracking the possible location of the lose-contact flight. Tomnod crowdsourcing platform[1] even releases the satellite images to find any possible clues of the flight by resort to the wisdom of crowds. Experts on Sina Weibo also provide strategies to solve tracking problem of MH370. For emergency agencies and government, it would be significant to identify such experts to form a specific team.

In this research, we propose a method to identify experts. One's social position is analyzed to seek for influential individuals who are important in social network. Centrality measures are used here. Besides, to guarantee individuals' expertise area and expertise level. One's post messages and social tags are compared with keywords relevant with problem solving in the emergency.

The rest of paper is organized as follows. Section 2 gives the related work about social network application in emergency management and experts finding in social network. Then the proposed method is illustrated in detail. A simple empirical analysis is present in Sect. 4. The paper concludes with the discussion of results. The last section also provides the drawbacks and future works of this research.

2 Related Work

2.1 Social Network Application in Emergency Management

According to [6], there are five disciplines of emergency management: mitigation, preparedness, communications, response and recovery. Previous researches study the role of social network in each discipline of emergency management. Reference [7] posits the importance of community for emergency management in achieving mitigation, a strongly knit social network is necessary to integrate sources to solve local problems. To satisfy the information communication in every discipline of emergency management, [8] provide guidelines to design an online social network, which can facilitate the cross-site collaboration and bring emergency domain experts together. Social network based community is also useful to [9] increase the predictive ability on preparedness component. Online communities of experts based on social network is significant to increase response ability of emergency management information system [3]. Social network has been proved efficient in every stage of emergency management through its function in gathering emergency information, building an information repository, information dissemination for recovery efforts, warning immediately as notification system [1]. Many examples are present to elaborate the relation between social network and emergency management [1].

In emergency disaster management, it's useful to integrate the scattered experts together by providing an online communication and collaboration system. However, little research is about expert finding in social network for emergency management.

2.2 Expert Finding in Social Network

Expert finding is to find users in social network with relevant expertise and a certain expertise level. The difference between social network based expert finding and

---

[1] http://www.tomnod.com/

traditional local expert finding is to analyze the social relationships between persons. Social network can be viewed as a graph by using nodes to denote person and lines (with or without direction) to denote their relationships. Density, degree centrality/prestige, closeness centrality/prestige and betweenness centrality/prestige are normal metrics to analyze social networks [10–14].

Social network provides a tool to detect relationship between experts and make the experts trustworthy, combining social network analysis and collaborative filtering can be efficient for expert profiling [15–17] proposes a propagation-based approach to combine one's local information and social relationships between individuals. Through Business Intelligence approach, [18] integrates social network PageRank algorithm and social link analysis for expertise finding. Reference [19] provides a method to identify the simulation rules to capture the dynamics in the social network for better expert finding.

3   Proposed Method

In this section, we propose a method to profile experts for specific emergency disaster by integrating an individual's social position in an online social network and one's domain expertise. The whole process is shown in Fig. 1. With the information overload in online social network, it would be less efficient to compute on the whole social network. Therefore, we first specify the network boundary by setting a list of constraints such as time, content, etc. The list of constraints is designed for a specific emergency disaster. Then, related data can be extracted from one's personal homepage. Social position analysis and domain expertise level are combined to profile an individual. Social position analysis measures an individual's degree of importance or

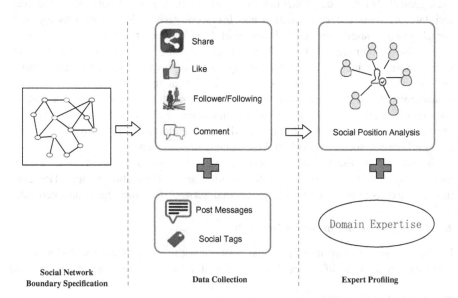

Fig. 1. The process of expert finding in online social network

prominence in online social network, which can be used to identify important or prominent individuals. The analysis of one's messages can also measure one's interest and attention on specific emergency disaster. For the specific purpose of seeking for professional experts to solve emergency problems, one's domain expertise is also an important feature to profile the individual. Individuals on the social network are profiled through our proposed method and professional experts are expected to be outperformed.

3.1 Boundary Specification and Data Collection

According to [20], positional, relational and event based approach are three main approaches to specify the network boundary. Event based boundary specification [21] means the network would be specified by a defined set of activities related to the specific event during a specific time. In this research, the set of activities is related with the specific emergency disaster. Keywords related with the occurrence of emergency disaster are given by an experienced expert as key-occurrence set: $\{key_1, key_2, ..., key_k\}$, and keywords related with problem solving of the emergency disaster are also given as key-solve set: $\{tec_1, tec_2, ..., tec_t\}$. Time period T is set from the start time when first emergency signal occurs to current time.

Post messages and social tags of an online individual are collected to profile one's domain expertise. Post messages are users' issued blogs and shared blogs with comments. For an individual in social network, one's post messages referring to occurrence of the emergency disaster are collected as one's attention to the disaster, while messages referring to problem solving of the emergency disaster are collected as one's attention to the unsolved problems of the emergency. Registered individuals are attached with social tags filled by themselves such as *student*, *cloud computing*, *data mining*, *cartoon* and so on to denote their interest, position as well as expertise area. Another kind of social tags is authorized by official such as certified *Academician*, *Actress*, etc. Certified tags can be more official and objective to denote one's expertise compared with self-filled tags.

Social information such as *share*, *like* and *comment* on one's emergency related messages, and followers/followings are collected to analyze one's position on the social network. Other users' *share, like* and *comment activities* of the individual's post messages referring to problem solving of the emergency disaster represent the importance degree of the message, which can be viewed as social acknowledgement to this individual's ability for the emergency. Besides, the individual's followers/followings are collected. Followers are fans of the individual while followings are users the individual pays attention to. If one user is both a follower and a following to the individual, then they are viewed as friends.

3.2 Social Position Analysis

According to [14], centrality and prestige are two generic metrics to measure the prominence of actors by analyzing relationship in social network. Centrality measures the involvement of an actor in the social network, regardless of senders or receivers.

On the other hand, prestige measures an actor who sent few relations but receive many relations. Prestige metric is more appropriate for expert identification, since experts are important actors with many admired fans. Prestige metric can dig out influential actors for the emergency. Similar with centrality measures, degree, closeness and between-ness are three main prestige measures [14]. Degree prestige is selected to measure the extent to which the individual infect others in social network. For binary networks, the degree prestige of note N_i can be computed as [12, 22]

$$P_D(N_i) = \sum_{j=1}^{g} x_{ji} \, (j \neq i) \tag{1}$$

where $x_{ji} = 1$ or 0 represent there is or no directed line from j to i and $\sum_{j=1}^{g} x_{ji}$ counts the number of incoming lines of node i.

In this research, we interpret the social network as valued graph. For each individual, the value on the directed line from j to i can be defined from four aspects: share, like, comment and follower. Shares, likes and comments are from other users to represent their attention to the individual's post messages related with the emergency disaster.

Post messages are first filtered by the keywords from key-occurrence set and key-solve set. Messages including keywords from both key-occurrence set and key-solve set are most likely about problem solving caused by the emergency disaster, and we name it as first-kind messages. Messages including keywords from only key-occurrence set reveal one's attention to the emergency, and we name it as second-kind messages. *Shares*, *likes* and *comments* for these two kinds of messages are given different weights. Information of whether the connected users are followers of this individual is also extracted. The *follow* activity shows one's respect, adoration and interest to the individual to a certain degree. Therefore, we can define the value on the directed line from other user j to this individual i as

$$v_{ji} = w_1 \cdot \left(cout_{sha,lik,com}^{1,j}\right)^n + w_2 \cdot \left(cout_{sha,lik,com}^{2,j}\right)^n + w_3 \cdot c_{fol}^{j}(j \neq i) \tag{2}$$

where $cout_{sha,lik,com}^{1,j}$ counts the number of user j's *share*, *like* and *comment* activities on first-kind messages, which are normalized as $\left(cout_{sha,lik,com}^{1,j}\right)^n = cout_{sha,lik,com}^{1,j}/cout_{sha,lik,com}^{1,max}$. $cout_{sha,lik,com}^{1,max}$ denotes the maximum number of one user's *share*, *like* and *comment* activities on first-kind messages to another user in the specified social network. Similarly, $cout_{sha,lik,com}^{2,j}$ counts the number of user j's *share*, *like* and *comment* activities on second-kind messages, which are normalized as $\left(cout_{sha,lik,com}^{2,j}\right)^n$. $c_{fol}^{j} = 1$, or 0 means user j follow or not follow individual i. The weight values order denotes the importance degree of three kinds of information from user j to individual i and can be decided by experienced managers.

Through this, the social network can be transformed into a valued directed network, as shown in Fig. 2. In the figure, nodes denote registered users on the online social network and the directed edge represents the like, comment, share and follow activities from user j to individual i. The acknowledgement and admiration degree from user j to individual i is computed as above: $0 \leq v_{ji} \leq 1$. v_{ji} is different with v_{ij}. For the directed valued network [12, 23], the degree prestige of note N_i can be computed as the summation of the incoming line values:

$$P_D(N_i) = \sum_{j=1}^{g} v_{ji} \ (j \neq i) \tag{3}$$

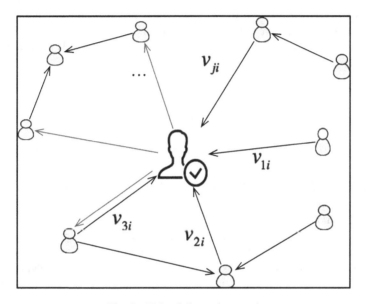

Fig. 2. Valued directed network

3.3 Expertise Level

Expertise level of the individual in online social network can be calculated from two aspects. For expertise finding problem, the critical issue is to find the relevant experts with high matching degree and expertise level. Keywords related with problem solving here are actually topics used to profile the expertise area, and can be used to query experts. On one hand, one's post messages reveal one's interest in and attention to the emergency disaster. If there are many professional terms about the knowledge related with the emergency disaster and problem solving in his/her original messages, then he/she probably be an expert in the knowledge domain of the emergency relief. Post messages used in this section are all original messages to reveal one's real expertise. Original messages are messages written by the individual but not shared from other sources. On the other hand, social tags are used to denote one's expertise. If the social tags match the specified professional keywords, or titles are present in the social tags, then the individual is probably the expert we expect. Detail algorithms are listed below.

Post Messages. For the purpose of identifying experts for emergency management, the individual's expertise area and expertise level are also profiled as important feature. As illustrated in data collection step, keywords related with problem solving are defined as key-solve set: $\{tec_1, tec_2, ..., tec_t\}$. Key-solve set involves professional terms defined by experienced managers, which can be relevant technology, method and strategy to solve the problems and support disaster relief operations. For example, for earthquake disaster, the key-solve set may include seismology, seismograph, earth crust, volcanic action, remote sensing, etc. as keywords; for wildfire disaster, the key-solve set may include other keywords like dry ice, aerosol fire extinguishing, etc. If these keyword co-occurred with sensitive words like *propose, method, solve* and so on, then the post messages are more likely about problem solving strategy for the disaster. Note that to measure the expertise level, all post messages are calculated here. For node N_i, frequency of the occurrence of the key-solve keywords in his/her all post messages are recorded as

$$Mes_i = \sum_{j=1}^{t} occur(tec_j) \tag{4}$$

where $occur(tec_j)$ represents the occurrence number of keyword tec_j in the individual's original post messages.

Social Tags. As aforementioned, there are two kinds of social tags for users in online social network collected: one is the self-filled tag, another is authorized tag. Our common notion is that authorized tags own more credibility than self-filled tags. Therefore, different weights are assigned for different kinds of tags. Besides the key-words included in key-solve set, information of the individual's expertise title occurring in social tags is also profiled. Expertise title information such as earthquake expert, cloud computing expert, academician, etc., so the key-title set can be defined as $\{manager, expert, engineer, researcher, professor, academician\}$. Different weights are given for different title. The social tags related information about key-solve set and information about title are separated here:

$$Tag_i = \sum_{j=s,a} w_j [w'_{kt} occur(key)^j + w'_t \cdot Title_t^j] \tag{5}$$

where $j = s$ or a represent the social tags are self-filled or authorized and weights are assigned for them, $occur(key)$ represents the overall occurrence number of keywords from key-solve set in social tags, and $Title_t$ represents the expertise title in social tags. w'_{kt} and w'_t are weights for information about key-solve set and information about expertise title.

At last, expertise level can be profiled through the integration of expertise from post messages and social tags after their normalization, which can be defined as

$$Exp_i = w_m Mes'_i + w_t Tag'_i \tag{6}$$

3.4   Expert Performance

Social position analysis profiles the prominence of the individual in a social network, and expertise level profiling one's expertise area and level. Both are indispensable for expert identification. So we define the expert performance in online social network by using the multiplication method:

$$P_i = P_D^*(N_i) \cdot Exp_i \tag{7}$$

4   Empirical Analysis

To validate the proposed method, we apply the method in the background of Sina Weibo to seek for experts for MH370 lose-contact emergency. Tracking problem is identified as the primary and critical problem for the emergency relief. Registered users as nodes on Sina Weibo constitute an online social network. Users publish micro-blogs and add social tags for themselves on Sina Weibo. Communications between different users are: *share*, *like* and *comment* on other's micro-blogs; *send* private letters; *follow* someone and become one's fan; *mention* someone, etc. The whole social network is huge with 5 hundred million registered users and more than 1 hundred million post micro-blogs on Sina Weibo, which makes the computation on the whole network difficult and inefficient. Keywords for key-occurrence set and key-occurrence set are first defined for network boundary specification. The time period is set from 8 March 2014 until now. Note that we only profile a small group for empirical analysis here. More research on the whole specified network is done in the future.

Experts who are experienced in such emergency disasters are invited to provide two kinds of keywords. One indicates the micro-blogs are about the MH370 lose-contact emergency, and another indicates the micro-blogs are about techniques or strategies for tracking problems of MH370. Keywords set are collected as: key-occurrence set {MH370, Malaysia Airlines Flight 370, Malaysia Airlines, etc.} and key-solve set {mathematic analysis, fuzzy mathematics, data mining, radar data analysis, satellite data analysis, supercomputing, trajectory prediction, information technology, flight route, black box detection, etc.}. Network is first specified by filtering micro-blogs through combination of above two keywords set. Post messages shared from other sources or tagged with news are not used to profile expertise level. Social tags with title like {engineer, data mining expert, research institution, professor, academician, etc.}. Research institution is adopted here because many scientific researchers' comments for this disaster are published on their research institution's homepage. A small group in the specified network is selected.

To simplify the process, weights are viewed as equal. By extracting their post messages as well as attached information (like, share, comment), social tags and followers, each individual is profiled according to our proposed method in this Section. An example of social position analysis is given in Fig. 3. Width of lines shows the communication frequency from one individual to another. For the user identified in center of the figure, its degree prestige is relatively high than others in the selected small group. By analyzing the individual's homepage, we can find that the original post messages related to MH370 include Big Data, black box, remote sensing, and wreckage salvage, satellite and so on with high frequency. The simple analysis shows the proposed method can be used to find experts for emergency in social network. However, more data are needed to give more detail analysis.

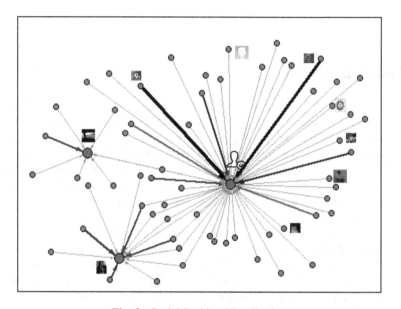

Fig. 3. Social Position Visualization

5   Conclusions and Future Work

With the development of Web 2.0, social network has changed the way of emergency management. The application of social network has increased the efficiency of emergency management. Nowadays emergency often needs lots of professionals with knowledge in certain domains. Social network is expected to be powerful in expert finding for emergency. In this research, we propose an innovativemethod combining social position analysis and expertise profiling for expertise finding in online social network for emergency management. Social position analysis captures the importance or prominence of an individual in social network, and expertise profiling measures one's expertise relevance and expertise level. Expert finding on online social can also capture one's interest in the specific emergency disaster.

However, there are several drawbacks of this research and some future works are needed. For the specified network analysis, only individual's degree prestige is measured. First, group-level degree prestige is needed to investigate whether the degree prestige in a social network is evenly or unevenly dispersed. Besides, keywords specified for the social network boundary and professional terms for expertise area determination are very important. These two keyword sets need to be changed and enrich dynamically to catch the latest development of the emergency. Third, for the empirical analysis, we only profile individuals from a selected small group. More data are needed here to profile the whole specified network. Weights and other parameters in the proposed method need to be decided by invited experts through scientific tools such as group decision analysis and AHP (Analytic Hierarchy Process). Still, experts will be invited to judge whether the identified experts in the whole specified network are appropriate for emergency problems solving. Questionnaires would be sent to selected individuals themselves to further test whether they want to contribute to the emergency problem solving. Last, the proposed model can be improved to expert team finding for the reason that collaboration is critically important.

Acknowledgment. This research work was partly supported by 973 Project (Grant No. 2012CB316205), National Natural Science Foundation of China (Grant No. 71001103, 91224008, 91324015), Beijing Social Science Fund (No. 13JGB035), Beijing Natural Science Foundation (No. 9122013), Beijing Nova Program (No. Z131101000413058), and Program for Excellent Talents in Beijing.

References

1. White, C., Plotnick, L., Kushma, J., Hiltz, S.R.: An online social network for emergency management. Int. J. Emer. Manag. **6**(3), 369–382 (2009)
2. Benbunan-Fich, R., Koufaris, M.: An empirical examination of the sustainability of social bookmarking websites. Inf. Syst. E-bus. Manag. **8**(2), 131–148 (2010)
3. Turoff, M., Chumer, M., Van de Walle, B., Yao, X.: The design of a dynamic emergency response management information system (DERMIS). JITTA **5**(4), 1–35 (2004)
4. Gao, H., Barbier, G., Goolsby, R.: Harnessing the crowdsourcing power of social media for disaster relief. IEEE Intell. Syst. **26**(3), 10–14 (2011)
5. Turoff, M.: Past and future emergency response information systems. Commun. ACM **45**(4), 29–32 (2002)
6. Haddow, G., Bullock, J., Coppola, D.P.: Introduction to Emergency Management. Butterworth-Heinemann, Burlington (2007)
7. Berke, P.R., Kartez, J., Wenger, D.: Recovery after disaster: achieving sustainable development, mitigation and equity. Disasters **17**(2), 93–109 (1993)
8. Plotnick, L., White, C., Plummer, M.M.: The design of an online social network site for emergency management: A one stop shop, p. 420. In: Amcis (2009)
9. Kirschenbaum, A.: Generic sources of disaster communities: a social network approach. Int. J. Sociol. Soc. Policy **24**(10/11), 94–129 (2004)
10. Brandes, U.: A faster algorithm for betweenness centrality. J. Math. Sociol. **25**(2), 163–177 (2001)
11. Freeman, L.C.: A set of measures of centrality based on betweenness. Sociometry **40**, 35–41 (1977)
12. Opsahl, T., Agneessens, F., Skvoretz, J.: Node centrality in weighted networks: Generalizing degree and shortest paths. Soc. Netw. **32**(3), 245–251 (2010)
13. Haythornthwaite, C.: Social network analysis: An approach and technique for the study of information exchange. Lib. Inf. Sci. Res. **18**(4), 323–342 (1996)
14. Wasserman, S.: Social Network Analysis: Methods and Applications, vol. 8. Cambridge university press, Cambridge (1994)

15. Kautz, H., Selman, B., Shah, M.: Referral Web: combining social networks and collaborative filtering. Commun. ACM **40**(3), 63–65 (1997)
16. Balog, K., De Rijke, M.: Determining expert profiles (With an Application to Expert Finding). In: Ijcai 2007, pp. 2657–2662 (2007)
17. Zhang, J., Tang, J., Li, J.: Expert finding in a social network. In: Kotagiri, R., Radha Krishna, P., Mohania, M., Nantajeewarawat, E. (eds.) DASFAA 2007. LNCS, vol. 4443, pp. 1066–1069. Springer, Heidelberg (2007)
18. Kardan, A., Omidvar, A., Farahmandnia, F.: Expert finding on social network with link analysis approach. In: 2011 19th Iranian Conference on Electrical Engineering (ICEE), pp. 1–6. IEEE (2011)
19. Zhang, J., Ackerman, M.S., Adamic, L.: Expertise networks in online communities: structure and algorithms. In: Proceedings of the 16th International Conference on World Wide Web, pp. 221–230. ACM (2007)
20. Laumann, E.O., Marsden, P.V., Prensky, D.: The boundary specification problem in network analysis. In: Freeman, L.C., White, D.R., Romney, A.K. (eds.) Research Methods in Social Network Analysis, pp. 61–87. George Mason University Press, Fairfax (1989)
21. Marsden, P.V.: Recent developments in network measurement. In: Carrington, P.J., Scott, J., Wasserman, S. (eds.) Models and Methods in Social Network Analysis, pp. 8–30. Cambridge University Press, Cambridge (2005)
22. Wasserman, S., Faust, K.: Social Network Analysis. Cambridge University, Cambridge (1994)
23. Barrat, A., Barthelemy, M., Pastor-Satorras, R., Vespignani, A.: The architecture of complex weighted networks. Proc. Nat. Acad. Sci. U.S.A. **101**(11), 3747–3752 (2004)

The 2nd International Workshop on Big Data Management on Emerging Hardware (HardBD 2014)

Wear-Aware Algorithms for PCM-Based Database Buffer Pools

Yi Ou[1]([⊠]), Lei Chen[2], Jianliang Xu[2], and Theo Härder[1]

[1] University of Kaiserslautern, Kaiserslautern, Germany
{ou,haerder}@cs.uni-kl.de
[2] Hong Kong Baptist University, Hong Kong, China
{lchen,xujl}@comp.hkbu.edu.hk

Abstract. PCM can be used to overcome the capacity limit and energy issues of conventional DRAM-based main memory. This paper explores how the database buffer manager can deal with the write endurance problem, which is unique to PCM-based buffer pools and not considered by conventional buffer algorithms. We introduce a range of novel buffer algorithms addressing this problem, called wear-aware buffer algorithms, and study their behavior using trace-driven simulations.

1 Introduction

Phase Change Memory (PCM) is a promising next-generation memory technology with a range of interesting properties: it is *non-volatile*, *bit alterable*, and *byte addressable*. Instead of being used as an external storage solution (like flash memory), PCM is more likely to be used in the main memory system, for two major reasons. First, similar to DRAM, bytes on PCM are *directly addressable* by the processor. Second, the *read latency* of PCM is close to that of DRAM. PCM has the potential to greatly impact core database technologies in the near future.

Similar to flash memory, PCM can endure only a limited number of writes (i.e., *limited write endurance*) and writes have a higher latency than read accesses (i.e., *read-write asymmetry*). However, PCM can endure about 10^7–10^8 writes per cell [12] (even up to 10^{12} by projection [7]), whereas flash memory can only endure about 10^5–10^6 erase cycles per block [5]. Furthermore, PCM allows in-place update and does not have the erase-before-write constraint of flash memory, which further negatively and significantly impacts performance and lifespan of flash devices [11]. In terms of access latency, PCM is two to three orders of magnitude faster than flash memory.

Compared with DRAM, PCM offers *higher density* and, therefore, potentially much lower cost per gigabyte. PCM is also *more energy-efficient* than DRAM in idle mode[1]. However, PCM suffers from *two critical problems*: higher

---

[1] DRAM consumes, independent of its utilization, the lion's share of energy for a typical computer system and, with growing memory capacities, this situation gets even worse.

© Springer International Publishing Switzerland 2014
Y. Chen et al. (Eds.): WAIM 2014, LNCS 8597, pp. 165–176, 2014.
DOI: 10.1007/978-3-319-11538-2_16

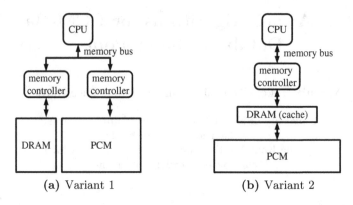

Fig. 1. Architectural variants of a hybrid main memory system based on DRAM and PCM, based on [2]

write latency and limited write endurance. Especially for data-intensive DBMS applications, it is therefore reasonable to consider a *hybrid main memory system* which consists of both PCM and (a relatively small amount of) DRAM, because such a design can overcome the capacity limit and energy issues of a conventional DRAM-based main memory without significant performance degradation.

1.1 Architectural Consideration

There are two architectural variants of such a hybrid main memory system, shown in Fig. 1. The first variant places PCM directly on the memory bus, side-by-side with DRAM [3]. The second variant places PCM below DRAM, as another layer in the memory hierarchy [12]. Their difference is that the first variant gives software explicit control over both types of memory (volatile and non-volatile), whereas the second variant manages DRAM as a hardware cache transparent to software developers [2].

The first architectural variant is more attractive for DBMS designers, because it allows the database software to take advantage of PCM's byte addressability and non-volatility. Our study follows such an architectural design and studies the management of a PCM-based database buffer pool, where the high-traffic, dynamic buffer management data structures, e.g., for indexing the buffer pool and for supporting efficient page replacement, are maintained in DRAM and the buffer pages are stored in PCM. By combining a relatively small amount of DRAM[2] with a large PCM buffer pool in such a way, the higher (write) latency can be partially hidden from the processor.

Although PCM's write endurance as compared to flash memory is improved by about two orders of magnitude, the write traffic to a buffer pool is expected to be significantly higher than to a secondary storage (e.g., based on flash memory).

---

[2] The DRAM can even accommodate a small number of hottest pages, depending on its available capacity.

Furthermore, page replacements also generate a substantial amount of write traffic to PCM. Therefore, special care has to be taken to prevent high-traffic writes (back from the processor's cache hierarchy) to the same PCM locations in the buffer pool. Minimizing such write traffic and distributing it uniformly across the PCM area may extend its lifetime long enough for practical DB use.

1.2   Goals

Conventional buffer algorithms are designed for DRAM-based buffer pools, which do not have the write endurance problem. Our study has three important goals:

1. Examine whether this problem can be effectively addressed by the database buffer manager.
2. Design buffer algorithms that address the endurance problem using wear-leveling techniques.
3. Study the behavior of such wear-aware buffer algorithms.

Our goals imply that, in addition to the hit ratio as the classical buffer pool metric, we shall be able to quantify (count or measure) the wear status of the PCM where the buffer pages are accommodated. Although PCM is managed by the buffer manager at a page level, other components of the DBMS can write to a page at a much smaller granule. Such writes can cause some parts of a page (more accurately, of the underlying PCM partition) to be worn out sooner than the other parts, unless the writes are uniformly distributed inside a page, which is a rare case. Therefore, studying the wear status at the page level, e.g., number of writes endured by each page, is not sufficient.

2   Wear-Aware Buffer Management

In this work, the smallest unit for which the wear is quantified is referred to as the *wear unit*. A candidate for the wear unit size is the lowest-level cache line size (e.g., 64 B), because that is the data transfer unit between CPU and the memory controller. For a page size of 8 KB and a wear-unit size of 64 B, we would have 128 wear units per page.

To stay with the conventional terminology when discussing DB buffer management, we denote the buffer replacement units as pages. For brevity, the term *page* either refers to a logical page or to a buffer page. A buffer page is a PCM partition having the size of a logical page, i.e., a buffer pool of B pages consists of B such partitions. Where it is unclear from context, we use the explicit terms *logical page* and *buffer page* to avoid ambiguity.

Considering the distinguished properties of PCM, some hardware optimizations have already been proposed, e.g., data comparison writes [14] or partial writes [7], to reduce the number of bits written to PCM. These hardware optimizations would be integrated into the conventional memory controller or even implemented in a dedicated memory controller for PCM. In both cases, the unit of data transfer between memory controller and PCM can be as small as a word (4 B) [7]. Due to these hardware optimizations, software optimizations at a granule smaller than the cache line size, e.g., those of [2], can also significantly improve wear leveling and performance. Therefore, we choose to examine the wear of PCM at the word level, i.e., the wear unit size is 4 B in our study. The wear unit size is a parameter for some of the algorithms that we will discuss. However, the logic of those algorithms are not specific to the parameter value.

2.1   Problem Analysis

For a wear unit u, we denote its *wear count* as $w(u)$, which is the number of times u was written. A wear unit u is *worn out* (i.e., it *fails*), if $w(u) = L$, where L is the *(wear) endurance limit*. Furthermore, we define *page wear* as the total wear count of all wear units of a buffer page and *total wear* as the total page wear of all buffer pages.

Assuming the wear count follows the normal distribution, we have $w(u) = \mu + \sigma \cdot Z$, where μ is the expected value of $w(u)$, σ the standard deviation, and Z the standard normal random variable. After a large number of writes to the PCM, the probability that u is not worn out is:

$$P[w(u) < L] = P[\frac{L - \mu}{\sigma} > Z] \tag{1}$$

If a PCM device with M wear units is considered worn out (i.e., the device fails) as long as one of its wear units fails, the probability of no device failure is (assuming that the wear units fail independently):

$$P[\text{no device failure}] = \left(P[\frac{L - \mu}{\sigma} > Z] \right)^{M} \tag{2}$$

There are two cases where the wear count increases. In both cases, we say that the corresponding buffer page is *modified*:

Page replacement When a buffer fault occurs, the buffer manager has to evict a victim page from the buffer pool to make room for the missing logical page, which will then be written to the buffer page where the victim resided. This means the wear count for all the wear units of that buffer page will be *potentially* increased by one.

Page update Update operations on the buffer page (insert, delete, and update of its records). Updating a record will only increase the wear count for the related wear units (instead of for the entire buffer page).

These two cases correspond to the two sources of write skew to the PCM used for a buffer pool: *inter-page* skew and *intra-page* skew. Although a buffer algorithm can only directly influence inter-page skew through its page replacement decisions, its page replacement strategy shall consider the page wear status for an effective wear leveling. Note, the page wear status is co-influenced by page replacement and page update.

2.2   Replacement Strategies

Formula (2) implies that to extend the device lifespan, two options are possible. The first one is to reduce the number of writes to PCM. This impacts μ. This option is taken by the LWD algorithm introduced in the following. The second one, taken by the remaining algorithms to be introduced, is to distribute the writes uniformly across the wear units. This reduces σ.

LWD Algorithm. The LWD (least wear-unit difference) algorithm is our first attempt to address the write endurance problem. On a buffer fault, the algorithm compares the page p waiting to enter the buffer pool with each page q in the buffer pool and chooses the one that has the smallest *wear-unit difference (WD)* to p as the victim. If multiple pages have the smallest WD to p, the least recently used (LRU) one among them is selected as the victim.

The *wear-unit difference* of a page q to page p, denoted as $d(p,q)$, is computed by comparing each corresponding pair of wear units (i.e., the two wear units at the same offset in both pages) and count the number of pairs that differ. In other words, $d(p,q)$ describes the physical similarity between p and q at the granule of a wear unit.

LWD assumes that to replace q by p in the buffer pool, instead of writing an entire page to the PCM, only the wear units that differ in both pages need to be written. This kind of optimization can be easily implemented in the buffer manager and it is also offered by the afore-mentioned hardware approach (data comparison write [14]). Therefore, LWD can potentially reduce the *total wear*.

However, LWD has a few problems. First, although the algorithm considers the WD, it does not consider the current *wear status*, e.g., how often is a buffer page already written or how are the writes distributed on its wear units. Consequently, a nearly worn-out page can still be selected as the victim (and thus to be written again) as long as it is the most similar one compared with the page entering the buffer pool. Therefore, it could not effectively achieve wear leveling. Second, the probability that two pages are highly similar is very small. Considering a wear unit size of 4 B, the probability that two wear units are identical is $1/2^{32}$, if the binary value of a wear unit follows a uniform distribution. Therefore, it cannot substantially reduce the total wear.

LPW Algorithm. The LPW (least page wear) algorithm addresses the first problem of LWD. Instead of considering the physical similarity, LPW always selects the page having the smallest *page wear* as victim. Page wear can only *approximately* represent the wear status of a page, because it does not capture the distribution of the wear count inside the page. This means, even if a few wear units of a page have high wear counts (e.g., due to skewed updates), the page can still can have a low page wear relative to other pages and, therefore, be (repeatedly) selected as victim. However, this is a trade-off for simplicity, because the buffer manager only has to maintain a counter for each buffer page. For a page replacement, the counter is incremented by the number of wear units the page contains. For a page update, the counter is incremented by the number of wear units that are updated.

Heuristic Algorithms. The LFM (least frequently modified) algorithm further simplifies the LPW approach by using page modification (i.e., page replacement or page update) *frequency* as the heuristic for the page wear status. As the name suggests, LFM selects the least frequently modified page as the victim, similar to the classical LFU (least frequently used) algorithm. However, LFU typically maintains, for each logical page currently in the buffer pool, an access frequency counter, which is reset at each page replacement. In contrast, the modification-frequency counter in LFM is maintained for each buffer page and the value of the counter survives page replacements.

Similarly, the LRM (least recently modified) algorithm resembles the classical LRU (least recently used) algorithm, but uses the modification *recency* as victim selection criterion. Similar to LRU, LRM can be implemented using a linked list of buffer page pointers. Page update or page replacement moves the modified page pointer to the MRM (most recently modified) position. On page hits, in contrast to LRU, LRM does not change the page (pointer) position in the list.

2.3 Complexity

The LWD algorithm has the highest time complexity, which is $O(B \cdot P)$ in our current implementation, where B is the buffer pool size and P the page size. Both LPW and LFM have a complexity of $O(B)$. The LRM algorithm has a complexity of $O(1)$.

In terms of space overhead, LRM is similar to LRU: the only overhead is introduced by the linked list structure. The LWD algorithm does not introduce any space overhead, because it compares the content of the wear units on the fly. Both LPW and LFM maintain a counter (of 8 B) for each buffer page. But this space overhead is ignorable, e.g., it is 0.1 % for a page of 8 KB. Note, all these data structures and management information are maintained in DRAM, based on the architectural assumption given in Sect. 1.1. If, for practical considerations, these small amount of meta data need to survive a DB server restart or crash, it is sufficient to propagate them to a persistent storage in appropriate intervals, e.g., hours or even days are acceptable.

3   Experiments

We used trace-driven simulations to evaluate the afore-mentioned algorithms. As explained in Sect. 1.2, we are primarily interested in two aspects of the algorithms: wear-leveling effectiveness and hit ratio. To study the wear-leveling effectiveness, we have to use record-oriented traces, instead of the page-oriented workloads (i.e., page reference strings) typically used by the studies on conventional buffer management [4].

3.1 Workload

Most experiments reported in this section used a synthetic trace which simulates a typical workload for a database buffer pool. The trace is generated as follows. Each tuple of the trace is a record request, which consists of a page identifier $\in [0, 131072)$, a record identifier $\in [0, 64)$, and a type identifier $\in \{R, U\}$. The type identifier describes the operation to be performed on the record: read (R) or update (U). For a U record request, the tuple additionally contains a randomly generated record of 128 B. A trace contains one million such record requests. The workload follows the 80–20 rule: both the page identifier and the record identifier follow a 80–20 self-similar distribution within their respective ranges to simulate skewed accesses (particularly skewed updates). Moreover, the update ratio is 20 %, i.e., 80 % of the requests are R requests and the remaining 20 % are U requests. Our experiments used a typical database page size of 8 KB. The page layout follows a simplified N-ary storage model (NSM) [1]: each page consists of N equi-length records ($N = 64$).

3.2 Methodology

We implemented a database storage manager supporting all the algorithms under examination. The traces are processed by the storage manager as follows. Prior to each trace execution, the database file (1 GB) is initialized with 131072 pages and each page contains 64 randomly generated records. An R record request is processed by getting the corresponding page via the buffer pool and reading the record. A U record request additionally overwrites the corresponding record in place using the generated modification pattern (sequence of bytes) contained in the request.

In addition, we implemented a PCM simulator, which maintains a wear counter for each wear unit of the buffer pool. The wear counters are all reset prior to each experiment. For each buffer page modification during the trace execution, the simulator increments the wear counters of the affected wear units. The wear-leveling effectiveness of the algorithms is expressed by the *standard deviation* σ of the wear count after each trace execution. The *hit ratio* h achieved by the algorithms are also compared. For a simulated PCM of n wear units $\{u_i, \text{ for } 0 <= i < n\}$, the wear count standard deviation σ is computed as:

$$\sigma = \sqrt{\frac{\sum_{i=0}^{n-1} (w(u_i) - \overline{w})^2}{n - 1}}$$

where $\overline{w}$ is the mean of $w(u_i)$.

We included LPW, LFM, and LRM in the comparison. Additionally, LRU and RND (random) are used as baselines for hit ratio (LRU) and wear-leveling effectiveness (RND). We did not include LWD in the comparison due to its problems discussed in Sect. 2.2, which are confirmed by our experiments (omitted due to space limitations). To facilitate visual comparison, we present the *normalized* performance figures, i.e., their ratios *relative* to the corresponding figures of RND.

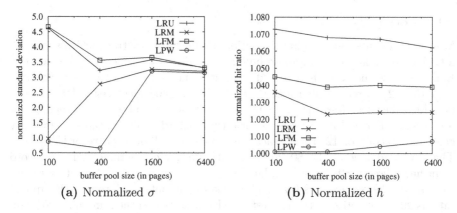

Fig. 2. Normalized σ and h, buffer pool size scaled from 100 to 6400 pages, typical workload

3.3 Typical Workload

Figure 2 compares the σ (wear count standard deviation) and h (hit ratio) of our algorithms (relative to RND) for the traces described above. Figure 2a confirms that buffer algorithms can have a great impact on wear leveling. For example, for a buffer pool of 100 pages, LRM improves the σ by nearly a factor of five compared with LRU.

To our surprise, the wear-leveling effectiveness of LFM is even worse than LRU for all buffer pool sizes. Our explanation is that the page modification frequency used by LFM is not a good approximation for the page wear status, because it can not distinguish between page replacement and page update. In contrast, this difference is captured by the page wear metric used by LPW, which, as a consequence, delivered the best σ for all buffer pool sizes. For a buffer pool of 400 pages, LPW reduced the σ by 24.2 % even compared with RND. The performance of LPW reveals that, although a buffer algorithm can primarily influence only the inter-page skew, considering the intra-page skew in the victim selection can help to improve wear-leveling effectiveness.

The hit ratios of LFM, LRM, and LPW are lower than that of LRU. LPW is even close to RND in terms of h. However, this is expected, because page reference statistics are not considered by LPW at all, while LFM (LRM) only partially considers the frequency (recency) of page updates in their victim selection decision. This implies that, if the workload is read-only, i.e., there is only page replacement and no page update, LFM and LRM would have no advantage over LPW in terms of h.

3.4 Read-Only Workload

This is confirmed by our experiment using a read-only trace, which shares all the parameters with the afore-mentioned trace except for the update ratio, i.e., the

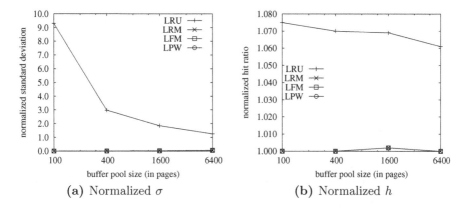

Fig. 3. Normalized σ and h, buffer pool size scaled from 100 to 6400 pages, read-only workload

read-only trace contains only R requests, which only trigger page replacements. The results of this experiment are shown in Fig. 3.

Under the read-only workload, all algorithms compared—LRM, LFM, and LPW—achieved nearly a perfect wear leveling (Fig. 3a), because there is no intra-page write skew (i.e., no record-level writes), but only a (inter-page) reference skew. In other words, the write is uniform within a page, but non-uniform among pages. The latter, inter-page write skew, can be effectively handled by our algorithms. This implies that if the write skew inside a page is less significant than the page reference skew, our algorithms are more effective in wear leveling. After all, the buffer algorithms have no direct influence on the update locations inside a page. This further implies that our algorithms would be even more effective in terms of wear leveling for a smaller page size. The price to pay for a nearly perfect wear leveling is a very low hit ratio which is close to that of RND, as shown in Fig. 3.

3.5   Total Wear

Figure 4 reports the total wear for the experiments corresponding to Figs. 2 and 3. The curves in Fig. 4b look (reversely) similar to the h curves in Fig. 3b. The same relation exists between Figs. 4a and 2b. This results from the fact that the total wear is dominated by the number of page faults. Each page fault requires a page replacement, which has a much greater impact on the total wear than a page update: the former overwrites an entire buffer page, whereas the latter only updates a record. Therefore, the total wear has a strong correlation with h.

4   Related Work

Qureshi et al. analyzed the write traffic to pages for two database applications and found that for both applications there is significant non-uniformity in which

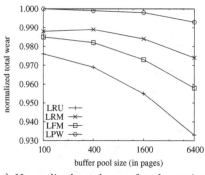 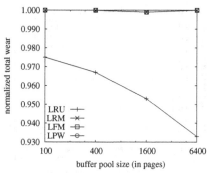

(a) Normalized total wear for the typical workload

(b) Normalized total wear for the read-only workload

Fig. 4. Normalized total wear, buffer pool size scaled from 100 to 6400 pages

lines (i.e., page partitions of cache line size) in the page are written back [13]. Their analysis confirmed the significance of the intra-page skew. As solution, the authors proposed a fine-grained wear-leveling technique, which randomly shuffles the lines when writing a page to PCM and restores the page layout when the page is written back to disk. Our approach to wear leveling is complementary to theirs, because our software approach manipulates the write traffic at a page level, whereas their hardware approach does this at the line level.

There have been some pioneer works on the use of PCM in database systems. For example, Chen et al. advocate that database algorithms should be adapted to PCM technology to improve performance, energy efficiency, and PCM's write endurance. For this purpose, they presented PCM-friendly algorithms for two core database techniques: B^+-tree index and hash joins [2]. Gao et al. presented a novel logging scheme that exploits the non-volatility and bit alterability of PCM for efficient transaction logging in disk-based databases [6]. However, the issues of using PCM for database buffer pool, i.e., the focus of our work, are not covered by these works.

Ou et al. identified the cold-page migration (CPM) problem related to the indirect use of flash memory for mid-tier buffer pool and proposed two effective solutions [11]. Our study has one thing in common with theirs: one of our goals is to address the write endurance problem and extend the device lifespan. However, the CPM problem studied by them is specific to flash memory, because it is rooted in the flash memory erase-before-write limitation, which is not present for PCM. Furthermore, wear-leveling techniques are not the focus of their study.

5   Conclusion and Future Work

In this work, we studied the write endurance problem of PCM-based database buffer pools. To attack this problem, we identified and classified the sources of write skew in such an environment and introduced a range of novel buffer algorithms, which examine page content or consider page wear status for their replacement decisions, to minimize and uniformly distribute the write traffic.

Using these algorithms, the database buffer manager can effectively address the endurance problem, as confirmed by our experiments. The experiments reveal that, although a buffer algorithm can primarily influence only the inter-page skew, precise page-wear status information can be used to improve its wear-leveling effectiveness.

Among the algorithms compared, LRM and LPW can effectively achieve wear leveling (e.g., up to factor five in one of the experiments or even a nearly perfect wear leveling for read-only workloads). However, they had a lower hit ratio compared to the conventional buffer algorithms represented by LRU. This suggests that wear leveling and hit ratio are two conflicting goals that must be considered by the buffer algorithm in a well-balanced fashion. Nevertheless, for a workload with lower page-level locality, e.g., 70–30 (or even 60–40) instead of 80–20, their penalty in terms of hit ratio would be smaller. For a hybrid main memory system, where the hottest pages are retained in the smaller DRAM, we can assume that the locality of page reference to the PCM would be lower than the 80–20 one.

Under the read-only workload, the wear-aware algorithms achieved the best wear-leveling effect but the lowest hit ratios. This is an issue requiring further algorithmic improvements. A possible approach is to dynamically adjust the behavior of the buffer manager based on workload statistics, similar to the approaches of [8–10].

Acknowledgement. Yi Ou's work is partially supported by the German Research Foundation and the Carl Zeiss Foundation. Jianliang Xu's work is partially supported by Research Grants Council (RGC) of Hong Kong under the grant no. G_HK018/11. The authors are grateful to German Academic Exchange Service (DAAD) for supporting their cooperation. They are also grateful to anonymous referees for valuable comments.

References

1. Ailamaki, A., DeWitt, D.J., Hill, M.D.: Data page layouts for relational databases on deep memory hierarchies. VLDB J. **11**, 198–215 (2002). http://dx.doi.org/10.1007/s00778-002-0074-9
2. Chen, S., Gibbons, P.B., Nath, S.: Rethinking database algorithms for phase change memory. In: CIDR, pp. 21–31 (2011). www.cidrdb.org
3. Condit, J., Nightingale, E.B., Frost, C., Ipek, E., Lee, B.C., Burger, D., Coetzee, D.: Better I/O through byte-addressable, persistent memory. In: SOSP, pp. 133–146 (2009)
4. Effelsberg, W., Härder, T.: Principles of database buffer management. ACM TODS **9**(4), 560–595 (1984)
5. Gal, E., Toledo, S.: Algorithms and data structures for flash memories. ACM Comput. Surv. **37**(2), 138–163 (2005)
6. Gao, S., Xu, J., He, B., Choi, B., Hu, H.: PCMLogging: reducing transaction logging overhead with PCM. In: CIKM, pp. 2401–2404 (2011)

7. Lee, B.C., Ipek, E., Mutlu, O., Burger, D.: Architecting phase change memory as a scalable DRAM alternative. In: ISCA, pp. 2–13 (2009)
8. Megiddo, N., Modha, D.S.: ARC: a self-tuning, low overhead replacement cache. In: USENIX FAST'03, USENIX (2003)
9. Ou, Y., Härder, T.: Clean first or dirty first? a cost-aware self-adaptive buffer replacement policy. In: IDEAS'10, Montreal, QC, Canada (2010)
10. Ou, Y., Jin, P., Härder, T.: Flash-aware buffer management for database systems. Int. J. Knowl. Based Organ. **3**(4), 22–39 (2014)
11. Ou, Y., Xu, J., Härder, T.: Towards an efficient flash-based mid-tier cache. In: Liddle, S.W., Schewe, K.-D., Tjoa, A.M., Zhou, X. (eds.) DEXA 2012, Part I. LNCS, vol. 7446, pp. 55–70. Springer, Heidelberg (2012)
12. Qureshi, M.K., Karidis, J., Franceschini, M., Srinivasan, V., Lastras, L., Abali, B.: Enhancing lifetime and security of PCM-based main memory with start-gap wear leveling. In: MICRO, pp. 14–23. IEEE (2009)
13. Qureshi, M.K., Srinivasan, V., Rivers, J.A.: Scalable high performance main memory system using phase-change memory technology. In: ISCA, pp. 24–33 (2009)
14. Yang, B.D., Lee, J.E., Kim, J.S., Cho, J., Lee, S.Y., Yu, B.G.: A low power phase-change random access memory using a data-comparison write scheme. In: ISCAS, pp. 3014–3017 (2007)

Accelerate K-means Algorithm by Using GPU in the Hadoop Framework

HuanXin Zheng$^{(\boxtimes)}$ and JunMin Wu

Department of Computer Science and Technology,
University of Science and Technology of China, Hefei, China
`291221622@qq.com`

Abstract. Cluster analysis, such as k-means algorithm, plays a critical role in data mining area, but now it is facing the computational challenge due to the continuously increasing data volume. Parallel computing becomes an efficient way to overcome the difficulty. In this paper, we use Graphics Processing Units (GPU) in the Hadoop framework to accelerate the k-means algorithm. As a result, our algorithm is about 10 times faster than the k-means implemented by Mahout.

Keywords: Hadoop · GPU · K-means

1 Introduction

The k-means is one of the most practical clustering algorithms. Under the assumption that the datasets are small and the research has traditionally focused on improving the quality of k-means algorithms, such as finding a better algorithm for centroids initialization [15]. However, many datasets now become large [1] and we need high-performance methods to deal with it.

K-means is an algorithm that can be easily partitioned into several independent tasks, so we can use GPU to overcome the challenge of the huge computational requirement and use Hadoop to resolve the limitation of main memory.

GPUs can be regarded as massively parallel processors with an order of the higher computation power and memory bandwidth than CPU [3]. Recently, several GPGPU programming frameworks have been introduced, such as NVIDIA CUDA [4]. However, it is a challenging task to develop general-purpose applications by using GPU. NVIDIA GPU cores are single-Instruction-Multiple-Data (SIMD), which discourages complex controlling flows. NVIDIA GPU's high bandwidth depends on the efficient manage on memory such as avoiding bank conflict [5].

Many researches focus on overcoming the huge computational requirement and designing a parallel k-Means algorithm for GPU [2, 6–8]. You Li [2] designed the k-means algorithm which simulates matrix multiplication and exploits GPU on-chip registers and also on-chip shared memory to achieve high bandwidth ratio. As a result, the algorithm is three to eight times faster than the best reported GPU-based algorithm [8].

Large-scale MapReduce clusters have become a popular solution to cope with the fast growth of big data. MapReduce programming model, originally introduced by

© Springer International Publishing Switzerland 2014
Y. Chen et al. (Eds.): WAIM 2014, LNCS 8597, pp. 177–186, 2014.
DOI: 10.1007/978-3-319-11538-2_17

google [9], allow programmers to focus on computational logic and get rid of complex details of parallelization. Hadoop is an open-source implementation of MapReduce programming model and becomes popular due to the flexibility and fault-tolerance. Mahout [10] focuses on machine learning and artificial intelligence areas, which is another open-source project depending on Hadoop. K-means, one of the most frequently used clustering algorithms, is implemented by Mahout [11].

Another important trend is to use accelerators on each node of a cluster to boost its total computational power. MarsHadoop [12] implements matrix multiplication on Hadoop and uses NVIDIA GPU to accelerate computation in each node. As a result, MarsHadoop is up to 2.8 times faster than the native Hadoop implementation. At the same time, Surena [14] is framework that allows employing both CPUs and GPUs for MapReduce-type applications. In particular, the authors found that by using simple scheduling optimizations, Surena is about 21 times faster than Hadoop.

Basing on Mars and Surena, we analyze the workflow of k-means and implement it on Hadoop accelerated by NVIDIA GPU. Because of the influence of data translation delay, we combine several map tasks to a huge computational task and send it to GPU. The GPU computation task is similar to matrix multiplication, we use partitioned matrix multiplication to exploit the shared memory to achieve high bandwidth ratio. The result shows that k-means is up to about 10 times faster than the implementation by Mahout.

2   Related Work

In this section, we review the CPU-based, GPU-based and Hadoop-based K-Means algorithm.

2.1   CPU-Based K-Means

Given a set of n points, $R = \{r_1, r_2,..., r_n\}$, where each point is a d-dimension real vector. K-means aims to partition the n points into k clusters $S = \{s_1, s_2,..., s_k\}$ $(k \leq n)$ such that $\sum_{i=1}^{k} \sum_{x_j \in s_j} \left\| x_j - u_i \right\|^2$ minimized, where u_i is the mean of points in s_i.

The algorithm for the standard k-means is given as following [11]:

1. Select k data points randomly as the initial centroids.
2. For each data point, find the centroid which it is closest to and assign the data point to this cluster.
3. Compute new centroids by taking the mean of all the data points in each cluster.
4. Iterate steps $1 \sim 3$ until the threshold or the number of iterations reaches the predefined value.

The whole process is shown in Algorithm 1 [2].

Algorithm.1. CPU-based k-Means

flag:shows whether it still needs to iterate;
iter:the current round of iteration
Max_iter: the maximum number of iterations
d(r,s):the distance between r and cluster s;
min_D:the temporal minimum distance;
index(r) ← s: r belong to cluster s

1 While flag && iter ≤ Max_iter
2   for each r in R
3   for each s in S
4   compute d(r,s);
5   end for
6   end for
7   for each r in R
8   min_D ← INF
9   for each s in S
10   if d(r,s) < min_D
11   index(r) ← s;
12   min_D ← d(r,s);
13   end if
14   end for
15 end for
16 if the change of centroid is less than threshold
17   flag←false;
18   iter←iter+1;
19 end of if
20 end of while

2.2   GPU-Based K-Means

In You Li's [2] paper, they observe the data dimension is an important factor that should be taken into consideration. For lower dimension data sets, they exploit GPU on-chip registers to decrease data access latency. However, Register Spilling will increase the reading latency when the data points cannot be loaded into the registers. For the high-dimensional data sets, they find lines $2 \sim 6$ in Algorithm 1 is similar to matrix multiplication and exploit the on-chip register and also shared memory to achieve high memory-access performance. Lines $7 \sim 19$ in Algorithm 1 are difficult to be fully parallelized due to the write conflict. The authors divide the data into group and get temp centers to decrease the write conflict. This way requires more memory and cannot be flexible to the increase of clusters or dimension.

2.3 Hadoop-Based K-Means

Hadoop is a popular open-source implementation of the MapReduce Model to cope with massive data. The algorithm for k-means implemented by Mahout is given by following:

1. Select k data points randomly as the initial centroids.
2. Partition the Map tasks: For each data point, a map task will find the closest centroid index. The index will be assigned to the Reduce task's key and the point to the reduce-task's value.
3. The Hadoop system classifies all point in accordance with the key.
4. Each reduce-task computing new centroids by taking the mean of all the data points with same key.
5. Iterate steps $2 \sim 4$ until the threshold or the number of iterations reaches the pre-defined value.

3 Design and Implementation

In this section, we focus on the implementation of the k-means algorithms by using GPU on Hadoop framework.

3.1 Overall System Design

According to Fig. 1, we add a GPU accelerator into each node. JNI is an interface between JVM and native program. The native program can control both the CPU and GPU resource. Hadoop's design aim is to combine many normal computer and sets up a powerful cluster, so we try to find a lower cost solution to resolve some

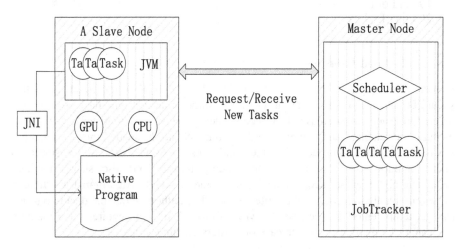

Fig. 1. Hadoop-GPU framework

computation-intensive problems. GPU GTS450 is a common hardware and some personal computers have already been equipped with a better GPU.

3.2 Expense of Data Transfer

Data transfer is an acute factor influencing the efficiency. If the expense of data transfer cover the time saved by GPU, all the work is meaningless.

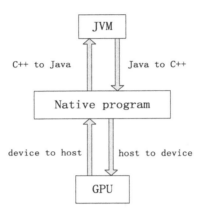

Fig. 2. Data transfer

As the Fig. 2 shows, the data will be translated frequently when using a GPU accelerator. The data should transfer from JVM to native program through JNI and then transfer from host to device. After GPU finish the task, the result should be copied back to JVM.

However, reducing the frequency of data transfer is the important point to Improve efficiency.

3.3 Using GPU to Accelerate the Hadoop-Based K-Means

The map-task and reduce-task are the processing units in the Hadoop framework, which can be accelerated by GPU.

As the k-means algorithm implemented by Mahout showing, a map-task only finds one point's closest centroid index. In order to reduce the frequency of data transfers, we can assign a map-task to find the closest centroid index of the points from $R' = \{r_{i_1}, r_{i_2}, \ldots, r_{i_k}\}(1 \leq i_1 < i_2 < \ldots < i_k \leq n)$, which means combining the small map-tasks into a big map-task and using GPU to accelerate the big one.

Algorithm.2. Big map-task

min_D: the temporal minimum distance;
d(r,s): the distance between r and the cluster s;
index(r) ← s: r belong to cluster s

```
1  transfer R' and S from Hadoop to native.
2  for each r in R'
3    for each s in S
4      compute d(r,s);
5    end for
6  end for
7  for each r in
8    min_D ← INF
9    for each s in S
10     if d(r,s) < min_D
11       index(r) ← d(r,s);
12       min_D ← d(r,s);
13     end if
14   end for
15 end for
16 transfer index information from navtive to Hadoop
```

As the Algorithm 2 show, line $2 \sim 6$ is simulate to the flow of matrix multiplication [2]. We implement it according to Algorithm 4 in You Li' paper and use partitioned matrix multiplication to exploit the shared memory to achieve high bandwidth ratio.

Line $7 \sim 15$ finds which cluster each point belong to. Because each point is independent, we can use one thread in GPU to calculate each point's closest cluster.

Each reduce-task computes new centroid by taking the mean of all the data points with same key. However, the reduce-task can't be combined into a big one due to the data independence. Only the computation complexity of one reduce-task was high enough, it is worth using the GPU accelerator to improve the efficiency.

4   Results

In this section, in order to show the efficiency of Hadoop-GPU framework, we compare k-means implemented by Hadoop-GPU with Mahout.

We use four nodes to set up the Hadoop platform and each node equips with 4.0 GB memory, one core-i3 CPU and one GeForce GTS 450 GPU. There is one map-task and one reduce-task in each node.

Our Hadoop's version is 0.20.203 and CUDA's version is 5.5.

In order to identify the algorithm's scalability, we use the control variable method to design experiments. As the following, we focus on four variables:

(1) $|R|$ represents the number of points.
(2) M represents the number of machines.
(3) $|R'|$ represents the number of map-tasks combine into a big one.
(4) K represents the points will be classified into k clusters.

Table 1. Control variable Method

Change of variable	$|R|$	M	$|R'|$	K
$|R|$	$3000 \sim 100000$	4	1000	1000
M	50000	$1 \sim 4$	1000	1000
$|R'|$	50000	4	$100 \sim 10000$	1000
K	50000	4	1000	$100 \sim 3000$

As Table 1 shows, in each experiment, we change one of the variables and keep another three constant.

The dimension is always 1000. Because the data is random and the iteration time may be very huge, we just regard the average time of the first five iterations as the running time. Then we get the following result.

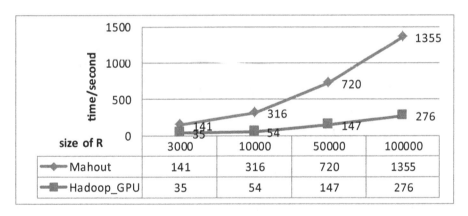

Fig. 3. Change variable $|R|$

As Fig. 3 shows, Hadoop-GPU always performs better than Mahout when $|R|$ is increasing.

As Fig. 4 shows, Hadoop-GPU performs better when the number of machine increases. At the same time, we can find only one machine equipts with an ordinary GPU performs better than four machines without GPU.

As Fig. 5 shows, combining small map-tasks into a big map-task contributes greatly to the performance. Here are two main reasons.

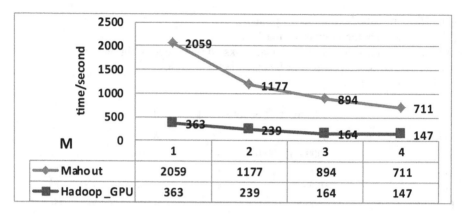

Fig. 4. Change variable M

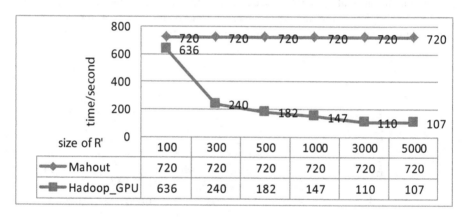

Fig. 5. Change variable $|R'|$

(a) As Sect. 3.2 discusses, reducing the frequency of data transfer is the important point to Improve efficiency. Combining small map-tasks is a useful way to reduce the frequency of data transfer. We design an experiment to identify this point. When $|R|$ is 5000, dimension is 1000 and K is 1000, We ignore all the computational tasks and just transfer the data from JVM to native program through JNI and then transfer it from host to device, and finally copy it to JVM. As Fig. 6 shows, transfer time is smaller when $|R'|$ is bigger.

(b) We use GPU to accelerate the big map-task, which can be about 100 times faster than the same computation assign to CPU.

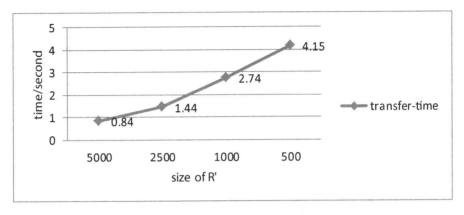

Fig. 6. Expense of data transfer

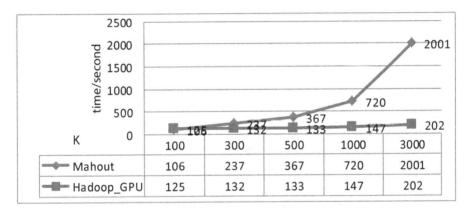

K	100	300	500	1000	3000
Mahout	106	237	367	720	2001
Hadoop_GPU	125	132	133	147	202

Fig. 7. Change variable K

As Fig. 7 shows, with the increase of the size of K, the running time increase quickly in mahout but not Hadoop-GPU. When K is 3000, Hadoop-GPU is about 10 times faster than Mahout.

5 Conclusion and Future Work

As the era of big data coming, data mining is not only faced with data-intensive problems, but also computation-intensive problems. Hadoop is the traditional way to resolve a data-intensive problem but is week at computation-intensive problem. We try to add GPU accelerator at every node and implement k-means algorithm on Hadoop-GPU framework, and the experiment indicates that Hadoop-GPU framework is powerful to resolve computation-intensive problem.

Fault tolerance, data redundancy and scalability are Hadoop's design goals, but at the same time they cause high delay problem. The better performance of GPU depends on higher IO speed. In the future, we try to impose distribute shared memory to decrease Hadoop's read-write disc operators, and then the utilization of GPU will become higher.

References

1. Farnstrom, F., Lewis, J., Elkan, C.: Scalability for clustering algorithms revisited. ACM SIGKDD Explor. Newsl. **2**(1), 51–57 (2000)
2. Li, Y., Zhao, K., Chu, X., et al.: Speeding up K-Means algorithm by GPUs. In: 2010 IEEE 10th International Conference on Computer and Information Technology (CIT), pp. 115–122. IEEE (2010)
3. Ailamaki, A., Govindaraju, N.K., Harizopoulos, S., et al.: Query co-processing on commodity processors. In: VLDB, vol. 6, pp. 1267–1267 (2006)
4. NVIDIA CUDA. http://www.nvidia.com/cuda
5. Harris, M., Sengupta, S., Owens, J.D.: Parallel prefix sum (scan) with CUDA. GPU Gems **3** (39), 851–876 (2007)
6. Fang, W., Lau, K.K., Lu, M., et al.: Parallel data mining on graphics processors. Hong Kong University of Science and Technology, Technical Report, HKUST-CS08-07, vol. 4 (2008)
7. Che, S., Boyer, M., Meng, J., et al.: A performance study of general-purpose applications on graphics processors using CUDA. J. Parallel Distrib. Comput. **68**(10), 1370–1380 (2008)
8. Wu, R., Zhang, B., Hsu, M.: Clustering billions of data points using GPUs. In: Proceedings of the Combined Workshops on UnConventional High Performance Computing Workshop Plus Memory Access Workshop, pp. 1–6. ACM (2009)
9. Dean, J., Ghemawat, S.: MapReduce: simplified data processing on large clusters. Commun. ACM **51**(1), 107–113 (2008)
10. Anil, R., Dunning, T., Friedman, E.: Mahout in Action. Manning, Shelter Island (2011)
11. Esteves, R.M., Pais, R., Rong, C.: K-means clustering in the cloud–a mahout test. In: 2011 IEEE Workshops of International Conference on Advanced Information Networking and Applications (WAINA), pp. 514–519. IEEE (2011)
12. Fang, W., He, B., Luo, Q., et al.: Mars: Accelerating MapReduce with graphics processors. IEEE Trans. Parallel Distrib. Syst. **22**(4), 608–620 (2011)
13. NVIDIA Corporation: Compute unified device architecture programming guide (2007)
14. Abbasi, A., Khunjush, F., Azimi, R.: A preliminary study of incorporating GPUs in the Hadoop framework. In: 2012 16th CSI International Symposium on Computer Architecture and Digital Systems (CADS), pp. 178–185. IEEE (2012)
15. Arai, K., Barakbah, A.R.: Hierarchical K-means: an algorithm for centroids initialization for K-means. Rep. Fac. Sci. Eng. **36**(1), 25–31 (2007)

An Efficient Hybrid Index Structure
for Temporal Marine Data

Dongmei Huang, Le Sun$^{(\boxtimes)}$, Danfeng Zhao, and Xiaoluo Zheng

Ocean University of Shanghai, Shanghai, China
{dmhuang,dfzhao,xlzheng}@shou.edu.cn,
josephsun1991@gmail.com

Abstract. Marine data is a typical big data that features multi-source, multi-class, multi-dimension and massiveness. Rapid query to big marine data is the fundamental request in vast marine applications. To improve query performance, we should devise a complete index structure. In this paper we propose a multi-layer index (ML-index, for short) with regarding to Time Interval B$^+$-tree and Hybrid Space Partition Tree (HSP-tree, for short). It employs Marine data value function that consists of data time length, data access frequency etc. to optimize the primary key index (i.e. B$^+$-tree). Moreover, we propose an adaptive space partition method on the basis of data characters, user query habits and data unit capacity particularly. Furthermore we build a secondary index, namely, the HSP-tree over the above partition result. We show the results of experiment that compares ML-index with two state-of-the-art index methods on the real marine data. These suggest that the ML-index enable user to perform marine data query in about 2/3 the time needed by the state-of-the-art tools.

Keywords: Marine Data Value Function · Time interval B$^+$-tree · Adaptive Space Partition · HSP-tree

1 Introduction

As technology makes inroads into our daily lives, massive data with high richness starts coming up, which is called big data [1]. According to what Meng and Ci [2] do with current Big Data resources, the notion of big data should be considered on the basis of its quantity (volume, variety, velocity) and its significant value property as well. Marine data is typical big data with property like multisource, multiclass, multi-dimensional, large-amount, real-time etc., which is also highly valuable. As China's work on marine proceeds, data acquisition method has evolved from traditional manual observation to informationalized observation by now, which results in exponential growth of data file sizes from GB, TB to PB [3, 4] and big challenge on data management. The traditional centralized Data Storage Management has some limitations in rapid access to target ocean data. The key to speeding up data access is to create a complete index mechanism. Meanwhile, data analysis and data space partition would have a direct impact on the performance [5]. In this paper, we comprehensively analyze the characteristics of marine data, reasonably partition the data and adopt multi-layer index (ML-index) flame to manage big marine data.

© Springer International Publishing Switzerland 2014
Y. Chen et al. (Eds.): WAIM 2014, LNCS 8597, pp. 187–199, 2014.
DOI: 10.1007/978-3-319-11538-2_18

Each index has its own features, including data types, data distribution and dimension size [6], which influence the performance of the index structure. In traditional disk database, B$^+$-tree has been widely adapted due to its simple structure, overall balance, etc., but has problems like lower storage utilization, low abilities on processing multi-dimension marine data. In addition, understanding of user requirements, data characteristics and data distribution helps to deal with data partition. Firstly, based on marine data characteristics, user query habits, we propose a multi-layer index (ML-index) out of time interval B$^+$-tree and hybrid space partition tree (HSP-tree), to improve query efficiency. The time B$^+$-tree as the main index supports fast querying tuples in a certain time interval. Secondly, we propose Adaptive Space Partition (ADSP) to partition marine sub-dataset and build HSP-tree as secondary index. Moreover, we define data time length, data access frequency, space unit capacity as foundation for time interval and subspace partition, and improve index structure.

Our Contributions. Our work makes the following contributions:

(1) We pre-process the monitoring data in a certain marine, analyze data characteristics and data distribution as well as define marine data value function to optimize data partition result.

(2) We propose a multi-layer index (ML-index) structure, time interval B$^+$-tree as the main index and HSP-tree as secondary index.

(3) We propose a partition strategy, ADSP, based on which, we build a secondary index that fits in rapid range query and process multi-dimensional data.

Our paper is organized as follows. We cover related work in Sect. 2 and provide an overview of data preprocessing work in Sect. 3. Section 4 describes how the whole index frame works. The experiment that demonstrates the efficiency of our algorithm is presented in Sect. 5. Finally, we conclude in Sect. 6.

2 Related Work

Multi-dimensional data, mostly spatial data, marine data, has promoted the study on multidimensional indexing. Recent study on multi-dimensional index has obtained many achievements, including quad-tree [7], K-D tree [8], R-tree [9] and grid files [10]. There are also a lot of new index algorithms that has appeared in recent years. Although the above algorithms are not widely used, they have distinctive features but share one thing in common — partition pattern. Multi-dimensional index structure can be basically divided into two catalogs: spatial and data partition. The first three ones are all based on space partition, while R-Tree is based on data partition. Grid file supports range query well, but as a kind of statistic index, it is hard to either expand data dynamically or expand dimension effectively. K-D tree is suitable for neighbor query, but information stored in leaf nodes is less and most of its space is wasted. Besides, K-D tree is unbalanced tree, which makes it hard to design an effective query algorithm, but it holds the advantage of partitioning data in order.

There is still another multi-dimensional index structure called composite index, which means users can build index on multiple columns. Innodb, MyISAM are the representative ones that are based on B$^+$-tree structure. Such engines have a primary index and secondary index structure, and make no difference in structure. Primary index requires a unique key while the key of secondary index can be repeated. The secondary index in above engines doesn't fit in quite well with marine data due to complicated and multiple marine attributes.

Above algorithms more or less exist some defects, like low responding in processing massive marine data by signal-dimensional index, difficulty in handling multi-dimensional data by composite index structure, huge I/O operations when querying by MyISAM and lack of understanding of data characters. From the perspective of marine data, we propose a more suitable index structure, overcoming low storage rate of grid file, and learn from K-D tree's well-ordered space partition and recognition.

3 Data Preprocessing

In this section, we preprocess marine dataset, present theoretical definition of marine data and define Marine Data Value Function to rectify time interval partition.

3.1 Marine Data Definition

Marine data is typical big data, where each tuple has time information, longitude and latitude as well. As an object in real world, if marine data loses its time, it will lose meaning. Hence, we define marine data scheme and marine dataset as follow.

Definition 1 (Marine Data Scheme). Marine data scheme is defined as R(A_1, A_2, $\cdots$, A_m), (A_1, A_2, $\cdots$, A_m) is its attribute set, where A$_1$ is time attribute.

Marine data scheme has lots of attributes, like temperature, visibility, and can be divided into three catalogs: water quality elements, meteorological elements and hydrologic elements. Thereby, we define marine dataset as follow:

Definition 2 (Marine Dataset). Marine dataset is a triple, A = (R, N, S), where R is marine data scheme; N is record amount in dataset A; S is file size of dataset A.

3.2 Time Interval of Marine Data

In marine data integration platform, data value has high correlation with data access frequency and date storage time. Marine data has timelessness, when new data enters, value of historical data changes. Users focus more on characteristic value. When data time length is shorter, data access frequency is higher, which renders faster query speed. Data time length and data access frequency are defined as follow:

Definition 3 (Data Time Length). Marine data has a consistent monitoring period and data falls into interval t_2-t_1, t_3-t_2, $\cdots$, t_i-t_{i-1} respectively. Such time length is marked as T_1, T_2, $\cdots$, T_n. We formalize Time Length T of marine dataset D as:

$$T = \sum_{i=1}^{n} T_i = T_1 + T_2 + \cdots + T_n \qquad (1)$$

Definition 4 (Data Access Frequency). F represents the frequency of data access. We mark data access frequency in interval T_k as f_k, and formalize frequency F of marine dataset D as:

$$F = \sum_{k=1}^{n} f_k = f_1 + f_2 + \cdots + f_n \qquad (2)$$

If data access frequency F grows in one interval, corresponding data importance gets higher and vice versa. From the above, marine data value is inversely proportional to data time length T, proportional to data access frequency F and inversely proportional to dataset file size S. Marine Data Value Function can be described as:

$$F(T) = \frac{1}{T_k} \times f_k \times \frac{1}{S} \qquad (3)$$

By computing $F(T)$, we can get the value of each interval. Given a threshold, we keep the original partition for intervals with high $F(T)$ and combine the intervals with low $F(T)$ [12]. In this way, we can render the results to be maximized to meet user's queries habits, and improve query efficiency as well. For example, we get the finally data time interval in 2012, T = {[6-24 8:00, 6-30 8:00), [6-30 8:00, 7-10 8:00), $\cdots$, [9-29 8:00, 10-5 8:00)}. The partition result is used to form time interval B^+-tree.

4 Multi-layer Index Structure

In this section, we describe the whole index structure, which fits well in marine business applications, and introduce implementation steps, algorithms and time complexity. In addition, we propose Adaptive Space Partition to classify marine data.

4.1 Multi-layer Index Frame

ML-index contains two layers: B^+-tree and HSP-tree, as showed in Fig. 1. Due to temporal continuity, we take time dimension as primary key to achieve overall control. After partitioning time into several intervals, we get some time series, which is called sub-region. In this way, adjacent data are stored in the same block. It shorts scanning regions when querying, which improves query efficiency.

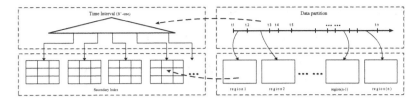

Fig. 1. Multi-layer Index Frame

We only index time interval and sub-region number other than data records, which reduces index update times when inserting records. We adopt B^+-tree to index time interval and insert marine data into intervals sequentially. Once inserted, former intervals cannot change. The records are inserted in ascending order.

4.2   Time Interval B^+-Tree

Marine data is extended by time, therefore, we partition the dataset into several non-overlap temporal dataset, $TIS = \{[t_0, t_1), [t_1, t_2), ..., [t_{i-1}, t_i)\}$. Each leafnode in B^+-tree corresponds to a HSP-tree, which indexes the sub-region. The leafnode stores its adjacent node's point to support range query. The principle of B^+-tree is as shown in Fig. 2.

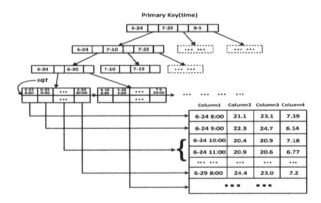

Fig. 2. B^+-tree index for Time Intervals

There are many records in one time interval T_n. To raise the query efficiency for massive points and multi-dimensional attributes, we should further decide a secondary key for indexing and whether to carry out the sub-region partition.

4.3   Adaptive Space Partition (ADSP)

Rational data space partition would have a direct impact on index performance [5]. In this section, we describe how ADSP works and adopt HSP-tree to index the partition result. Goil et al. [11] propose CLIQUE for data cluster, and we propose adaptive space partition based on CLIQUE and equidistant partition.

Definition 5 (Equidistant Partition). Given marine data scheme R with domains $\{D_1, D_2, D_3, \cdots, D_m\}$, defining $S = A_1 \times A_2 \times \cdots \times A_M$ is a d-dimension space. The space S is partitioned into a grid consisting of k^M non-overlapping rectangular units [11].

The result of 2-d space partition of WaterTemp Dataset is shown in Fig. 3. As dimension grows, blank cell grows exponentially, which results in lower storage and higher traversal time consuming. Hence, we propose ADSP to avoid above problem.

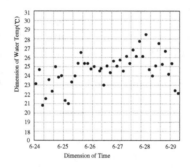

Fig. 3. A 2-d space equidistant partition among T_1

4.3.1   Adaptive Space Partition
We partition each dimension according to its correlated data scale and data distribution, which is called Adaptive Space Partition, as described below.

(1) Get domains form each marine scheme R and project into data space.

(2) Partition each dimension A_i into several equal intervals according to Definition 5.

(3) Compute the histogram for each unit of A_i and set the value of each bins.

(4) Given a threshold, from left to right merge two nearby units with the threshold and get the interval $R[i, j]$ (No. j interval in A_i).

(5) If number of bins is one, we have an equi-distributed dimension and equidistant partition it; else we take the result from step 4 as preliminary partition result.

(6) Get the partition and iteratively compute the capacity for each unit. If merged unit capacity $\theta \succ \mu$, it will be further partitioned according to Definition 5; Doing so repeatedly till all qualified sub-regions are done partitioning.

If there exists so many similar data in one unit after partition, data scanning will get harder. Therefore, we set a threshold for each unit and once it is over, we execute further partition. We define space unit capacity as follow:

Definition 6 (Space Unit Capacity). The capacity of partitioned space unit c_i, $\theta = \frac{S_i}{cs_i}$, in which S_i is the number of points in the unit c_i, cs_i represents the available storage of unit c_i. If unit capacity $\theta \succ \mu$, equidistant partition will be executed, μ is the threshold for space unit capacity.

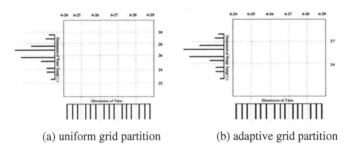

(a) uniform grid partition   (b) adaptive grid partition

Fig. 4. The result of different grid partition

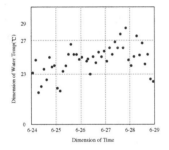
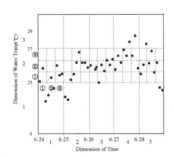

Fig. 5. The result with step 4 in ADSP   **Fig. 6.** The final result of ADSP among T_1

Figure 4 shows the 2d-space partition result. Figure 4(a) is the result of equidistant partition, which is a space-based partition. Figure 4(b) shows the result of adaptive partition, which is a data-based partition. The result in step 4 is shown in Fig. 5.

Suppose the upper data unit capacity θ is 3, we compute the capacity in each unit and further partition if necessary. Doing so iteratively, we get the final partition result that is shown in Fig. 6. Under same conditions, ADSP holds higher storage rate and less blank cells, which is a data-based and space-based partition way.

4.3.2   The Algorithm for ADSP

The algorithm is described as follow:

ALGORITHM 1. GetAdaptiveSP(A,k,th)

INPUT: A=(R, N), $|A_i|$(cardinality of A_i), threshold th_i

OUTPUT: final partition SP

   1: **for** each dimension A **do**

   2:      Divide $|A_i|$ into windows of some small fixed size r_m, m=1,2,$\cdots$, k_i;

   3:      Compute the histogram for each unit of A_i;

   4: **If** ($val_{m+1} \in [val_m (1- th_i) , val_m (1+ th_i)]$) **then** merge r_m and r_{m+1} into r_m;

   5: **else if** (number of bins ==1)

   6:      Divide the dimension A_i into equal partitions and set a threshold th_i for it;

   7:      Get the spatial interval partition SP and mark *Cellnumber* for each cell;

   8: **while** ($i < C_i$){

   9:    **for** ($;j < C_i; j$ ++){

  10:   **if** ($\theta = \frac{S_i}{cs_i} > \mu$)

  11:    **then** partition each dimension with a fixed size k_i/a;   /*a=1,2,3,$\cdots$*/

  12:    **else** break; **else** i^{++};} }

  13:   Get the final SP;

The algorithm is efficient to lower blank cells and avoid scanning and store it. In this paper, we propose ADSP method to divide marine dataset and the key is to compute the record number in each unit C_i. Hence, time complexity is $O(n)$, if number of non-empty cells is huge, marked as n, then time complexity will be $O(n^2)$.

4.4   Hybrid Space Partitioning Index Tree

From above method, we can get a relatively stable date partition that is suitable for data distribution in a certain marine. In this paper, we adopt HSP-tree to store the final data partition. As dimension amount grows, data units grow exponentially, thereby we only index non-empty cell.

Definition 7 (Hybrid Space Partitioning Index Tree). We define HSP-tree under marine data scheme R of Dataset A as follow:

(1) It has one root, and optimistically M + 1 layers.

(2) Each dimension corresponds to one layer in HSP-tree, and HSP-tree rotates with dimensions judged by Recognition.

(3) Aside from leafnode layer, all the non-leaf nodes contain two parts: (1) cell number c_i; (2) the pointer (nextLayer) that points to all the conjoint non-empty cells on the next layer. If the node is in the penultimate layer, it points to leafnode.

(4) The route from root to leafnode corresponds to a non-empty cell.

Figure 7 shows the HSP-tree that indexing the interval derived from space partition as shown in Fig. 6. The tree has five layers, and the first layer has five nodes that

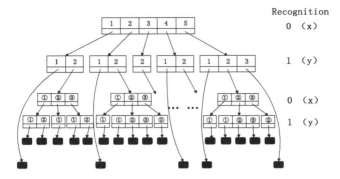

Fig. 7. The result with HSP-tree corresponding to Fig. 6

respectively represent its cell number. The leaf node stores the information such as the point amount and the point number. All the nodes have the same unit capacity.

4.4.1 HSP-Tree Algorithm

HSP-tree takes the preprocessing data partition as the reference, and inserts each point of Dataset A_i into tree in sequence. Step InsertPoint() searches corresponding *cell number* in the tree, if success, then insert the data point and plus one record amount; else if not, we should update the tree to create a new node by Step UpdateTree(). When the leaf node exists, the point will be inserted in ascending order.

ALGORITHM 2. CreateHSPtree(A, SP)

INPUT: A=(R,S,F), SP
OUTPUT: HSP-tree

1: root=NULL;
2: **for** each point (record) in A_i, i∈d **do**
3:      Mark the cell number C_i into which data falls;
4:      InsertPoint(root, C_i , point);
5:      UpdateTree();   */delete node and publish new node*/
6: **return** HSP-tree;

In the process of inserting, record amount keeps changing. For those leaf nodes whose unit capacity $\theta \succ \mu$, we should update the tree and split the node. The operations on HSP-tree (e.g. Range query, exact query) are basically based on tree traversal.

4.4.2 Time Complexity of HSP-Tree

(1) Insert Operation

Suppose the amount of sample data is N and the dimension is d, we partition each dimension into m intervals. After merging, each unit is intensive at worse, then we keep the partition and unit amount is m^d. Typically, the tree has *2d + 1* layers. Each point goes through *2d* layers when inserting and is compared $log_2 m$ times in each layer at most, then compare times is $2d log_2 m$ at most. The whole compare times for dataset are $N \times 2d log_2 m$. Hence time complexity for insert operation is $O(N)$.

(2) Range Query

Firstly, we partition dataset by key time to avoid searching extra redundant blocks; therefore, the key is to find the adjacent unit that overlaps most with the given unit. Each unit has $2d$ adjacent units. At worst, every unit is included and the search operation amount is $m^d \times 2d$. Time complexity for each search operation per unit is $O(d\log_2 m)$, so the whole time complexity is $O(2d^2 m^d \log_2 m)$, and maximum for non-empty units is N/a (a is unit capacity), so the complexity can be changed as follow:

$$O \leq O\left(\frac{2d^2 \log_2 m}{a} \times N\right)$$

As parameters a, d and m are constant, so $O \approx O(N)$.

(3) Exact Query

We also think of exact query as query for a certain point, so we only need to find the corresponding block and the only disk I/O operation is to read this block. At worse, the method would go through $2d$ layers, as well as compare $log_2 m$ times in each layer and $2d\log_2 m$ in the whole. So the whole time complexity is $O(N)$.

4.5   Range Query Algorithm for Multi-layer Index

Given a 2D range query $Q(Y_0, Y_i)$, Y_0 is a certain range in domain D_0 of dimension A_0, $Y_0 \in D_0$, Y_i is a certain range in domain $D_i (i \neq 0)$ of dimension A_i. First off, user can find the corresponding time interval through B$^+$-tree. Then for each time interval Y_0, user can find corresponding Y_i range. Furthermore, we find all needed data from Y_0, Y_i ranges. There exist two situations: (1) If given range happen to be the sub-range of (Y_0, Y_i), we directly scan this range. (2) If the overlap ratio between given range and (Y_0, Y_i) is over predefined threshold, we would scan the whole range and adopt another key to filter the data. The algorithm is described as follow:

ALGORITHM 3. Range Query

INPUT: $Q(Y_0, Y_i)$

OUTPUT: R_Q /*query dataset*/

1: $R_Q \leftarrow \emptyset$; $S_Q \leftarrow \emptyset$;
2: $T_Q \leftarrow getHSPtree(B^+\text{-tree}, Y_0)$;
3: **for each** HSP-tree T in T_Q **do**
4:    $S_Q \cup getRegionFromHSPtree(T, Y_i)$;
5: **end for**
6: **for each** region R in S_Q **do**
7:    **if** abstract(R)$\subseteq (Y_0, Y_i)$ **then** $R_Q \cup scanRegion(R)$;
8:    **else if** overlap(abstract(R), (Y_0, Y_i))$> \epsilon$ **then**
9:      **for each** record r in R **do**
10:        **if** r $\in (Y_0, Y_i)$ **then** $R_Q \cup$ r;
11:        **end if**
12:      **end for**
13: **end for**

The key to the paper is to build the multi-dimension secondary index. In terms of Innodb and MyISAM, time complexity for primary index B$^+$-tree and response time keeps the same. The performance evaluation will be presented in experiment part. When we execute range query, we set the initial threshold ϵ a little higher, which is a heuristic search method so as to avoid restart scanning due to the lower ϵ.

5   Experiment

Queries for marine scenario reproduction are mainly range query and exact match query. Exact query can be considered as a special type of range queries by setting search range to zero. In this paper, we evaluate the performance of range query among ML-index, Innodb and MyISAM in three aspects: response time, CPU consuming time and number of I/O. All environment configuration are given as follows: (1) Hardware: Intel Core i5 CPU, 4G RAM, 1 TB dish; (2) Operation system: XP Professional; (3) Database: MySql; (4) Programming environment: C++ Compiler.

We respectively experiment ML-index, Innodb, MyISAM on MySql platform. Figure 8 shows the query performance for range queries. We can get from the figure that when the search radius is relatively short, ML-index plays better in search time and CPU time against Innodb and MyISAM, but needs more I/O counts due to the heuristic principles. The higher initial threshold ϵ results in lower data filter ability. With the search proceeding, the threshold will be declined to fit in well. When the search radius is zero, ML-index filters over half points by primary key (time) at first. At this time, we pick a second key to further filter data and get a small intensive unit; thereby query efficiency is quite better. As dataset size rises, ML-index plays much better in response time and consuming time. As search radius grows, Innodb and MyISAM perform worse in data filter due to their single-dimensional secondary indexes.

As can be seen from the performance curve, ML-index need less CPU time and three indexes get close as search radius grows. Unfortunately, ML-index needs more I/O operation times initially, because HSP-tree stores the primary key (time) information like Innodb. Moreover, although such mature indexes as MyISAM, Innodb performs better in I/O counts and CPU time, they are weak at response time especially facing with marine date with great similarity. For region query, ML-index has a better query performance and is a stable index that fits in well with marine data.

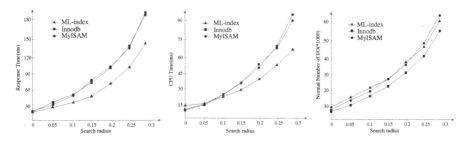

Fig. 8. Test with different methods in response time, CPU time, I/O cost respectively

The experiment verifies that data space partition would have a direct impact on the index performance [5]. Meanwhile, under marine business platforms, almost 80 % of queries focus on 20 % data objects. Hence, analyzing features of data and planning on better data partition helps to build a complete index structure for marine data.

6   Conclusions and Outlook

The index is the key to raise the query speed for marine business platform, while index performance is mostly determined by rational data space partition. In this paper, we study on a better index structure that fits in well with marine data. The experiment suggests that as dataset size grows, ML-index performs better in query speed and consuming time. Furthermore, HSP-tree improves query efficiency for data hotspots with characteristic values, and guarantees access speed for data with low data value function. In conclusion, the structure shows robustness and stability in the whole.

In this paper, we formalize marine data value function to keep data partition make great sense. According to data unit capacity (See Definition 6), we can get that the threshold for data unit capacity is related to the entire system's storage utilization and access efficiency, and we will still work on optimizing data unit capacity. In the future we will study on query problems upon several different computer nodes as well as combine data migration [12] under cloud platform to work out a better solution.

Acknowledgments. This work is Supported by National Natural Science Foundation of China (61272098), Natural Science Foundation of Shanghai (13ZR1455800) and Scientific Research Foundation for Ph.D. Of shanghai Ocean Univ.

References

1. Mayer-Schönberger, V., Cukier, K.: Big data: a revolution that will transform how we live, work, and think. Houghton Mifflin Harcourt (2013)
2. Meng, X.-F., Ci, X.: Big data management: concepts, techniques and challenges. J. Comput. Res. Dev. **50**(1), 146–169 (2013)
3. Shi, S., Lei, B.: Theory and Practice on China Digital Ocean, pp. 1–16. Ocean Press, Beijing (2011)
4. Liu, X.-S., Zhang, X., Chi, T.-H., et al.: Study on China digital ocean prototype system. In: Proceedings of the 2009 WRI World Congress, pp. 466–469. IEEE, Piscataway (2009)
5. Zhou, X.-M., Wang, G.-R.: Key dimension based high-dimensional data partition strategy. J. Softw. **15**(9), 1360–1374 (2004)
6. Chen, J., Fang, B.-X., Tan, J.-L., et al.: Index filtering algorithm based on minimum enclosing circle partition. J. Comput. **35**(10), 2139–2146 (2012)
7. Weber, R., Schek, H.J., Blott, S.: A quantitative analysis and performance study for similarity-search methods in high-dimensional spaces. VLDB 98, pp. 194–205 (1998)
8. Robinson, J.T.: The KDB-tree: a search structure for large multidimensional dynamic indexes. In: Proceedings of 1981 ACM SIGMOD on Data Management, pp. 10–18. ACM (1981)

9. Beckmann, N., Kriegel, H.P., Schneider, R., et al.: The R*-tree: an efficient and robust access method for points and rectangles, pp. 322–331. ACM (1990)
10. Nievergelt, J., Hinterberger, H., Sevcik, K.C.: The grid file: An adaptable, symmetric multikey file structure. ACM Trans. Database Syst. (TODS) **9**(1), 38–71 (1984)
11. Goil, S., Nagesh, H.: MAFIA: efficient and scalable subspace clustering for very large data sets. In: Proceedings of SIGKDD on Data Mining, pp. 443–452 (1999)
12. Huang, D.-M., Du, Y.-L., He, Q.: Migration algorithm for big marine data in hybrid cloud storage. J. Integr. Plant Biol. **2014**(01), 199–205 (2014)

On Massive Spatial Data Retrieval
Based on Spark

Xiaolan Xie[(⊠)], Zhuang Xiong, Xin Hu, Guoqing Zhou,
and Jinsheng Ni

Institute of Information Science and Engineering,
Guilin University of Technology, Guilin, China
xie_xiao_lan@foxmail.com

Abstract. In order to search more efficiently for rapidly growing spatial data, cloud quad tree and R-tree is adopted in spatial index for the non-relational databases of cloud HBase, by which data can be retrieved successfully. By comparison of retrieval efficiency between cloud quad tree and R-tree, that how different parameters acted is tested on data index and retrieval efficiency and put forward an relatively more reasonable solution. Subsequently, we verify the validity of index calculation when there is a giant one.

Keywords: Spatial data · Spark · Quad tree · R tree · Index · Cloud computing

1 Introduction

With the growing development of computer technology, Geographic Information System (GIS) have been developing rapidly, with the expansion and augment of its application and demand. For faster processing and better spatial index, cloud computing is applied to retrieve spatial data prove to be a leading subject of resolving massive spatial data and geographic distribution [1].

R-tree (Rectangle Tree) is a data structure which is widely used in spatial retrieval. The depth of the tree is relatively smaller than the quad-tree, therefore the query efficiency is much higher. R-tree as an index can achieve a great faster speed in less changing spatial data. Besides, the controllable section for the user to control is much less and therefore it is more convenient to use. However, the add or delete operation on R-tree, which has already been generated, will decrease the quality of R-tree index, while it will consume a huge amount of computing if the index is repeatedly rebuilt (especially at present a lot of mutation of R tree, such as R+ tree, R* tree, are at the expense of insert and modification to improve the performance of query). As for the demand for this type, the quad-tree as represented in segmentation rule is a better choice. In this paper, we achieve a quad-tree and R-tree algorithm for parallel query based on Spark, to lay the foundation for the massive spatial data query and processing [2].

© Springer International Publishing Switzerland 2014
Y. Chen et al. (Eds.): WAIM 2014, LNCS 8597, pp. 200–208, 2014.
DOI: 10.1007/978-3-319-11538-2_19

2 Quad-Tree Index Algorithm on Hbase

The other paper entitled On Massive Spatial Data Cloud Storage and Quad-tree Index Based on the Hbase design a linear quad-tree spatial data index structure based on Hbase, but MapReduce is a framework of batch to process massive data, and it is not appropriate for the query operation. As for this type of operation, the memory computing framework Spark is a better choice.

Spark index quad-tree is divided into three steps: preparation in advance, computing, and consolidation. The Hbase query interval data will be read into a RDD block before Spark index, and then the data block will be calculated. Compared to calculation and operation on the Region of MapReduce, the same operation in memory RDD block is much faster and more stable. The index procedure for Spark on quad-tree cloud is as following program flow chart [3] (Fig. 1):

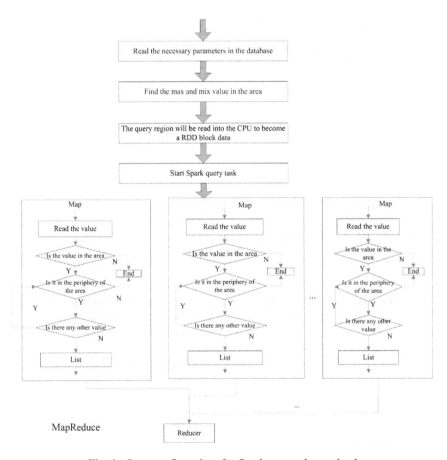

Fig. 1. Program flow chart for Spark on quad-tree cloud

Map operation output of regional spatial data may have duplicate values, therefore, Reduce operation must delete the duplicate values, and ultimately the same spatial data only output once. The nonexistent duplicate value after the Reduce operation of spatial data sets is the query results.

3 R Tree and Variants

Traditional sort tree, B tree appeared to be inadequate in the spatial data index. Because the traditional index tree is mostly linear index tree, the spatial data are basically more than two-dimensional spatial data, which does not have a very clear linear structure. In order to solve this problem, Guttman proposed R-tree, and then the advantages of dynamic balance is extended to the multidimensional space [3]. As a result the sort tree can be used in lots of new area, such as spatial data.

The directory node of the R tree is mainly composed of Mbr and sub node array, where Mbr is the smallest rectangle containing the node, and the rectangle can accommodate all the child nodes of the node. Data node of the R tree is mainly composed of Mbr and spatial data array, R tree by itself generally does not preserve spatial object itself, but to save space in the location of the database object. The biggest problem of R-tree is the brother node Mbr overlap. If the query region contains exactly several nodes of coincident area, the index of these nodes under the same root of the subtree must search. In order to minimize overlap, domestic and foreign scholars have proposed many R-tree variants, such as R+ trees, R* tree, X trees [4]. These improvements have advantages and disadvantages, and R * tree is generally considered as the most efficient R-tree variants.

4 R-Tree Index Algorithm on Spark

4.1 Traditional R-Tree Index Algorithm

Traditional R tree and its various variants of index algorithms can be roughly divided into three steps [3–5]:

- If the root node R overlaps with the query area S, and R is a directory node, it is used as an ordinary node T, and then carry out the second step. If the root node R overlaps with the query area S, and R is the data node, and then it is treated as a ordinary node T, and then carry out the third step.
- If T is not a leaf node, every child node of T will be checked to determine whether the child node overlaps the S. For all overlapping sub nodes, the index algorithm will be used to search for subtree under the root of sub nodes.
- If T is a leaf node, every child node of T will be checked to determine whether the child node overlaps the S. If the data Doverlaps S, the data D is one of the search condition in accordance with the data.

The main steps are highly parallel computing, and do not affect each other. Therefore, the algorithm will be directly converted into cloud algorithm, which is theoretically feasible. But it is found that in actual operation the direct conversion of these steps to the cloud operation cannot bring too much performance as Spark comparison algorithm running intersection is relatively more rapid, and the distribution and aggregation of data blocks is relatively slow. Reduction operation is frequently used and this will consume a lot of network bandwidth and transmission time, especially during the iteration, the bandwidth will become a significant bottleneck. Therefore, we can consider the query K layer subtree in one time of iteration, Reduce operation will be processed after each K layer query, then re-allocate computing tasks.

4.2 Cloud R-Tree Index Algorithm and Optimizing

Cloud R tree query algorithm is also divided into three steps, which can be subdivided into the following steps [5–8]:

(1) If the root node R intersect with the query region S, which is a directory node, put the root node into Dir List, and carry out the second step. If the root node R intersect with the query region S, which is a data node, then put the root into DataNode List, and carry out the third step.

(2) Recursive query is used for each node of Dir List until all the nodes are up to leaf nodes.

- Convert Dirlist to RDD data block, denoted Rd, is the value of k is default to 0;
- Perform the Map operation of Rd data block, if Rdi intersects with S, then output all of its nonempty nodes as an array, k value plus 1;
- If k < K and the output is not empty and the child node of output node is not a leaf node, continue to process 2.2 until k = K or the output is empty or the child node of output node to the leaf node;
- Perform the Reduce operation of the Ra data block RDD after the Map operation, combine all the outputs with a large array, and convert the array into List, denote Li.
- If the elements of Li are Li leaf node, perform the third step, otherwise, make Li as the new DirList, continue to process 2.1 until Li is a leaf node.

(3) Determine whether the data contained in DataNode List overlap S, output the overlapping data.

- Convert The Data Node List into a RDD data blocks, denoted by Rn;
- Perform the Map operation on data block Rn. If Rni intersects with S, then output all if its data to an array.
- Perform the Reduce operation on data block Ra after the Map operation, combine all the outputs into a large array, and convert the array into List, denoted Ld.
- Convert Ld into the RDD data block, denoted by Re;
- Perform Map operation on Re. If Re (i) intersects with S, Re [i] data is one of the index criteria that meet the requirement;
- Perform the collect operation on the Ra data block after the Map operation, and obtain the final result.

The value of K is determined by the number of the spatial data and the order of the R tree itself. The smaller the order is, the deeper depth of the tree. While the data for each node is relatively fast, K can be set to a larger value. Meanwhile, if the network bandwidth is small, Reducing operating costs is relatively large. It can increase the K value in order to reduce the Reduce operation frequency.

5 Simulation

5.1 Experimental Platform

Hardware: Master Node: 3.40 GHz i5-3570 CPU 4G RAM;
Slave nodes: four 3.30 GHz i3-3220 CPU 4G RAM;
Platform: Fedora18 x86_64 Hadoop-1.0.3 HBase-0.94.2 mesos-0.12.1 spark-0.7.3;
Java environment: jdk-7u40;
Development and debugging tools: Eclipse J2EE Galleo;

Experimental data: about 90 % of them with text describing properties, 50 % with image description attributes, 30 % with audio description attributes, 10 % with video description attribute.

5.2 Comparison of Performance Between Spark and Hadoop Retrieval

In the Spark and Hadoop clusters are retrieval operations on spatial data of different sizes of space, record retrieval time as shown in Fig. 2 (unit: Ms):

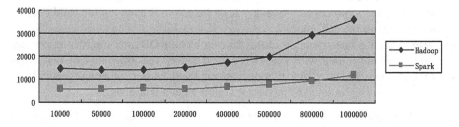

Fig. 2. Comparison of different regional spatial index time

It can be seen from the table that the efficiency of Spark query is significantly higher than conventional MapReduce. On one hand, the code of Spark is more compact and efficient, on the other hand, the computational model and HDFS coupling of MapReduce is too high, and most of the convergence of computing operations are required to read and write HDFS, while Spark are to be treated as a block of memory to process the data and process results, it not only improves the data processing speed and fault tolerance, but also reduces the procedures for hard disk and network IO requirements.

5.3   The Performance of Spark Spatial Data Retrieval

Using the Collect method to the RDD data block instead of the Reduce operation, the Spark cluster on the space point different size of data operation space retrieval operation, retrieval time record retrieval time and with regional data were compared, as shown in Fig. 3 (unit: Ms).

It is can be seen from the Table 2 that after the elimination of the step. Reduce operation of the spatial point cloud quad-tree query operation, the index is increased by nearly 10 times compared to Reduce operation. To improve the index efficiency, operations like space plotting should be processed by using as many as point data.

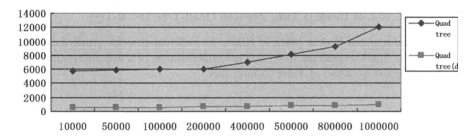

Fig. 3.   Different types of spatial data index time

5.4   Cloud Quad Tree for the Uneven Distribution of Spatial Data Processing

To test the cloud quad tree unevenness of the distribution of the data processing, Java program is employed to distribute 50 percent of the data to around 10 % space with equality, and the other half of the data is still evenly distributed in the entire space. The test is divided into two steps: intersection with the search range and intensive space; or vise versa. The index time is recorded as shown in Table 1 (unit: ms, OM: table representative of a Java out of memory error):

Table 1.   Non uniform spatial data index time

Test	The index range and space intensive intersection				The index range and space -intensive disjoint			
Volume data sheet	100,000	200,000	400,000	500,000	100,000	200,000	400,000	500,000
Level = 6	17809	18347	OM	OM	6065	6015	7276	8125
Level = 8	10250	10544	14628	17244	6102	6003	7256	8044
Level = 10	7125	7049	7536	8247	6097	6115	7179	8062

When Level is small, the imbalance data has had a great impact on the quad-tree. When the search range and dense space intersects, the level is too small to trigger the out of memory error. Because the division of cloud quad-tree space is completed before the data is inserted, data amount of local area is increasing greatly, which has a particular similar effect on cloud quad-tree and the time under large global data amount. A large amount of individual data item for key causes a longer time to judge the space data communication when searching the data, and thereby it leads to the bottleneck for the inquiry. However, the bottleneck of this local query performance only appears in the query region which contains or intersects data dense areas, but it does not have any impact on query performance of sparse region. The solution for this problem is same with the problem of global intensive data, which is to increase the Level depth of cloud quad tree. When Level in Table 3 is big the increasing data of local data does not cause too much reduction in query efficiency, which also illustrates this point.

5.5   Comparison of Reading Time Between Quad-Tree and R-Tree

Creating quad-tree and R -tree index in HBase for different sizes of spatial data, and then using the Spark to read the index and the establishment of a tree query structure, time consumption is shown in Table 2 (unit: ms):

Table 2. Time setting for quad-tree and R-tree

Amount of data	10,000	50,000	100,000	200,000	400,000	500,000	800,000	1,000,000
Quad-tree	80	82	81	95	102	134	183	227
R Tree	1147	1453	2139	4412	6435	7946	9532	12453

In order to create the directory node, R-tree needs to repeatedly read HBase database, causing the result that the index time of spark R-tree is much greater than R-tree. In addition, all the index of R-tree must be entirely read to CPU and creates a tree structure. Compared to quad tree which may contain parts of the data read into the memory space, the flexibility of the R-tree seemed to be insufficient. Therefore, when compared to the query efficiency of the quad tree and R tree, the query time will only be calculated, without considering the creation of R-tree and cloud quad-tree time required to read from HBase. R-tree is mostly used for static data index, while, when retrieval operations are very frequent, you can save time by using the generated R tree shared by multiple retrieval operations.

5.6   Comparison of the Retrieval Time of R Tree and Space Quad Tree under Different Parameters

The value M of R tree and value K of the index program are set to different values, and perform retrieval operations for spatial data of different sizes in the Spark cluster. The comparison of time between record index time and cloud quad tree index time are as shown in Table 3 (unit: Ms):

Table 3. Comparison of index time between quad tree and R tree

The amount of data	10 thousands	50 thousands	100 thousand	200 thousand	400 thousand	500 thousand	800 thousand	1 million
Quad tree(point)	579	606	593	632	759	820	883	965
Quad tree(region)	5786	5946	6082	5986	7043	8146	9237	12103
M = 100 K = 4	1090	1786	4450	6670	8753	9548	23436	26690
M = 100 K = 8	1850	1927	4197	6375	8032	8113	13042	17890
M = 500 K = 4	1136	1534	1942	5236	7526	7845	8932	11279
M = 500 K = 8	1627	2179	3045	5124	7372	6239	9768	15367

It is shown that, optimized space quad-tree points query index is much faster than other index methods. While the value M of R tree and value K of the index program are taken to the appropriate time, the query efficiency of spatial data is much higher than the quad tree. But if both of them do not match well, the query efficiency of R-tree will be greatly affected, especially in a large amount of data. The choice of smaller K and value M will cause frequent transfer of Reduce, which could greatly decrease the efficiency of R-tree index. When value M is larger, the depth of R tree is smaller, while if value K is set to a larger value, unbalanced load will be caused. Meanwhile, M value will be bound to the use of memory (out of memory error occurs when the value of M in the table above), and the larger value M will make the MBR of R-tree node overlap rate increase, so the value of M and K must be adjusted according to the dynamic demand: when value M increases, K values should be reduced accordingly.

6 Summary

Due to the quality of cloud storage by columns, smaller computing needs larger IO, data arranged according to Key, etc., there are many differences between the cloud spatial indexing and retrieval with single selection. As changes in less space area data, query efficiency is slightly higher than R tree spatial quad tree, and cache memory sharing technology to reduce the generation time R-tree, and after each data changes, you must re-build the R-tree index. If the data changes greatly, you can use the space for better overall performance quad-tree. For processing point data, the algorithm efficiency is much higher than the R-tree following the space-optimized quad-tree search. Therefore, it's better to use quad-tree spatial point data retrieval in the cloud.

Acknowledgments. This research work was supported by National High Technology Research and Development Program 863 under Grant No. 2013AA12A402, Natural Science Foundation of Guangxi Provincial under Grant No. 2013GXNSFAA019349.

References

1. Zhu, Q., Zho, Y.: Distributed spatial data storage object. Geom. Inf. Sci. Wuhan Univ. **31**(5), 391–394 (2006)
2. He, R.Y., Li, Q.J., Fu, W.J.: Guide to Developing and Application With Oracle Spatial Database, pp.1–48. Surveying and Mapping Press (2008)

3. Hu, J.: Research and implementation of parallel clustering algorithm in cluster environment. Master's degree thesis of East China Normal University, pp.15–29 (2012)
4. He, X.Y., Min, H.Q.: Hilbert R- tree spatial index algorithm based on Clustering. Comput. Eng. **35**(9), 40–42 (2009)
5. Zhang, Z.B., Wang, Y.Z., Li, H.: Research on the cost model for spatial query based on R-Tree. Micro Comput. Syst. **24**(6), 1017–1020 (2003)
6. Cary, A., Sun, Z., Hristidis, V., Rishe, N.: Experiences on processing spatial data with MapReduce. Sci. Stat. Database Manage. **2009**(6), 302–319 (2009)
7. Wang, L., Zhang, H.H., Li, K.S., Ju, H.B.: Region matching algorithm for DDM based on dynamic R-tree. Comput. Eng. **34**(3), 56–58 (2008)
8. Apache. Spark Handbook (2013). http://spark.incubator.apache.org/docs/0.7.3/

HDStore: An SSD/HDD Hybrid Distributed Storage Scheme for Large-Scale Data

Zhijie Feng[1,2], Zhiyong Feng[1,2], Xin Wang[1,2(✉)], Guozheng Rao[1,2],
Yazhou Wei[1,2], and Zhiyuan Li[1,2]

[1] School of Computer Science and Technology, Tianjin University,
Tianjin, China
{fengzhijie,zyfeng,wangx,rgz,weiyz,
lizhiyuan0718}@tju.edu.cn
[2] Tianjin Key Laboratory of Cognitive Computing and Application,
Tianjin, China

Abstract. Traditional data storage schemes are primarily based upon Hard Disk Drives (HDD). However, with the appearance of large amount of data on the Web, the read/write performance based on HDD has reached a bottleneck. Thus the emerging of Solid State Drives (SSD) has provided an opportunity for the storage of the Web of data. In this paper, we propose an SSD/HDD hybrid distributed storage scheme, called HDStore, for large-scale data, in which the single fix-sized journal file using the append-only mode is stored on SSD to support efficient read and write, while several segment files focusing on read are stored on HDD. Through a series of operations *build*, *split*, *move*, and *merge* between the journal and segment files, we constructed HDStore storage scheme based on JS-model. The experimental results show that HDStore obtains an efficient optimization of data read/write, especially the write performance has increased by 15 % compared to the traditional HDD-based scheme.

Keywords: JS-model · HDstore · SSD · Hybrid distributed storage scheme

1 Introduction

With the development of modern Web, tremendous information growth has led us to the age of big data [1], which poses a great challenge to the traditional data management methods, and for which we get big data technology [2, 3]. The big data technology is built on a variety of technologies, such as parallel computing, distributed file system, distributed database, scalable storage system and so on, among which the key issue is how to store and manage big data effectively and efficiently [4].

To solve the aforementioned problem, there are several choices for the data storage management. We can classify the data according to their significance, or make reasonable arrangements for the data processing, or utilize hybrid storage technology to improve the data read/write performance [5–7].

The traditional high performance database system is primarily based on HDD [8], whose performance bottleneck is mainly concentrated on the I/O rates. SSD is a data storage device using integrated circuit assemblies as memory to store data persistently.

© Springer International Publishing Switzerland 2014
Y. Chen et al. (Eds.): WAIM 2014, LNCS 8597, pp. 209–220, 2014.
DOI: 10.1007/978-3-319-11538-2_20

Compared with the HDD, SSD has outstanding read/write performance, while it has low-capacity and high-priced [9]. Considering the characteristics and taking advantage of both devices to constitute a hybrid storage will effectively improve the performance of big data management, which is a new attempt for the large data storage [10].

In this paper, we take advantage of a JS-model to achieve the hybrid SSD/HDD storage scheme under the framework of a distributed system for big data storage management. A single journal file is stored on the SSD for fast read/write operations, and the corresponding multiple segment files are stored on HDD for data access operations. Thus we make full use of the advantages of both devices, and avoid not only the deficiencies of SSD such as low capacity, but also the deficiencies of HDD such as low performance.

The main contributions of our work include:

– We build efficient storage model, called JS-model, for large-scale data. Based on the existing big data storage model, we abstract JS-model that is targeted to store and access data quickly.
– We devise a new SSD/HDD hybrid distributed storage scheme, called HDStore, which is designed to optimize data read/write by managing journal and segment files on diffident kinds of disk drives.
– We perform comprehensive experiments to examine the performance of our JS-model and HDStore scheme for large-scale data under the distributed environment.

The rest of this paper is organized as follows:

In Sect. 2, the data storage JS-model is defined. In the JS-model, the journal file is focused on fast data write and the segment files are focused on the persistent data storage. In Sect. 3, the HDStore storage scheme is described in details, which is based on the JS-model for SSD/HDD hybrid distributed storing large-scale data. In Sect. 4, we conducted a series of experiments. The results show that our HDStore storage scheme is more efficient than the traditional storage scheme based on the HDD, and the write performance under the HDStore is increased by 15 %. Finally, we conclude this paper and outlook the future work.

2   Data Storage Model

In this section, we summarize the existing big data storage model and abstract the hybrid data storage model, called JS-model, for large-scale data in the distributed environment. The JS-model includes (i) the definition of *JS-model*, (ii) a series of operations under the *JS-model*.

2.1   JS-Model

Unlike traditional data storage schemes, our hybrid distribute storage scheme is based on JS-model. The files which stores data persistently is divided into journal file and segment files under the JS-model. First, we give the definition of *JS-model*.

Definition 1. Assume that there is a data item i, where i is a tuple of r and t, r is a data record and t is the time stamp of the record submitted. Then the journal file J is a finite set of data item i.

$$J = \{i | i = \langle r, t \rangle\} |J| \leq N, \text{where } N \text{ is an integer.}$$

J refers a journal file, while i refers a data item. In a journal file J on a node in a cluster, there are a number of data items i composed by original record r and the submitted time t. The i is the basic unit of the data, which appends to a J after being processed to realize initial storage in this model. The size of the J is limited.

Definition 2. Let I be a tuple of R and T, where R is a data record and T is the time stamp of the record submitted. The segment file S is an infinite set of data item I.

$$S = \{I | I = \langle R, T \rangle\}.$$

S refers a segment file, while I refers a data index. The segment file S use B-tree structure to store the data index I persistently. The leaf node of the B-tree is composed by I. It is worth mentioning that data item i in J is unordered and append-only, while I in S is ordered by B-tree.

Definition 3. JS is a set of the union of journal file J and a series of segment files $S_1, S_2, \cdots, S_n$.

$$JS = J \bigcup S_1 \bigcup, \cdots, \bigcup S_n.$$

JS refers one single journal file J corresponding to multiple segment files $S_1, S_2, \cdots, S_n$. The size of J is limited. It is used as a data buffer during the loading data process. There are zero or more S which store data persistently. The number of segment files n depends on the size of data set.

2.2   Data Operation

We have defined the JS-model as the collection of files. There is a single journal file J, whose elements are finite. While there are a lot of segment files $S_1, S_2, \cdots, S_n$, whose elements are infinite. In order to describe the data storage scheme, we also need to define a series of basic operations on J and S. Figure 1 shows the data operations.

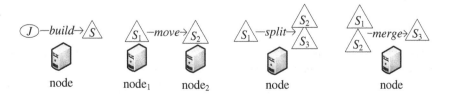

Fig. 1. Data operations

The operation *build* builds a segment file S from a journal file J. In the initial state, there is a single J and zero S on each node in the cluster. With the number of the data item i increase in J, the J reaches its capacity limitation. Thus, a new S is built by J pouring the old i.

The operation *split* splits a segment file S into several segment files $S_1, S_2, \cdots S_n$, where $n \geq 2$. In order to adapt to the disk space, file system limits the size of file. With the number of the data index I increase in S, the S reaches its capacity limitation. Thus, it is split into $S_1, S_2, \cdots S_n$ to prepare for the data storage.

The operation *move* moves a segment file S from one node to another in the cluster. To achieve the load balancing in the distributed system, the S need to be moved from large node where more $S_1, S_2, \cdots S_n$ are stored to tiny one where less $S_1, S_2, \cdots S_m$ are stored. But the journal file J stays local.

The operation *merge* merges several segment files $S_1, S_2, \cdots, S_n$ into one segment file S. With the data load process proceeding, a number of $S_1, S_2, \cdots, S_n$ with small size is existed. Considering the data compactness and the space utilization, *merge* operation is very important.

3 Hybrid Storage Scheme Based on JS-Model

3.1 Characteristics of HDD and SSD

The hardware of modern computers is developing rapidly, while the HDD is an exception [11]. The physical structure of HDD has limited the I/O rate. Due to the transfer speed of hard disk not changed, there is always a bottleneck for the improvement of the whole computer performance according to the bucket effect.

The basic working principle of mechanical hard disk is as follows. When application needs to read and write data, the HDD will receive instructions. Then a series of actions take place, such as the movement of the head and the spinning of the disk. As a result, a few milliseconds is wasted because of the movement of mechanical equipments [12]. Avoiding the mechanical structure and adopting new structure are the fundamental ways to improve the hard disk read and write performance.

SSD is made of an array of solid-state electronic memory chips. The interface specification and definition, functionality and usage of SSD are exactly the same as HDD's. While SSD has more advantages, such as faster I/O rate, anti-shock, low power consumption, noiseless, lightweight and so on. However, the capacity of SSD is limited. Moreover, SSD has a short lifetime because of the restrictions of erase cycles and a higher price than HDD in unit storage [13, 14].

3.2 HDStore Storage Scheme

To manage and store the big data effectively, we put forward and implement HDStore hybrid storage scheme under a distributed environment based on JS-model.

The Fig. 2 shows that our scheme is based on a distributed cluster. There are different kinds of servers running on each node in the distributed cluster. Data server is in charge of building data indexes, and store data in the file system persistently.

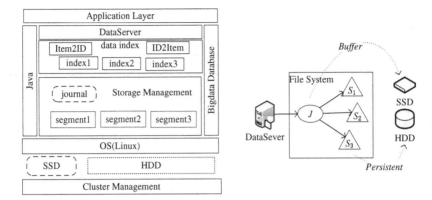

Fig. 2. HDStore storage scheme

On each node in the cluster, smaller capacity SSD and larger capacity HDD are set. The SSD is used for the fast read/write, while the HDD is used for the persistent storage of large-scale data.

In our storage scheme, several indexes are constructed from the source data in the data server, which stores the data record in the journal file J and segment files $S_1, S_2, \cdots S_n$, where one J corresponds s series of $S_1, S_2, \cdots S_n$.

The journal file J is used as a data buffer and write-once read-many, in which all historical consistent states can be accessed. It is maintained over a reference to the data record r and the time stamp t. It is loaded and store temporarily. However, the J is different from the buffer in the memory, for the J is used as an immortal store in which all data items i are persistent on disk. The data item i is out of order and appends to J logically. The J supports the concept of overflow. When it exceeds a threshold extent, overflow is triggered and a segment file S is activated. During the overflow processing, the i in the J is moved onto S and the date item R in S is modified. A journal file J is corresponds to an index file in one data server.

The segment file S stores data persistently and relies on B-tree to provide a persistent mapping from keys to values. The B-tree is used as an object store and based on the identifier of data record R. The Clusters of B-tree serialize object within the leaves naturally and provides IO efficiencies. The journal file J corresponds to zero or more S in every data server. Among different nodes, the data server shares the same metadata of the R.

At the very beginning, there is one single J and no S on each node. The data loading process begins by randomly choosing a J in the cluster. With the data loading, *build* operation is happened as soon as the chosen J reaches its max capacity, before when the size of S is set. Whenever one S reaches its max capacity, it *split* into $S_1, S_2, \cdots, S_n$. The first time n is the number of the nodes in the cluster, in the later n is two. The next step is that the S_i *move* to each node in the cluster. Thus there is an S on every node respectively. Then, the data loading is proceeding parallelly in the cluster.

After that, along with the data is loading, if S reaches its max capacity, it *split* into two in each own node. If size of S is too small, the *merge* operation happens periodically. In order to balance the performance of the nodes in the cluster, S can *move* among the nodes in the cluster. Algorithm 1 shows the procedure for managing data with the HDStore scheme.

Algorithm 1. management process of HDStore

Input: a cluster C and the number of node n
Output: a journal file J and segment files $S[n]$
1:   data_loading(C):
2:   choose J from C
3:   **if** J is overflow **then**
4:     S is *built*
5:   **end if**
6:   **if** S reaches limit first **then**   // the segment file is split for the first time
7:     S is *split* into $S[n]$
8:     S_i is *move* to every node
9:   **end if**
10:   **for** each S **in** C **do**   // it reaches a steady state
11:     **if** J is overflow **then**
12:       S is *built*
13:     **end if**
14:     **if** S reaches limit **then**
15:       S is *split* into S_1,S_2
16:     **end if**
17:     S is *move* for load balancing//the move operation maintains system balance
18:     S is *merge* periodically   // it optimizes the system performance
19:   **end for**

In the HDStore scheme, the operation times between the SSD and HDD dynamic change based on the size of data sets. What we concern is that the ratio of the read/write times on SSD and the read/write times on the HDD affects the overall system performance for the same data set. In the process of loading data, the journal has more read/write operations and the segments only provide batch write. The HDStore optimizes the read/write latency of the journal to improve overall performance. It stores one single journal file J on the SSD to support data buffer. The fast read/write characteristic of SSD is suitable for the frequently read and write operations on J. Besides, because of its low capacity, J is limited in size which is suitable for the SSD. While considering the size of S to reduce the cost of storage, it stores corresponding $S_1, S_2, \cdots, S_n$ on HDD for persistent data storage.

By using HDStore storage scheme, journal file J and segment files $S_1, S_2, \cdots, S_n$ are reasonably placed with the use of the characteristics of SSD and HDD. In this way, it compromises between effective data management and cheap storage, which leads to a higher cost performance.

4   Experiments

The experiments are based on an open source system called Bigdata [15]. Bigdata is a scale-out architecture for persistent, ordered data storage. Based on Bigdata, we have implemented HDStore storage scheme for large-scale data. In our experiments, we compared 3 different storage schemes, which are based on HDD, SSD and HDStore respectively.

4.1 Settings

All experiments are carried out by using a 4-node cluster as the underlying storage system. The operating system is Ubuntu 12.04. and the database is Bigdata 1.3. Each node uses Dell PowerEdge R820 Rack Server that has 4 6-cored CPUs @ 2.2 GHz supporting hyper-threading and 64 GB memory. The storage devices are composed of two SSDs and a HDD disk array. Eight HDDs (ST91000640SS, 1 TB, 2.5″, SAS, 7200 rpm) composes the HDD disk array, in which seven of them form a RAID5 disk array under the PERC H710P RAID controller, and one of them is used as a hot backup. The SSD is Dell LB406 M (400 GB, SAS, 8C38W).

On a node in the cluster, a single HDD fulfills the I/O by mechanically rotating the disk which leads to a low I/O rate. But under the RAID5 technology, the read and write performance of the HDD disk array is significantly improved. The PERC H710 Controller of the Dell Server is a medium-class RAID controller card which has a 512 MB cache. The random read and write performance of RAID5 is further improved under the controller.

The system performance differs according to the different ratios of SSD and HDD. The higher proportion of SSD, the faster data read/write. If we all use SSD, it can achieve the best performance while it has the highest cost. Thus the HDStore balances of the performance and cost. Theoretically, the SSD should contain all journal file on the DataServer. One DataServer has five indexes and each index corresponds to one journal file. All journal files do not exist at the same time, because the corresponding indexes are constructed sequentially. So it only needs a total of five journal files. In order to ensure the journal batch transforming into the segment file, each one journal file also needs a backup. Thus, the number of journal files is at most ten, and the limit of one journal file size is 200 M. The size of one single journal file limits 200 M. In this way, the capacity of SSD is only 2 GB, while no limitations are placed on HDD since it depends on the size of data set.

The detail configuration is shown in Fig. 3. There are four servers in the Bigdata distribute system. They are in charge of the whole cluster. The manager server will start other services on the local host. The jini server performs service discovery. The zoo-keeper server registers current nodes which represent their existence as part of logical service. The metadata server is used to manage the life cycles of scale-out indices. Besides, there is a data server running on each node in the cluster.

4.2 Datasets and Queries

In order to evaluate the performance of data loading, our datasets are provided by the LUBM generator [16]. We generated 5 datasets in increasing size, which are listed in Table 1. Although we only use the LUBM, the HDStore storage scheme is not opti-mized for the dataset. In other words, our scheme is general to be used for other datasets.

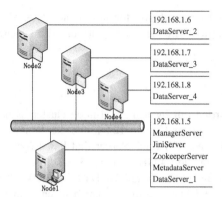

Fig. 3. Distributed cluster configuration

Table 1. Characteristics of lubm data sets

Data set	size(Mb)	Triples	Uris	Literals
lubm_1	3,162	19,946,347	3,243,555	1,669,166
lubm_2	6,344	39,874,257	6,483,710	3,334,970
lubm_3	9,575	78,352,162	9,771,773	5,028,070
lubm_4	12,767	97,328,377	13,022,107	6,699,946
lubm_5	15,956	116,611,812	16,268,151	8,369,852

For the reasons that (i) large-scale data will be successfully loaded in low memory and (ii) data is loaded fast in the distributed parallel environment, we have to pre-process into a number of small files whose size is around 20 MB, to adapt to the distributed environment.

The query sets in our experiments are taken from LUBM SPARQL queries [17]. There are 4 queries randomly selected (See Appendix A).

4.3   Results: Data Loading

We performed three experiments on measuring the loading data performance according to three different storage schemes based on HDD, SSD, HDStore. To achieve an accurate experimental result, all data sets are experimented 3 times to get an average evaluation. The results are shown in Table 2.

The data loading on the HDD is first analyzed. All persistent storage of the server is located on the HDD. The journal file in the Data Server is naturally placed on the HDD, so are the segment files. From the results, we observe that the time of loading data on HDD is nearly linear growth with the size of data. It takes the most time, compared with the other two schemes under the same size of data.

Then, we analyze the cost of the data loading on the SSD. Similarly, the journal file and the segment files are both persistently stored on the SSD, which should achieves the best performance theoretically. It is experimentally proved. It is not the best choice to put both journal and segment files on the SSD, considering the cost of SSD.

Table 2. Data loading performance

Data set	Files	HDD(s)	SSD(s)	HDStore(s)
lubm_1	150	975	727	820
lubm_2	300	1,770	1,381	1,587
lubm_3	450	2,863	2,148	2,799
lubm_4	600	3,571	2,999	3,305
lubm_5	750	4,428	3,908	4,187

We care the most about the performance of data loading under the HDStore hybrid storage scheme. The single journal file on the SSD can be quickly read and write due to the hight I/O performance of the SSD, and segment files on the HDD can be massively stored. The cost of HDStore is lower than SSD's while the performance is better than the HDD's.

With the data set size increase, data loading time is basically a linear growth. While the experimental results show that the Bigdata database system on the SSD has the absolute advantages over the other two schemes, but the cost will be 10 times over HDD. The performance under the hybrid scheme is close to that under the SSD scheme, but the hybrid scheme is impressive considering the loading time and cost issues. It is the best choice to put the single journal file on the SSD and put the multiple segment files on the HDD.

Notice that the hybrid storage scheme has increased the performance just by 15 % compared to the HDD scheme, which is far below our expectation of the HDStore performance. This is because of the RAID technology and the Dell PERC H710P mentioned above. We have analyzed the storage device of the experimental environment and obtain the statistics as follows: The read and write rate of the HDD disk array under RAID5 are 550 MB/s and 270 MB/s respectively, while the read and write rate of the SDD are 630 MB/s and 320 MB/s respectively.

4.4 Results: Data Querying

The querying experiments are conducted with data set of lubm_1 on different storage scheme based on HDD, SSD and HDStore. Four queries (see the Appendix A) are executed. Every result of the query test is experimented 5 times to get an average evaluation. The result is shown in Table 3 which refers to the hot query.

The results draw from the querying experiments shows that the time of data querying under SSD and HDStore both have advantages over the HDD considering the data sets with smaller size. This is not surprising since we add SSD on the distribute cluster, so the performance based on SSD and HDStore are better than the performance based on HDD.

We can see that the time query1 takes is much more than the other 3 queries, because most of the time of query1 is spent on data index. We have analyzed the reason for it, and find that a longer time of data index means a longer interaction times with the disk, that's when the advantages of the SSD appears. While the result set is small, there

Table 3. Data querying performance (time in milliseconds)

Query	Data	HDD(ms)	SSD(ms)	HDStore(ms)
query_1	lubm_1	921	662	779
query_2	lubm_1	272	260	263
query_3	lubm_1	259	238	247
query_4	lubm_1	281	268	273

is not much difference among the 3 schemes. As is shown in the results, the query performance is not ideal because of the same reason mentioned in the data loading experiment part.

5 Related Work

Since SSD develops rapidly in recent years, users expect to make full use of it. For this purpose, lots of hybrid storage systems based on both SSD and HDD have been proposed For example, the HB-Storage [18] considers the characteristics of SSD and HDD, and builds a HDD write buffer to optimize the SSD write requests. It can maintain a high read performance and significantly reduce the write requests on the SSD. However, our HDStore scheme focuses on the storage of large-scale data, and builds SSD buffer to optimize the HDD batch read/write requests. It has higher overall read/write performance for the big data storage. The FlashLogging [19] is a logging solution that exploits multiple (USB) flash drives for synchronous logging. It handles the large variance of write latencies because of device erasure operations. Nevertheless, our HDStore scheme is suitable for the fast indexing and achieves optimization of the HDD batch write.

6 Conclusions and Future Work

Traditional storage schemes based on HDD are unable to deal with big data efficiently. However, the storage solution based on the SSD is too expensive. So currently it is the best choice to use the HDStore hybrid distributed storage scheme for large-scale data. In this paper, we put forward a hybrid storage scheme that is based upon the JS-model that we abstract from various present storage schemes for large-scale data. The HDStore puts the one single journal file on the SSD that support fast read and write and puts a number of segment files on the HDD that could support massive storage. Benchmark experimental results show that the time of data loading increased by 15 % and the querying time is also improved.

For the future work, to improve the performance of the HDStore significantly, we will count the times of the merge or split between SSD and HDD and optimize these operations. We will also try to put the journal files on the SSD that uses RAID technology. Besides, to optimize the querying performance, we will also take data indexes into account. The Bigdata system constructs several kinds of indexes for the data. What we will do is to consider the placement of those indexes onto the SSD or HDD to increase the querying performance.

Acknowledgments. This work is supported by the National High-tech R&D Program of China (863 Program) (2013AA013204) and the National Natural Science Foundation of China (61100049, 61373165), Special Fund for Fast Sharing of Science Paper in Net Era by CSTD (No.2012008).

Appendix A: The Query of LUBM Data Set

```
# Query1
    PREFIX rdf: <http://www.w3.org/1999/02/22-rdf-syntax-ns#>
    PREFIX ub: <http://www.lehigh.edu/~zhp2/2004/0401/univ-bench.owl#>
    SELECT ?X WHERE
    {?X rdf:type ub:GraduateStudent .
     ?X ub:takesCourse http://www.Department0.University0.edu/GraduateCourse0}
# Query2
    PREFIX rdf: <http://www.w3.org/1999/02/22-rdf-syntax-ns#>
    PREFIX ub: <http://www.lehigh.edu/~zhp2/2004/0401/univ-bench.owl#>
    SELECT ?X, ?Y1, ?Y2, ?Y3 WHERE
    {?X rdf:type ub:Professor .
     ?X ub:worksFor <http://www.Department0.University0.edu> .
     ?X ub:name ?Y1 .  ?X ub:emailAddress ?Y2 .  ?X ub:telephone ?Y3}
# Query3
    PREFIX rdf: <http://www.w3.org/1999/02/22-rdf-syntax-ns#>
    PREFIX ub: <http://www.lehigh.edu/~zhp2/2004/0401/univ-bench.owl#>
    SELECT ?X WHERE {?X rdf:type ub:Student}
# Query4
    PREFIX rdf: <http://www.w3.org/1999/02/22-rdf-syntax-ns#>
    PREFIX ub: <http://www.lehigh.edu/~zhp2/2004/0401/univ-bench.owl#>
    SELECT ?X WHERE
    {?X rdf:type ub:Person .
     <http://www.University0.edu> ub:hasAlumnus ?X}
```

References

1. Lohr, S.: The age of big data. New York Times **11** (2012)
2. Davis, M.: Specialized data management method: U.S. Patent 5,423,038. P. 6 Jun. (1995)
3. Minelli, M., Chambers, M., Dhiraj, A.: Big data technology. Big Data, Big Analytics: Emerging Business Intelligence and Analytic Trends for Today's Businesses, pp. 61–88 (2013)
4. McAfee, A., Brynjolfsson, E.: Big data: the management revolution. J. Harvard Bus. Rev. **90** (10), 60–68 (2012)
5. Fienberg, S.E.: The Analysis of Cross-Classified Categorical Data. Springer, New York (2007)

6. Dean, J., Ghemawat, S.: MapReduce: simplified data processing on large clusters. J. Commun. ACM. **51**(1), 107–113 (2008)
7. Porcarelli, D., Brunelli, D., Magno, M., et al.: A multi-harvester architecture with hybrid storage devices and smart capabilities for low power systems. In: 2012 International Symposium on Power electronics, Electrical Drives, Automation and Motion (SPEEDAM), PP. 946–951. IEEE (2012)
8. Tsirogiannis, D., Harizopoulos, S., Shah, M.A., et al.: Query processing techniques for solid state drives. In: Proceedings of the 2009 ACM SIGMOD International Conference on Management of Data, pp. 59–72. ACM (2009)
9. Agrawal, N., Prabhakaran, V., Wobber, T., et al.: Design tradeoffs for SSD performance. In: USENIX Annual Technical Conference, pp. 57–70 (2008)
10. Madden, S.: From databases to big data. J. IEEE Internet Comput. **16**(3), 4–6 (2012)
11. Hitchcock, R., Smith, G.L., Cheng, D.D.: Timing analysis of computer hardware. J. IBM J. Res. Dev. **26**(1), 100–105 (1982)
12. Sarajlic, E., Yamahata, C., Cordero, M., et al.: Electrostatic rotary stepper micromotor for skew angle compensation in hard disk drive. In: IEEE 22nd International Conference on Micro Electro Mechanical Systems, 2009, MEMS 2009, pp. 1079–1082. IEEE (2009)
13. Chen, F., Koufaty, D.A., Zhang, X.: Understanding intrinsic characteristics and system implications of flash memory based solid state drives. ACM SIGMETRICS Perform. Eval. Rev. **37**(1), 181–192 (2009)
14. Rizvi, S.S., Chung, T.S.: Flash SSD vs HDD: High performance oriented modern embedded and multimedia storage systems. In: 2010 2nd International Conference on Computer Engineering and Technology (ICCET), vol. 7. IEEE V7-297–V7-299 (2010)
15. SYSTAP, L.: Bigdata. http://www.systap.com/bigdata.htm
16. Guo, Y., Pan, Z., Heflin, J.: LUBM: A benchmark for OWL knowledge base systems. Web Semant. Sci. Serv. Agents. World Wide Web **3**(2), 158–182 (2005)
17. Prud'Hommeaux, E., Seaborne, A.: SPARQL query language for RDF. W3C recommendation, 15 (2008)
18. Yang, P., Jin, P., Wan, S., Yue, L.: HB-Storage: optimizing SSDs with a HDD write buffer. In: Gao, Y., Shim, K., Ding, Z., Jin, P., Ren, Z., Xiao, Y., Liu, A., Qiao, S. (eds.) WAIM 2013 Workshops 2013. LNCS, vol. 7901, pp. 28–39. Springer, Heidelberg (2013)
19. Chen, S.: FlashLogging: exploiting flash devices for synchronous logging performance. In: Proceedings of the 2009 ACM SIGMOD International Conference on Management of data, pp. 73–86. ACM (2009)

International Workshop on Data Management for Next-Generation Location-Based Services (DaNoS 2014)

Rating Aware Route Planning
in Road Networks

Junqiang Dai, Wei Jiang, Guanfeng Liu$^{(\boxtimes)}$, Jiajie Xu, Lei Zhao, and An Liu

School of Computer Science and Technology, Soochow University, Suzhou, China
{gfliu,xujj,zhaol,anliu}@suda.edu.cn

Abstract. Nowadays, with the increasing popularity of mobile Web (e.g., mobile social network), geo-positioning technologies and smart devices, it enables users to generate large amounts of location information and corresponding descriptive activities. Location-based services (LBSs) have been widely studied and applied into many real applications. During LBSs selection, the users want to do multiple activities on the route, which has many demands (e.g., time, site, service, etc.). However, the existing approaches do not consider the ratings of activities in route planning. In this paper, we propose a novel route planning method, called Rating Aware Route Planning (RARP). Given a set points of interest (POIs) in road network, where each point belongs to a specific category (e.g., resturant, gas station, bank, etc.) with several properties (rating, geo-position, etc.), a starting point S and an ending point T, our route planning method retrieves the best route with the constraint of rating specified by users that starts at S, passes through at least one point from each of the category in order, and ends at T. In addition, we propose two algorithms for the problem and conduct the experiments on a synthetic dataset in a real road network. The experimental results demonstrate that our propose method can plan a route having the shortest distance and high ratings with good efficiency.

Keywords: Road networks · Route planning · Rating aware · Location based service

1 Introduction

Nowadays, many websites provide location-based services (LBSs) that allow users to take advantage of information from the web about points of activities (e.g., shopping, restaurants, etc.), such as Foursquare, Facebook Place and Flickr and so on. With these services, people can record their location and activity information and then upload them on to these websites, so that they can share their experience with their friends. It is necessary for people to plan a route (called route planning) when they want to go from one place to another place and do several activities in the route. Despite these efforts, the route plannings have been considered on tags [1], the shortest path [2] and spatial keyword queries [3,4], which describe some features of the activities that people want to

© Springer International Publishing Switzerland 2014
Y. Chen et al. (Eds.): WAIM 2014, LNCS 8597, pp. 223–235, 2014.
DOI: 10.1007/978-3-319-11538-2_21

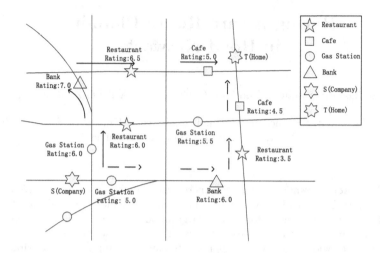

Fig. 1. A rating aware route planning

take part in. In these points of interest (POIs) based route planning, people are willing to select the POIs that have high ratings (reputations) as they want to enjoy the good services, so that they are more concerned about the rating of the POIs (e.g., coffee shop, restaurant, tennis hall, etc.).

For example, as shown in Fig. 1 there are some POIs that are divided into different categories (e.g., gas station, bank, restaurant, cafe) and each POI has an average rating given by other people who have visited these points. Assume that Jack wants to go to home (T) from his company (S), in his way, he plans to stop at a Gas Station, a Bank, a restaurant to have lunch, a coffee shop, and then go back to home. He wants to conduct these activities in order and the average ratings of the activities (denoted as λ) should be not be less than 6.0 ($0 \leq \lambda \leq 10$). From Fig. 1 we can find there are two routes from S to T. One is represented by the dotted line denoted as *route* 1, the other is represented by the solid line denoted as *route* 2. Now, we know that the distance of *route* 1 is 20 km, and the distance of *route* 2 is 23 km. We can see that *route* 1 is shorter than *route* 2 that can across all the categories. But the average rating of the route is 4.75 that can not satisfy the user's requirements. Then the *route* 2 could satisfies the demands of the average rating of the path (6.125). Therefore an advanced trip planning that has the shortest distance and high rating of POIs in the route need to be supported for the queries.

In our daily life, people usually conduct several activities in order. It is a hard work for people to plan an efficient route that considers both the distance of the route and the rating of the activities. Therefore it's significant and challenge to plan a route that meets multiple requirements of users, like the requirements of distance of the route and the rating of POIs located on the route.

In the literature, the existing work for route planning in road networks can be categorized into two types. The first type is the route planning for the existing online web mapping service (e.g., Google Map, Bing Map, Baidu Map) and spatial keyword query [3,4]. This type of route planning usually plans the best possible route which considers distance, traffic and road conditions, etc. The other type route planning method is for user based on order constraints and multi-tags [1,2]. However, this type of route planning method does not consider the rating of POIs. All these existing researches focus on finding a route with the shortest distance.

In this paper, we aim to provide an efficient and high-quality route planning method for users in road networks. Our main contributions are summarized as follows.

(1) We first propose a new type of route planning method, Rating Aware Route Planning, taking the distance of the route and the average rating of POIs in the route (e.g., the service, the atmosphere, etc.) as attributes to ensure the quality of the planned route.

(2) We propose two algorithms for solving the Rating Aware Route Planning problem, where we consider both the distance and the average rating of POIs in the route.

(3) We conduct extensive experiments on both synthetic and real datasets in road networks. The experimental results demonstrate that our method outperforms the existing methods.

The organizations of our paper is as follows. Section 2 introduces the problem definition. Section 3 introduces the related work. Section 4 introduces our new algorithms. Section 5 is the experiment results and analysis. Section 6 is the conclusion and future work.

2   Problem Formulation

In this section we define formally the problem and introduces the basic notation that will be used in the rest of paper.

A set of definitions are presented below to specify the problem we address in this paper:

Definition 1 (A road network). A road network is represented as a set of intersecting polygonal lines. We consider the road network as an undirected connected graph $G = <V, E>$, where V is the set of vertices representing road ends or intersections, and $E = \{(v_i, v_j)|v_i, v_j \in V\}$ is the set of edges representing road segments. For each edge $e = (v_i, v_j) \in E$, the weight of the edge represents the distance. We use $dist(v_i, v_j)$ to denote its distance.

Definition 2 (Points of Interest). A point of interest, or POI, is a specific point location that someone may find useful or interesting, such as restaurant or hotel. Each POI $o_i = <p, r, c>$ has three attributes, where $o_i.p$ represents its geographical location on two dimensional space; and $o_i.r$ ($0 \leq o_i.r \leq 10$) represents the rating of the POIs given by users according to their experience; also $o_i.c$ represents the category that o_i belongs to (e.g., restaurant, bank, gas station, etc.). We assume that each POI is mapping to a vertex in the road network.

Definition 3 (Route Query). In an undirected road network, a route query is defined as $Q = <S, T, q, \lambda>$, where S is the starting point and T is the destination point specified by users, and the ordered sequence $q = \{c_1, c_2, ..., c_m\}$ contains a set of user-input categories of POIS that the user want to do one by one. Otherwise, λ is a constraint given by users that the average rating η of the POIs should not be less than λ.

Problem Definition. We target to solve the problem of rating aware route planning which is formally defined as: Given a set of POIs O and a route query $Q = <S, T, q, \lambda>$ on road network $G = <V, E>$, we process the query for an optimal valid route $R = <path, O_R>$ on this road network. A valid route R must be able to satisfy the following constraints: 1. the route path starts from S and end at T; 2. O_R covers all the POIs of required categories in sequential order specified by q, such that $\forall i \in [1, m]$, $v_{t_i} = q_i$; 3. the average rating η of O_R is no less than the given threshold such that $\eta \geq \lambda$. A valid route is the one that can satisfy all the requirements specified by user. Among all of the valid routes, we targets to find the one that has minimum total distance in this paper, so as to reduce the travel cost of users.

3   Related Work

Recently, many researches are proposed for planning an optimal route and the related works can be generally divided into three categories: the first category is distance based shortest route, which arms to plan a route with some constraints; another category is spatial keyword query, which aims to plan a group of spatial objects related to users input keywords; the last category is to plan an optimal route, but not related to covering categories or keywords.

Many efforts have been made to handle the shortest route with some constraints. The traveling salesman problem (TSP) [5] is the well known problem that searches for a route of minimum distance in a graph. In [5], the general TSP was first studied by Karl et al. in the 1930s. TSP asks for the minimum distance round-route drom a point passing through a given set of points. In addition, Li et al. [2] propose Trip Planning Query (TPQ) which adopts fast approximation algorithms to find the shortest route constrained by passing at least one point from each category of place. The optimal sequenced route (OSR) query, proposed by Sharifzadeh et al. [6], is to find the shortest route constrained by visiting at least one point of each catgory. An additional constraint is that the route must contain categories in the same sequence as the categories appear in the sequence. The problem of asking for a route constrained by partial ordering between categories is also studied in [1,7,8]. Different from these existing works, we target to find the route with minimum distance and the average rating that meets user's demands.

Researches in spatial database also address applications in spatial networks, many studies on spatial keyword search have been proposed recently [9–13]. In [13], Li et al. proposed a direction-aware spatial keyword search method, which finds the k nearest neighbors to the query that contain all input keywords and satisfy the direction constraint. For the perspective of covering keywords, the collective spatial keyword querying asks a group of spatial objects rather than a route. However, these researches do not consider the rating of the objects. Therefore, we cannot apply their methods to the rating aware route planning directly.

There are some other approaches which are related to finding a travel route, but not related to covering categories or keywords. Work [14] studies how to continuous answer route planning according to the updated road condition. In addition, in [15] Roy et al. focus on planning an itinerary based on the user's preferences and a time budget that considering the user's feedbacks. Lu et al. [16] devised a problem of finding the optimal trip with respect to a travel score, which is the sum of ratings of places to visit, and a travel time. Next, several researches consider a user's historical data such as trajectories when recommending a travel route [17–21]. These works do not consider the coverage of all categories or keywords. Therefore, we cannot apply them to the rating aware route planning directly.

All the above researches cannot solve the rating aware route planning problem, because we consider both the POIs that the users need to visit and the average rating of the POIs in the route.

4   Algorithms

In this section we propose two kinds of algorithms for the rating aware route queries. In Sect. 4.1 we propose the local optimization based algorithms first, called *Greedy Heuristic Algorithm*. The Minimum Distance Heuristic Algorithm is introduced in Sect. 4.2 afterwards.

Suppose we have an RARP query with a starting location, an ending location, a set of POIs O and a sequence. The brute-force search algorithm for finding the optimal solution of RARP query must compute all sequenced routes that follow the sequence. Considering the definition of RARP query, the number of these routes is n^m, n is the number of POIs in each category and m is the number of the searching categories. Both n and m can be very large, so RARP is an NP-Complete problem.

4.1   Greedy Heuristic Algorithm

The rating aware route planning is basically subject to several parameters, such as the distance on spatial network, rating of POIs, and some heuristic search methods. In this paper, we propose a Greedy Heuristic Algorithm (GHA) to plan a short valid route. The algorithm finds a POI for one category from the sequence q in order (Line 3), which must satisfy the following constraints: (1). the average rating η of the POIs is no less than the given threshold λ such that $\eta \geq \lambda$ (Line 6–8); (2). The POI has minimum network distance from current location $curr$ using Dijkstra algorithm in road networks (Line 5). Add the shortest path between the POI and current location $curr$ to the route (Line 9) and set the

POI as the current location (Line 11); Repeat the above operations until all the categories in q are processed (Linc 3 to 12); At last, it returns a route R consisted of a series of shorter distance that starts from S, goes through the POIs found above one by one, and reaches the ending T (Line 13 to 15). Details of the GHA are shown in Algorithm 1.

Algorithm 1. Greedy Heuristic Algorithm

Input: source location S, target location T, a data set O.
Output: *Route*
 1: $curr = S$;   //** the current location */
 2: $Route = null$;
 3: **for** each $c \in q.c$ **do**
 4:     **repeat**
 5:         Find the POI o in category c having minimum network distance from $curr$;
            //** using Dijkstra Algorithm */
 6:         If $\rho < \lambda$ //**ρ represents the current rating in the path when choose o*/
 7:            $c = c - o$;   //** ignore the POI o from the category c */
 8:     **until** $\rho < \lambda$
 9:     $pth = shortestpath(curr, o.p)$; //** pth represents the route from $o.p$ to $curr$ */
10:     Add pth to *Route*;
11:     $curr = o.p$;
12: **end for**
13: $pth = shortestpath(curr, T)$;
14: Add *path* to *Route*;
15: Return *Route*;

We give an example in Euclidean space to illustrate how GHA works in Fig. 2. A user wants to find a route R that starts from S, contains the categories of POIs the user wants to take in are the sequence $q = \{A, B, C\}$, and the average rating constraint λ is 6.0. The algorithm will first find the nearest point o_1 ($o_1.c = A$) and the average rating η is 4.5 ($\eta < \lambda$). But if we select o_1, then we cannot find a valid route eventually, then the algorithm ignores the point o_1 and chooses another the nearest point o_2 ($o_2.c = A$) and checks the average rating $\rho = 7.0$ ($\eta > \lambda$). Repeating above operations, it will get the route $S \rightarrow o_2 \rightarrow o_3 \rightarrow o_5 \rightarrow T$ that the average rating $\eta = 6.17$ ($\eta > \lambda$).

This algorithm can find a route that satisfies the average rating constraint λ, and the route has the local shortest distance. However, it cannot make sure that the route is the global optimal.

4.2 Minimum Distance Heuristic Algorithm

GHA can not find the shortest route with the average rating that can meet the requirement specified by a user. In processing a query, we need to search m categories and each category has n POIs. So there are n^m route combinations of

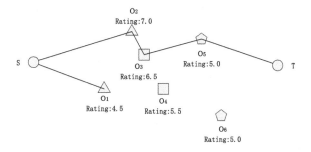

Fig. 2. How GHA works

the categories, and additional time cost is necessary for route planning in road networks by A* algorithm [22]. Therefore, it is not realistic to find an optimal route when the value of m and n is very large. So it is necessary to reduce the value of n in order to narrow the search space. In this section, we propose a novel heuristic algorithm for the rating aware route planning problem, called Minimum Distance Heuristic (MDH) algorithm, which is shown in Algorithm 2. In the algorithm, we locate k points from each category in sequence q; Then we compute the shortest path of the adjacent categories of POIs in q to initialize the necessary partial route information; At last, we find the shortest route that meets user's demand among all the possible routes.

Basic Candidate Selection. In this section, we make a basic candidate selection. Similar to the generalized traveling-salesman problem (GSTP), we know that when the POIs is clustered between the starting point S and the ending point T that we can choose an optimal route. So we choose k POIs for each category that are corresponding to the categories in the sequence q and the k POIs represents the *top-k* minimum traveling distance from S to T through the POIs (Line 3 to 5). We use bidirectional Dijkstra algorithm to find the point efficiently, as it can achieve much better pruning effect than traditional approaches. A bidirectional Dijkstra algorithm works by keeping track of two Dijkstra search scopes: one from the source, and one from the destination working on the reverse graph. When the two search scopes reach the same node, the shortest route pass through a POI that has been reached from both the source and the destination ([23], p. 30).

To select k points of each category in sequence q, the key idea is to find closest POIs and their distance based on Dijkstra's algorithm separately from both S and T concurrently. When they find the same POI, the POI has the shortest distance from S to T if there is no other POI in the distance range of the POI. For example, in Fig. 3, array L_1 stores the POIs extending from S and array L_2 stores the POIs extending from T; When each array adds a new POI, it need to search another array to find that the POI has been added; In Fig. 3, we see that array L_2 adds the POI C and finds that the POI C has been in array L_1; Then the sum of the distance $dist(S, C) + dist(C, T)$ is 23 km, continue to search array L_1 until the distance is greater than 16 Km and search

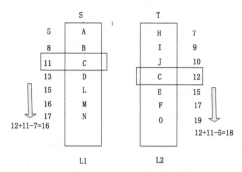

Fig. 3. How to find the point

array L_2 until the distance is greater than 18 km. If it does not find another pair of identical POIs, the sum of the distance $dist(S,C) + dist(C,T)$ is minimum; The POI that belongs to the each category we need before will be added in the arrays. It needs to do m bi-direction search when locates all the points of each category. To make our algorithm more efficient, we combine the categories that specified by an user together so that we only need to do bi-direction search once to find k points for each category in sequence q.

In order to find a valid route, we must make sure that we find a valid route from these POIs. Then we can find a valid route with shortest distance and the average rating that meets user's demand. Here we use a method to ensure at least one valid route can be covered in search space. The average rating of the highest rating among k points of each category in q should not be less than $\beta*\lambda$ $(\beta > 1)$. If not, the algorithm need to expend k twice. Obviously the lower bonder of the average rating η is the constraint average rating λ.

Afterwards, we can calculate the adjacent route $dist(o_i, o_j)$, $\forall\, t\in[1,\, m\text{-}1]$ $(o_i.c = q_t,\ o_j.c = q_{t+1})$ according to the sequence $q = \{q_1, q_2,..., q_m\}$. we can avoid accessing some hopeless routes, if the expected highest average rating (chooses the highest rating θ of other categories in q) of the route is lower than the constraint rating λ when we choose the route contains two POIs.

Route Search. After the search space has been reduced, we need to find a route with the minimum distance and proper average rating. The route R starts from S and adds the POIs among the points which are selected according to the sequence q. In order to improve the algorithm more efficient, we can delete some combinations of the POIs. Before we add a POI into the route R, we assume a route that if we choose the POI and then choose the other POIs with the highest rating of each categories. If the average rating of the assumed route is less than λ, then we delete the path, otherwise, add the POI to the route and continue to search the POI of other categories (Line 11–24). Among all the combination, there are several routes that can satisfy all the requirements specified by users. At last, the algorithm will return the route with the minimum distance among all the routes that satisfy user's demands (Line 8 to 9).

Algorithm 2. Minimum Distance Heuristic Algorithm

Input: source location S, ending location T, a data set O.

Output: *Route*;

1: $curr = S$; //** the current location */
2: $Route = null$;
3: $L = null$;
4: **for** each $c \in q.c$ **do**
5:     Find *top* k POIs o having minimum network distance $(dist(S,o) + dist(o,T))$ in each category c;
6: **end for**
7: $j = 0$;
8: $COMBINATION(Route, curr, j, L)$;
9: Find the Route which has the *mindistance(L)*; //** the *Route* with minimum distance in L; */
10: Return *Route*
11:
12: **Subroutine** $COMBINATION(Route, curr, j, L)$
13: **if** $j > m\text{-}1$ **then**
14:     $pth = shortestpath(curr, T)$; //** pth represents the route from T to $curr$ */
15:     Add path to *Route*;
16:     $L \leftarrow Route$;
17: **end if**
18: $curr = S$; //** the current location */
19: **for** $i=1$ to k **do**
20:     chooses a POI o from the k points of one category;
21:     θ=the expected highest average rating of the route; //**chooses o with the highest rating of each category in the remaining categories */
22:     **if** $\theta \geq \lambda$ **then**
23:         $pth = shortestpath(curr, o.p)$; //** pth represents the route from $o.p$ to $curr$ */
24:         Add path to *Route*;
25:         $curr = o.p$;
26:         call $COMBINATION(Route, curr, j+1, L)$
27:     **end if**
28: **end for**

For analyzing the complexity of the algorithm, we note that it takes one time with Dijkstra's algorithm to locate the top-k points for all categories, we need to do $k*(m\text{-}1)$ times Dijkstra's algorithm to calculate the distance among adjacent points. At last,the combination for calculate the possible routes costs less than $O(k^m)$, Thus, the time complexity of MDH is $O((km - k + 1)n^2 + k^m)$. Where k is the size of candidate set of each category, m is the number of categories in query, n is the number of points in road networks.

5   Experiments

This section presents a comprehensive performance evaluation of the proposed techniques for rating aware route planning in spatial databases. We used synthetic

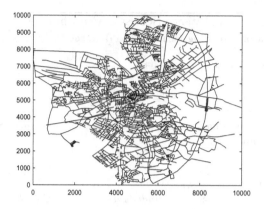

Fig. 4. Road network in OL

datasets generated on road networks. All experiments are conducted on a PC equipped with a Xeon processor 2.39 GHz, 3 GB of main memory and Windows 7 Profession operating system.

5.1 Experimental Setup

To generate synthetic datasets we obtain a real road network, the city of Oldenburg(OL) with 6105 nodes and 7035 edges, which is shown in Fig. 4. In the datesets, we generated uniformly at random a number of points of interest on the edges of the network. The number of categories is 20, while each category has 60 POIs, and the rating of the POI is in the range $r \in [0, 10]$. Then we combine the datasets of POIs with the real road network.

5.2 Query Performance Analysis

In this section we study the performance of the two algorithms in the road network. First, we study the effect of the number of categories in query. We set

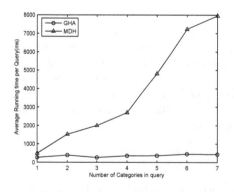

Fig. 5. Runing Time of algorithms

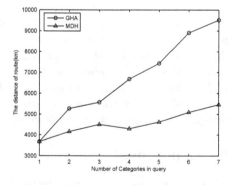

Fig. 6. The route distanceof algorithms

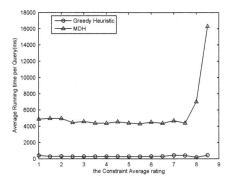

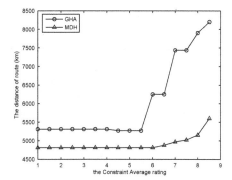

Fig. 7. Runing Time of algorithms

Fig. 8. The distance of the route of algorithms

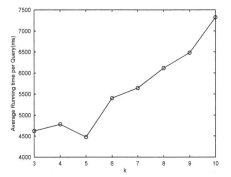

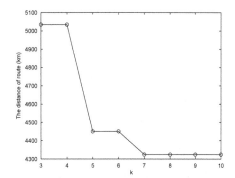

Fig. 9. Running Time of MDH

Fig. 10. The distance of the route of MDH

k=5, λ=7. When the number of categories in query increases, Fig. 5 represents the change of running time, Fig. 6 represents the change of the route distance and Fig. 7 represents the average rating of the route found by the two algorithm. In Fig. 5 we can see that the running time increases when the number of categories in query increases, as it needs to access more space and calculate more distances in the road network. In Fig. 5, GHA costs less time than MDH, but in Fig. 6, the distance of the route it finds is longer.

First we set k=5, m=5. In Fig. 7, it presents the running time when the average rating constraint λ increases. We can see that GHA costs less time. Figure 8 presents the average distance of the route of the two algorithms when λ increases. We can see that when λ is larger than 5.5, the distance of the route found by GHA increases fast, while the distance of the route found by MDH is far smaller than the route by GHA. When λ is larger than 5.5, there are many

POIs that cannot satisfies user's demand and it may find the POIs with high rating that is far away from the destination.

First we set $m=5$, $\lambda=6.5$. In Fig. 9, it presents the running time when the *top-k* increases. We can see that the running time increases rapidly when the search space increases, so we can hardly search the whole space. Figure 10 presents the average distance of the route when the *top-k* increases. We can see that when the *top-k* is larger, the distance of the route will be shorter. (Here we set $m=5$, $\lambda=6.5$)

6   Conclusions and Future Work

We first propose a new type of route planning, Rating Aware Route Planning, taking the distance of the route and the average rating of POIs in the route (e.g., the service, the atmosphere, etc.) as attributes to ensure the quality of route that we provide. We aim to find a route with shortest distance and proper rating of the route that satisfies the demands of users. Local optimization algorithms run quickly but can not find a route with shorter distance. The MDH can find a route with possible shortest distance.

In the future, we plan to study the route planning in a dynamic road network, also we can consider the time constraint that the user need to arrive a location.

References

1. Zhang, J.-Z., Wen, J., Meng, X.-F.: Multi-tag route query based on order constraints in road networks. Jisuanji Xuebao (Chin. J. Comput.) **35**(11), 2317–2326 (2012)
2. Li, F., Cheng, D., Hadjieleftheriou, M., Kollios, G., Teng, S.-H.: On trip planning queries in spatial databases. In: Medeiros, C.B., Egenhofer, M., Bertino, E. (eds.) SSTD 2005. LNCS, vol. 3633, pp. 273–290. Springer, Heidelberg (2005)
3. Hariharan, R., Hore, B., Li, C., Mehrotra, S.: Processing spatial-keyword (sk) queries in geographic information retrieval (gir) systems. In: 19th International Conference on Scientific and Statistical Database Management, SSBDM 2007, pp. 16–16. IEEE (2007)
4. De Felipe, I., Hristidis, V., Rishe, N.: Keyword search on spatial databases. In: IEEE 24th International Conference on Data Engineering, ICDE 2008, pp. 656–665 (2008)
5. Bigg, N.L., Lloyd, E.K., Wilson, R.J.: Graph Theory, pp. 1736–1936. Oxford University Press, Oxford (1976)
6. Sharifzadeh, M., Kolahdouzan, M., Shahabi, C.: The optimal sequenced route query. VLDB J. **17**(4), 765–787 (2008)
7. Chen, H., Ku, W.-S., Sun, M.-T., Zimmermann, R.; The multi-rule partial sequenced route query. In: Proceedings of the 16th ACM SIGSPATIAL International Conference on Advances in Geographic Information Systems (2008)
8. Li, J, Yang, Y., Mamoulis, N.: Optimal route queries with arbitrary order constraints (2013)
9. Cong, G., Jensen, C.S., Wu, D.: Efficient retrieval of the top-k most relevant spatial web objects. Proc. VLDB Endow. **2**(1), 337–348 (2009)

10. Cao, X., Cong, G., Jensen, C.S.: Retrieving top-k prestige-based relevant spatial web objects. Proc. VLDB Endow. **3**(1–2), 373–384 (2010)
11. Rocha-Junior, J.B., Nørvåg, K.: Top-k spatial keyword queries on road networks. In: Proceedings of the 15th International Conference on Extending Database Technology, pp. 168–179 (2012)
12. Cao, X., Cong, G.,Jensen, C.S., Ooi, B.C.: Collective spatial keyword querying. In: Proceedings of the 2011 ACM SIGMOD International Conference on Management of Data, pp. 373–384 (2011)
13. Li, G., Feng, J., Xu, J.: Desks: Direction-aware spatial keyword search. In: 2012 IEEE 28th International Conference on Data Engineering (ICDE), pp. 474–485 (2012)
14. Malviya, N., Madden, S., Bhattacharya, A.: A continuous query system for dynamic route planning. In: 2011 IEEE 27th International Conference on Data Engineering (ICDE), pp. 792–803 (2011)
15. Basu Roy, S., Das, G., Amer-Yahia, S., Yu, C.: Interactive itinerary planning. In: 2011 IEEE 27th International Conference on Data Engineering (ICDE), pp. 15–26 (2011)
16. Lu, E.-C., Lin, C.-Y., Tseng, V. S.: Trip-mine: An efficient trip planning approach with travel time constraints. In: 2011 12th IEEE International Conference on Mobile Data Management (MDM), vol. 1, pp. 152–161 (2011)
17. Zheng, Y., Xie, X.: Learning travel recommendations from user-generated GPS traces. ACM Trans. Intell. Syst. Technol. (TIST) **2**(1), 2 (2011)
18. Chen, Z., Shen, H.T., Zhou, X.: Discovering popular routes from trajectories. In 2011 IEEE 27th International Conference on Data Engineering (ICDE), pp. 900–911 (2011)
19. Ge, Y., Liu, Q., Xiong, H., Tuzhilin, A., Chen, J.: Cost-aware travel tour recommendation. In: Proceedings of the 17th ACM SIGKDD International Conference on Knowledge Discovery and Data Mining, pp. 983–991 (2011)
20. Kurashima, T., Iwata, T., Irie, G., Fujimura, K.: Travel route recommendation using geotags in photo sharing sites. In: Proceedings of the 19th ACM International Conference on Information and Knowledge Management, pp. 579–588 (2010)
21. Cheng, A.-J., Chen, Y.-Y., Huang, Y.-T., Hsu, W.H., Liao, H.-Y.M.: Personalized travel recommendation by mining people attributes from community-contributed photos. In: Proceedings of the 19th ACM International Conference on Multimedia, pp. 83–92 (2011)
22. Hart, P.E., Nilsson, N.J., Raphael, B.: A formal basis for the heuristic determination of minimum cost paths. IEEE Trans. Syst. Sci. Cybern. **4**(2), 100–107 (1968)
23. Schultes, D.: Fast and exact shortest path queries using highway hierarchies. Universität des Saarlandes, Saarbrücken, Master-Arbeit (2005)

Optimal RSUs Deployment in Vehicular Networks

Chunyan Liu[1], Hejiao Huang[1,2(✉)], Hongwei Du[1,2], and Xiaohua Jia[1]

[1] Harbin Institute of Technology Shenzhen Graduate School, Shenzhen, China
[2] Shenzhen Key Laboratory of Internet Information Collaboration, Shenzhen, China
hjhuang@hitsz.edu.cn

Abstract. In order to transfer an alert message from the accident site to the control center as soon as possible, the Roadside Units (RSUs) act as the most important roles in vehicular ad hoc networks (VANETs). However, we always hope to deploy minimum number of RSUs because of the low market penetration of VANET-enabled vehicles and the deployment cost of RSUs. In this paper, the optimal RSUs deployment problem is formulated as a set cover problem. We propose a heuristic greedy algorithm such that the number of RSUs is minimum. Meantime, alert messages would be propagated to RSUs within a time constraint, whenever an accident happens in anywhere of a road. Simulation results show that the scheme is efficient and the time complexity is better than those existing algorithms.

Keywords: RSUs deployment · Delay · VANETS

1 Introduction

Nowadays, the number of vehicles on the roads increases every year, and road safety becomes a great risk in our social life. Although, governments are establishing restrictive regulations to improve safety on roads, the number of accidents still increases every year all over the world [4].

A close looking at the accidents shows that many of the deaths occurred during the time between the happening of the accident and the arrival of the medical assistance in a road. This period of time is so called golden period to save lives. Generally, a high traffic would happen on the road. For saving time, other vehicles should avoid to select this road. All these management and coordination need an intelligent transportation system (ITS). Above all, alert messages from vehicles should be propagated to the ITS.

Vehicular ad-hoc networks (VANETs) is one of the most important part of ITS, and mainly consist of On-Board Units (OBUs) and Roadside Units (RSUs) [19].

This work was financially supported by National Natural Science Foundation of China with Grants No.11371004 and No.61100191, and Shenzhen Strategic Emerging Industries Program with Grants No. ZDSY20120613125016389, No. JCYJ20120613151201451 and No. JCYJ20130329153215152.

© Springer International Publishing Switzerland 2014
Y. Chen et al. (Eds.): WAIM 2014, LNCS 8597, pp. 236–246, 2014.
DOI: 10.1007/978-3-319-11538-2_22

OBUs are installed on vehicles to provide wireless communication capability and RSUs are deployed along roadside as internet access point for VANETs. Therefore, there are two main wireless communication paradigms, namely vehicle-to-vehicle (V2V) and vehicles-to-RSU (V2R). VANETs could communicate with ITS using RSUs. Generally, RSUs connect to internet directly, therefore, RSUs could communicate with ITS instantaneously. When accident happens, alert messages should be propagated to ITS within half of the golden period. The delay is mainly caused in propagating messages to RSUs. Hence, RSUs location is very important. Usually, it is very difficult to decide where to place RSUs [15]. Current work concerning RSUs placement considers either to serve maximum number of vehicles [21], or to minimize the communication delay [1, 20, 23], or to minimize the placement costs with the limited number of RSUs [2, 5].

In this work, we propose a RSU deployment scheme that alert messages could be propagated to RSUs within a given delay bound, wherever road accidents happen. Considering the cost to deploy RSUs, our optimization objective is to use the minimum number of RSUs to achieve that alert messages would be propagated to RSUs within a delay bound, whenever an accident happens in anywhere of a road.

The remaining part of the paper is organized as follows. Sections 2 and 3 introduce related works and the network model, respectively. The RSU deployment problem is formulated in Sect. 4 and a heuristic greedy algorithm is presented in Sect. 5. Section 6 presents comparative experiments and results. Finally, Sect. 7 is a conclusion.

2 Related Works

Most work for wireless access point and base station placement so far consider it as a continuous infrastructure radio coverage [21]. Trullols O et al. [21] proposes a maximum coverage approach for the information dissemination problems. The base station placement problem in WISE [9] is only one of a wide variety of optimization problems that may arise in wireless systems. Wright [11] proposes a variant of the Nelder-Mead "simplex" method for finding optimal base station placement. Li et al. [18] provide an optimal placement of gateways to minimize the average number of hops from RSUs to gateways in vehicular networks. Barrachina J et al. [3] presents a density-based approach for RSUs deployment problem in urban scenarios. It deploys RSUs according to an inverse proportion to the expected density. Yongping Xiong et al. [24] transforms the RSUs deployment problem as a vertex cover problem to provide the desired connectivity performance. Lee J et al. [15] considers the optimal placement of RSUs by analyzing the number of the reported locations per minute by taxis to a telemetric system. Recently, Karp [14] proposes a RSU placement scheme based on genetic algorithm. However, this scheme needs manual setting of candidate set. Recently, Lochert C et al. [16] proposes a RSU placement scheme based on genetic algorithm. However, this scheme needs manual setting of the candidate set. Trullos O et al. [22] presents a max coverage formulation to maximize the number of vehicles that get in contact with RSUs. It could be formulated as a maximum coverage problem (MCP), and could be solved by using heuristic algorithms. A α−coverage notion for vehicles was proposed in [25], which guarantees the interconnection gap of RSUs. For better performance, it would work with networking protocols and applications. Aslam B et al. [2] presents two different optimization methods, BIP and BEH, to minimize the average reporting time with a limited number of RSUs in an urban environment.

3   Network Model

We consider an urban road topology area. There are some intersections and roads in the area. Initially, all vehicles appear randomly in the area. Information dissemination time for each roads would be computed through vehicle mobility model or statistic data.

Since [21] has proved by experiments that intersections are more better locations than road segments for the deployment of RSUs, in terms of information dissemination potential, we consider the case that RSUs are deployed at intersections. However, there may be some long roads in which information dissemination would take very long time from one end to another. In this paper, if a road whose transmission time from one end to another is larger than the delay bound, then the road will be segmented and the cut-off points are also viewed as intersections.

Generally, RSUs deployment problem is a coverage problem from the viewpoint that RSUs are placed to cover vehicles on roads. Our goal is to deploy as less RSUs as possible such that alert messages would be transmitted to RSUs within the given delay bound wherever there has a crash or other accidents.

Consider the road topology as an undirected graph $G = (V, E)$ with each intersection as a vertex and each road segment as an edge, V is a set of all vertices, and E is a set of all edges. Obviously, $E \subseteq V^2$. Define a delay function $delay(e), \forall e = (p, q) \in E$ to denote the transmission time from p to q, and a constant α denote the delay bound. Based on the assumption that long roads would be cut-off, $\forall e \in E, delay(e) < \alpha$. Generally, signal propagation speed in the air is much larger than signal forwarding speed and vehicle speeds. Furthermore, signal forwarding speed is greater than vehicle speeds. For each $e \in E$, let $length(e)$ denote the length of road e, λ_e denote the number of vehicles for each kilometer in the road, $speed_e$ denote the vehicles average speed in road e, and constant fw denote the signal forwarding period. For simplicity, we default OBUs transmission range r equals to RSUs'. Then $delay(e)$ has a lower bound($\frac{length_e}{r} fw$) and an upper bound($\frac{length_e}{speed_e}$), where $\frac{length_e}{r} fw$ means that there are enough vehicles ($length(e) * \lambda_e \geq \frac{length_e}{r}$) that information would be forwarded by vehicles-to-vehicles with multi-hops; $\frac{length_e}{speed_e}$ means that there are so few vehicles ($length(e) * \lambda_e < \frac{length_e}{r}$) that information cannot be propagated until the vehicle with the alert message reaches a RSU.

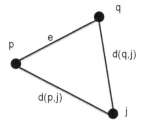
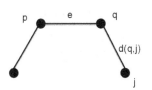

Fig. 1. Edge e can be covered by j using two paths by crossing $q \to j$ and $p \to j$

Fig. 2. Edge e can be covered by j using one path by crossing $q \to j$

4   Problem Formulation

We call a road e is covered by a RSU j, if an alert message from any point of e can be successfully propagated to j within the given delay bound α. Our goal is to cover all the roads with minimum number of RSUs.

4.1   Edge Cover

Let $S_j, j \in V$ denote the set of edges covered by j. We find those edges S_j covered by a vertex j within a delay α. The steps to find cover edges for each vertex are as follows:

> *Step 1:* For each vertex pair $(p, q) \in E$, compute the shortest delay $d(p, q)$.
> *Step 2:* For each vertex j, find those edges covered by j
>   for all $e = (p, q) \in E$
>   for all $j \in V$
> $S_j = \{e | delay(e) + d(p, j) + d(q, j) \leq 2 * \alpha\}$
> $\cup \{e | delay(e) + d(q, j) \leq \alpha\} \cup \{e | delay(e) + d(p, j) \leq \alpha\}$

Step 2 is to find those edges covered by a vertex where two cases are considered as Figs. 1 and 2 shown.

Figure 1 shows the case that $delay(e) + d(p, j) + d(q, j) \leq 2 * \alpha$. In this case, alert message in any point of e will be propagated to j either by crossing p or by crossing q. Figure 2 shows the case that $delay(e) + d(q, j) \leq \alpha$. In this case, any alert message in e will be propagated to j by crossing q.

4.2   Problem Transformation

Let $S = \{S_j | j \in V\}$. Then S is a collection of edge sets. Obviously, $\bigcup_{j \in V} S_j = E$. Our problem is to select as few as possible subsets from S such that their union equals E. This problem is a classical set cover problem and it is NP-hard [14]. Results of [7] show the hardness of approximation within a ratio of

$(\log_2 n)/2 \approx 0.72 \ln n$ and U. Feige [23] proves that $(1 - o(1)) \ln n$ is a threshold below which set cover cannot be approximated efficiently, unless NP problem has slightly super polynomial time algorithms.

4.3   Set Cover Problem

Set cover problem is a historic classical problem and it could be solved with optimal solution using $0 - 1$ integer liner programming when the problem scale is not too large. For our problem, we could formulate it to a $0 - 1$ integer liner programming problem. Let A denote a $|E| * |V|$ matrix and

$$A(e, j) = \begin{cases} 1, \text{ if } e \in S_j \text{ for all } j \in V \\ 0, \text{ otherwise} \end{cases} \tag{1}$$

Therefore, the problem can be formulated as:
Objective:

$$\text{Minimum } z = \sum_{j \in V} x_j \tag{2}$$

Subject to:

$$Ax \geq 1 \tag{3}$$

$$x_j = \{0, 1\}, j = 1, 2, \ldots |V| \tag{4}$$

where x_j means whether the vertex $j \in V$ is selected as a RSU. If vertex $j \in V$ is selected, $x_j = 1$, otherwise, $x_j = 0$. Constraint (3) guarantees that each road is covered by at least one of RSUs.

5   Heuristic Greedy Algorithm

In this section, we propose a heuristic greedy algorithm to solve the RSUs deployment problem. As the key factor of the heuristic algorithm, a heuristic greedy function satisfying some comprehensive strategies is constructed. Our heuristic greedy function is a polymatroid function. Therefore, the heuristic greedy algorithm is a polymatroid solution for set cover problems.

5.1   Heuristic Function

S_j is the set of edges covered by vertex $j \in V$. Let C denote a subset of V and $N(C) = \bigcup_{j \in C} S_j$. Specially, $N(V) = E$. Define $F(C) = |N(C)|$ and $c(C) = |C|$. $F(C)$ (resp., $c(C)$) is polymatroid functions [13] on 2^E (resp., 2^V) since the following three factors are satisfied [17] for F (resp., c):

- F is increasing;
- F is submodular, that is for any two subsets C_1 and C_2 of V, $F(C_1)+F(C_2) \geq F(C_1 \cup C_2) + F(C_1 \cap C_2)$;
- $F(\emptyset) = 0$.

So the problem is to find $\min c(C) : F(C) = F(V), C \subseteq V$ on 2^V.

Comprehensive strategies are constraints that could reduce the complexity of solving problems, and the simplified problem has the same optimal solutions with the original problem. In this problem, there are some comprehensive strategies:

(1) If $\exists j \in V, N(j) = S_j = N(V)$, then j is an optimal solution, and the optimal solution only contains one RSU j;
(2) If $\exists i, j \in V, N(i) \subset N(j)$, then i is not in the solution.

Based on the above comprehensive strategies, we construct the heuristic function: $F(N(C \cup \{j\})) - F(N(C))$.

Then, the greedy strategy is:

select $j \in V$, with $\max(F(N(C \cup \{j\})) - F(N(C)))$.

5.2   Algorithm

The heuristic greedy algorithm is described in Algorithm 1.

Algorithm 1. Heuristic Greedy Algorithm(HGA)

Input:
    $G = (V, E)$, N, $C = \emptyset$
Output:
    C: minimum subset of V which covers E
1: $EE \leftarrow E$
2: $VV \leftarrow V$
3: **while** $EE \neq \emptyset$ **do**
4:      **if** $\exists j \in VV, N(j) \backslash N(C) = N(VV) \backslash N(C)$ **then**
5:          select j
6:      **else**
7:          Select $j \in VV$ with $\max(F(N(C \cup \{j\})) - F(N(C)))$
8:      **end if**
9:      $C = C \cup \{j\}$
10:     $EE = EE \setminus N(j)$
11:     $VV = VV \setminus \{j\}$
12: **end while**
13: **return** C

The "while loops" will not stop unless the set EE is empty, in other words, all edges are covered. In line 4, if the remaining edges could be covered by a vertex j, then select vertex j and end the algorithm. Line 4 satisfies the comprehensive strategies (1). Line 7 achieves the greedy strategy.

5.3   Algorithm Analysis

Initially, $C = \emptyset, VV = V$. In the first iteration of the algorithm, if $\exists j, N(j) = N(VV)$ (line 4), then select j (line 5) and $EE = EE \setminus N(j) = \emptyset$ (line 7), the "while loops" stops. Therefore, HGA satisfies the comprehensive strategies (1).

If $\exists i, j \in V, N(i) \subset N(j)$, then, $N(C \cup \{i\}) \subseteq N(C \cup \{j\})$. While $N(C \cup \{i\}) = N(C \cup \{j\})$, whether i or j is selected, there is a same impact for the next iteration. While $N(C \cup \{i\}) \subset N(C \cup \{j\})$, $F(N(C \cup \{i\})) < F(N(C \cup \{j\}))$, and $F(N(C \cup \{i\})) - F(N(C)) \subset F(N(C \cup \{j\})) - F(N(C))$, our algorithm prefers to select j rather than i. Therefore, if $\exists i, j \in V, N(i) \subset N(j)$, our greedy strategy prefer to select j rather than i unless i and j have the same affect for the next iteration.

Cormode G et al. [8] provides a proof that the greedy algorithm produces a solution within a factor $1 + \ln n$ of optimum for set cover problem. Similarly, the HGA could also be proved as [8] that it can produce a solution within a factor $1 + \ln n$ of optimum solution, where $n = |E|$.

Let σ denote the number of optimal solutions. Then greedy algorithms need at least $1 + \sigma \ln n$ iterations. Each iteration gets an element of the final solution. In each iteration, the judgement in line 4 would be executed, and there are at most $|V|$ operations for the judgement. Then the computational complexity of HGA is $(3 + O(|V|))(1 + \sigma \ln n)$. The computational complexity of transformation is $O(|V||E|)$. The process of finding shortest delays for each pair of vertices need $O(|V||V|)$ operations. Let $m = |V|$. The total computational complexity is $O(|V|)(1 + \sigma \ln n) + O(|V||E|) + O(|V||V|) = O(\sigma m \ln n + mn + m^2)$ and $\sigma \leq m$.

6   Simulation Results and Discussion

In order to evaluate the RSU deployment methodology outlined above, we apply it to a random city scenario model. The model covers an area of $10 \text{ km}^2 \times 10 \text{ km}^2$, and there are 50 intersections in the model.

Our experiment consists of two parts. The first is to transform the city scenario as roads topology graph G and to find those edge cover sets $S_j, \forall j \in V$. The second part is to solve RSUs deployment problem.

The road topology is shown in Fig. 3. Each dot means a intersection. Without loss of generality, roads are represented by straight line segments. Intersections and road segments are labeled with Arab numbers, respectively.

It is proved that CPLEX is the best method for solving small-scale set cover problem with optimal solution [6], we present the simulation results comparing our greedy algorithm (HGA) to cplexbilp of cplex and bintprog of matlab respectively. Table 1 shows the number of RSUs and running time of them, respectively while the given delay bound α is from 1.8 to 4.0.

Both cplexbilp of cplex and bintprog of matlab could get optimal solutions for this set cover problems. Moreover, cplexbilp costs less time. Our heuristic greedy algorithm could get an approximate optimal solution with less time than that of cplexbilp. Specially, heuristic greedy algorithm would get an optimal solution while α equals 1.8 unit time (Fig. 3).

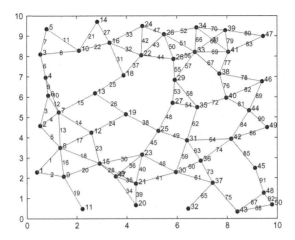

Fig. 3. 10 km^2 × 10 km^2 area with 50 intersections, and $\forall e \in E, delay(e) < 1.8$ unit time

Table 1. Results for cplex, bintprog and HGA algorithm

Given delay	Cplex		HGA		Bintprog	
α	RSUs	Time	RSUs	Time	RSUs	Time
1.8	27	0.016	27	0.014475	27	15.5056
2.0	23	0.016	25	0.012126	23	0.7956
2.2	20	0.016	21	0.011459	20	0.546
2.4	16	0.016	19	0.009049	16	0.0468
2.6	16	0.016	18	0.007372	16	0.4524
2.8	13	0.016	15	0.006636	13	0.078
3.0	11	0.015	13	0.005008	11	0.0468
3.2	8	0.015	11	0.004821	8	0.0936
3.4	7	0.015	10	0.003791	7	0.0468
3.6	6	0.015	8	0.003219	6	0.0468
3.8	6	0.015	8	0.002853	6	0.0936
4.0	6	0.015	7	0.002668	6	0.1248

Set cover problem is historic classical NP-hard problem. Approximation algorithms cannot get a less $1 + \ln n$ optimal solution. Recently, study prefers to the variant problems of set cover problem. Reference [10] proved that the neural network algorithm (NN) is the best algorithm for set cover problem than Greedy, Thresh, SortLP, RR, Tgreedy and Altgreedy algorithms. The comparison of NN algorithm and our heuristic greedy algorithm is shown in Tables 2 and 3, by running the examples of [12].

Table 2. Computational results for HGA and NN algorithms

Examples	mygreedy	NN	Examples	mygreedy	NN	Examples	mygreedy	NN
scp41	42	43	scp58	39	37	scpb5	24	23
scp42	41	39	scp59	41	38	scpc1	46	46
scp43	43	40	scp510	38	37	scpc2	47	45
scp44	44	43	scp61	23	22	scpc3	46	47
scp45	41	41	scp62	22	21	scpc4	47	46
scp46	42	41	scp63	23	21	scpc5	47	46
scp47	43	43	scp64	22	22	scpd1	26	26
scp48	43	40	scp65	23	23	scpd2	27	26
scp49	44	42	scpa1	41	41	scpd3	26	26
scp410	44	42	scpa2	43	42	scpd4	27	26
scp51	38	37	scpa3	42	41	scpd5	27	26
scp52	37	37	scpa4	41	40	scpe1	6	5
scp53	37	37	scpa5	42	41	scpe2	5	5
scp54	38	37	scpb1	23	24	scpe3	5	5
scp55	36	37	scpb2	23	23	scpe4	5	5
scp56	38	37	scpb3	24	23	scpe5	5	5
scp57	38	35	scpb4	24	24			

Table 3. Cpu time in seconds for HGA and NN algorithms

Examples	HGA	NN
scp41	0.4	6.6
scp51	0.7	12
scp61	0.2	4.4
scpa1	1.4	21
scpb1	0.8	17.2
scpc1	2.5	30.8
scpd1	1.4	26.8
scpe1	0	1

Table 2 shows that the solution of HGA is close to the solution of NN. The performance of NN is a little better than HGA in most cases. However, NN costs more time than HGA as shown in Table 3.

7   Conclusion

In this paper, we present a scheme to deploy RSUs such that alert messages could be transmitted to RSUs within a given delay bound wherever there is an accident.For future work, we will study information dissemination delay according to the real traffic information.

References

1. Aslam, B., Zou, C.C.: Optimal roadside units placement along highways. In: 2011 Consumer Communications and Networking Conference (CCNC), pp. 814–815 (2011)
2. Aslam, B., Amjad, F., Zou, C.C.: Optimal roadside units placement in urban areas for vehicular networks. In: 2012 IEEE Symposium on Computers and Communications (ISCC 2012), pp. 423–439 (2011)
3. Barrachina, J., Garrido, P., Fogue, M., et al.: D-RSU: a density-based approach for road side unit deployment in urban scenarios. In: International Workshop on IPv6-based Vehicular Networks (Vehi6), Collocated with the 2012 IEEE Intelligent Vehicles Symposium, pp. 1–6 (2012)
4. Barrachina, J., Garrido, P., Fogue, M., et al.: Road side unit deployment: a density-based approach. IEEE Intell. Transp. Syst. Mag. **5**(3), 30–39 (2013)
5. Lin, P.-C.: Optimal roadside unit deployment in vehicle-to-infrastructure communications. In: 12th International Conference on ITS Telecommunications (ITST), pp. 796–800 (2012)
6. Caprara, A., Toth, P., Fischetti, M.: Algorithms for the set covering problem. Ann. Oper. Res. **98**(1–4), 353–371 (2000)
7. Lund, C., Yannakakis, M.: On the hardness of approximating minimization problems. J. ACM **41**(5), 960–981 (1994)
8. Cormode, G., Karloff, H., Wirth, A.: Set cover algorithms for very large datasets. In: Proceedings of the 19th ACM international conference on Information and knowledge management. pp. 479–488 (2010)
9. Fortune, S.J., Gay, D.M., Kernighan, B.W., Landron, O., Valenzuela, R.A., Wright, M.H.: Wise design of indoor wireless systems: practical computation and optimization. Comput Sci. Eng. **2**(1), 58–68 (1995)
10. Grossman, T., Wool, A.: Computational experience with approximation algorithms for the set covering problem. Eur. J. Oper. Res. **101**(1), 81–92 (1997)
11. Wright, M.H.: Optimization methods for base station placement in wireless applications. In: 48th IEEE Vehicular Technology Conference, pp. 387–391 (1998)
12. Beasley, J.E.: An algorithm for set covering problem. Eur. J. Oper. Res. **31**(1), 85–93 (1987)
13. Liu, J., Huang, H., Du, H.: Greedy algorithm for least privilege in RBAC model. In: Wang, W., Zhu, X., Du, D.-Z. (eds.) COCOA 2011. LNCS, vol. 6831, pp. 537–545. Springer, Heidelberg (2011)
14. Karp, R.M.: Reducibility among combinatorial problems. In: Complexity of Computer Computations, Plenum Press (1972)
15. Lee, J., Kim, C.M.: A roadside unit placement scheme for vehicular telematics networks. In: Kim, T., Adeli, H. (eds.) AST/UCMA/ISA/ACN 2010. LNCS, vol. 6059, pp. 196–202. Springer, Heidelberg (2010)
16. Lochert, C., Scheuermann, B., Wewetzer, C., Luebke, A., Mauve, M.: Data aggregation and roadside unit placement for a vanet traffic information system. In: Proceedings of the fifth ACM international workshop on VehiculAr Inter-NETworking, pp. 58–65 (2008)

17. Wan, P.-J., Du, D.-Z., Pardalos, P., Wu, W.: Greedy approximations for minimum submodular cover with submodular cost. Comput. Optim. Appl. **45**(2), 463–474 (2010)
18. Li, P., Huang, X., Fang, Y., Lin, P.: Optimal placement of gateways in vehicular networks. IEEE Trans. Veh. Technol. **56**(6), 3421–3430 (2007)
19. Peng, Y., Abichar, Z., Chang, J.M.: Roadside-aided routing (RAR) in vehicular networks. In: IEEE International Conference on Communications, pp. 3602–3607 (2006)
20. Sun, Y., Lin, X., Lu, R., et al.: Roadside units deployment for efficient short-time certificate updating in VANETs. In: 2010 IEEE International Conference on Communications (ICC), pp. 1–5 (2010)
21. Trullols, O., Fiore, M., Casetti, C., et al.: Planning roadside infrastructure for information dissemination in intelligent transportation systems. Comput. Commun. **33**(4), 432–442 (2010)
22. Trullols, O., Barcelo-Ordinas, J.M., Fiore, M., et al.: A max coverage formulation for information dissemination in vehicular networks. In: IEEE International Conference on Wireless and Mobile Computing, Networking and Communications (WIMOB 2009), pp. 154–160 (2009)
23. Feige, U.: A threshold of ln n for approximating set cover. J. ACM **45**, 634–652 (1998)
24. Xiong, Y., Ma, J., Wang, W., Tu, D.: Roadgate: mobility-centric roadside units deployment for vehicular networks. Int. J. Distrib. Sens. Netw. **2013**, 1–10 (2013)
25. Zheng Z, Sinha P, K.S.: Alpha coverage: Bounding the interconnection gap for vehicular internet access. In: INFOCOM 2009. pp. 2831–2835 (2009)

Cache-Aware Query Processing
with User Privacy Protection
in Location-Based Services

Zhengang Wu[1,2,3], Liangwen Yu[1,2,3], Huiping Sun[1,2,3](✉),
Zhi Guan[1,2,3], and Zhong Chen[1,2,3]

[1] Institute of Software, EECS, Peking University, Beijing, China
[2] MoE Key Lab of High Confidence Software Technologies (PKU), Beijing, China
[3] MoE Key Lab of Network and Software Security Assurance (PKU), Beijing, China
{wuzg,yulw,sunhp,guanzhi,chen}@infosec.pku.edu.cn

Abstract. Location-Based Service (LBS) providers can send geo-tagged data to mobile users who report their current location information. Location-related user privacy is a vital concern since untrusted LBS can monitor and misuse user location information. This paper aims at how to efficiently protect location privacy in query processing. In proxy-based location anonymizers, location cloaking algorithms need heavy bandwidth for query processing after blurring exact locations into a region. To alleviate it, the proposed query framework can integrate location cloaking and data caching. Next, we devise a novel location anonymization algorithm which takes advantage of the caching feature. Finally, we evaluate this solution experimentally for showing its improvement.

Keywords: Caching · Cloaking · Location privacy · Location Bases Services · Query processing

1 Introduction

Recent advances in mobile Internet and low-cost smartphones with global positioning capability facilitate Location-Based Services (LBSs). Generally, LBS providers manage massive Points of Interest (POIs) in spatial databases. A POI object represents a real-world location entity which involves its geographical coordinates and related semantic information. In a location-based query, a remote LBS server sends on-demand POI data to mobile users. However, this useful function implies a latent privacy threat that mobile users leak sensitive location information to risky LBS providers.

The location anonymizer is a middleware which achieves user privacy protection between clients and servers of LBS. A proxy server as one of common architectures can run a location anonymization algorithm. However, this proxy-based method generally needs heavy bandwidth with low efficiency because blurring an exact user location into a cloaking region implies massive additional POI data

© Springer International Publishing Switzerland 2014
Y. Chen et al. (Eds.): WAIM 2014, LNCS 8597, pp. 247–258, 2014.
DOI: 10.1007/978-3-319-11538-2_23

which have to be downloaded. For this, we aim at how to improve performance of the location anonymizer proxy by caching requested POI data.

K-Anonymity means that at least k entities are undistinguishable in a set. This metric is readily applied into location-related user privacy protection. Spatial cloaking techniques supporting location k-anonymity generally construct an anonymous group that generalizes exact locations of at least k users into a shared spatial region and thus need to catch multiple queries of registered clients via proxies or P2P networks.

Contributions. With improving performance of location anonymizing proxies, the proposed solution provides three-fold contributions. First, we present a framework *CachedCell* which achieves a useful integration of location anonymization and caching POI data to reduce overhead of spatial cloaking in LBS, and existing schemes fail to study the impact of caching in proxy-based location anonymizers. Next, this paper devises a novel cache-aware location anonymization algorithm *GCC* using location k-anonymity. Finally, this solution is evaluated experimentally and compared with different parameters.

Outline. The rest of the article is structured as follow. The next section summarizes related works on the topic and Sect. 3 describes overview of system. Next, we introduce the proposed solution described in Sect. 4 and its performance is evaluated through a set of experiments in Sect. 5. In the last section we sum the solution up and point out future works.

2   Related Works

A wide range of research works cover user privacy preservation in LBS. In general, location-related user privacy is two-fold. First, query privacy means the linkable ability that which user sends a query in which location. Second, exact location information also is sensitive.

Pseudonym and Mix Zone [1] aims at how to break the user-location relationship. This mechanism involves removing or replacing real identities but fails to alter original location information. Location k-anonymity can achieve both query privacy and location privacy. Sweeney proposed k-anonymity [2] for protecting sensitive data in 2002 and then Gruteser et al. [3] in 2003 adopted k-anonymity into the location privacy protection field. Besides, some recent novel metrics [4–6] defined location privacy according to different background knowledge of adversaries.

Many existing schemes rely on the location k-anonymity because of its concision and usability. Their system structures fall into two major types generally. First, a proxy-based scheme [7] needs a middle server between clients and servers of LBS to handle requested clients under k-anonymity. Second, in a P2P-based scheme [8], multiple clients can group themselves via peer-to-peer (P2P) overlay networks to achieve k-anonymity.

An important focus is spatial cloaking in continuous queries of LBS. Gruteser et al. [9] firstly discussed location privacy in continuous LBSs and devised disclose-control algorithms for it in 2004. Xu et al. in 2007 introduced K-Anonymity Area (KAA) [10] to minimize the size of the cloaking area and measure the area's entropy. In 2009, Pan et al. designed the distortion-based method [11] to trade off location privacy and Quality of Service(QoS) and Yiming Wang et al. [12] studied how to preserve possible future locations by hiding direction and speed of user movement. In 2011, Pingley et al. designed DUMMY-Q [13] using perturbation to protect query privacy. DUMMY-Q does not need an additional trusted middle server since it runs in client-side devices. In 2012, Lee et al. proposed a cloaking region construction method GCAA [14] to improve KAA in the continuous LBS setting. GCAA focuses how to rapidly generate a compact cloaking region that is small but enough for privacy guarantee under the k-anonymity metric. A scheme from Wang et al. [15] clusters some similar users into a cloaking region by calculating similarity of velocity and acceleration with movements of mobile users.

In addition, cryptographic methods can provide a strong privacy guarantee unlike the above methods that achieve only a limited privacy. Private Information Retrieval (PIR) [16], a useful cryptographic primitive, helps a client to obtain his desired data without leaking the index number of the data entry to the database holder. So PIR-based location privacy protection schemes can provide a proven privacy assurance and resist Access Pattern Attack, but they are of low performance or rely on specialized hardware, constrained by specific PIR protocols. Existing PIR-based schemes involves range queries [17], NN/kNN queries [17,18] and so on.

However, the above-mentioned works fail to explore the caching feature in proxies which support location privacy preservation.

3   System Overview

According to existing location cloaking schemes, our system employs the three-tier location anonymization architecture which involves three major entities, clients, location anonymizers and LBS servers. First, mobile users reports their real-time locations for receiving on-demand POI data from remote LBS servers and the corresponding clients are GPS-equipped mobile devices installed by LBS client software. Second, LBS servers maintain POI databases and respond POI data to clients. Third, for enhancing user privacy, a location anonymizer, which

Fig. 1. Proxy-based location anonymization architecture

is a trusted intermediate server, can seamlessly forward requests of clients and hide user locations to untrusted LBS servers (Fig. 1).

Without loss of generality, a location anonymizer executes the following steps: first, it blurs an exact location into a cloaking region by location anonymization algorithms and sends this region as a required parameter to LBS servers; next, it receives the region's POI data from LBS server and filters and dispatches exact POI results to corresponding clients.

4   Location Anonymizer Proxy Supporting Caching

The data cache can improve processing performance of proxy servers intuitively. Therefore, we introduce the following framework to integrate a location anonymizer with the POI cache feature. More importantly, the cache feature also improve location cloaking mechanism and for this we introduce a novel algorithm that can trade off location privacy and query efficiency with POI cache.

For simplicity, we assume that a single LBS server manages all POIs and a single location anonymizer handles all user queries. Locations of both POIs and user queries are pined in a rectangle region of a map and so they can be readily indexed by a full quad-tree whose leaf nodes represent cells in this rectangle region. Each cell is labeled by a unique string $CellID$.

In the location anonymizer LA, a user query request can be denoted by a tuple $req = (uid, loc, arg)$ which involves the user identifier uid, the location coordinates loc and the query arguments arg. By the quad-tree index, LA maintains an inverted list $RequestDB$ where a row represents a tuple $(CellID, REQs)$ where $REQs$ is the set of user queries in this cell $CellID$ to handle user queries in a custom run-time snapshot in an on-demand manner.

The POI Database $POIDB$ of the LBS server store POI entries each of which can be denoted by a tuple $poi = (pid, loc, info)$ where three elements represent the unique identifier, the coordinates and other semantic information of this location point respectively. After receiving a requested cell $CellID$, the LBS server can response this cells data denoted by a tuple $Data(CellID) = (CellID, POIs)$ where $POIs$ represents the set of POIs in this cell.

4.1   Query Framework with POI Caching and Location Cloaking

The proposed query framework in Algorithm 1 integrates location anonymization with caching POI data. The main idea involves two major aspects: first, for a user query, LA gives priority to fetch cached results; second, for an uncached query, LA runs a location anonymization algorithm to hiding the exact location of this query when requesting demand results from an untrusted LBS server.

The net gain of caching is twofold. First, for a cached query, cell cache is stored in a trusted middle server LA, so untrusted LBS learns no information about the cached query and the privacy risk is reduced naturally. Second, cache in the proxy improves communication performance by cutting down the amount of duplicate requests and users can share cached resources.

Algorithm 1. CachedCell Framework: Caching POI Data by Cells

Input: A query request $q = (uid, loc, arg)$ of a user uid;
Output: The user uid obtains POI data of $CellID_q$

1 $CellID_q \longleftarrow Parse(q)$;
2 **if** $Cache.check(CellID_q)$ **then**
3    $\quad \lfloor\ Data(CellID_q) \longleftarrow Cache.load(CellID_q)$;

4 **else**
5    $\quad$ /*Calling a location anonymization algorithm*/
6    $\quad CR \longleftarrow GCC(CellID_q)$;
7    $\quad$ **foreach** $CellID \in CR$ **do**
8    $\quad\quad$ downloads $Data(CellID)$ from POI databases of LBS;
9    $\quad\quad$ **if** $CellID = CellID_q$ **then**
10   $\quad\quad\quad \lfloor\ Data(CellID_q) \longleftarrow Data(CellID)$;
11   $\quad\quad \lfloor\ Cache.save(CellID, Data(CellID))$;

12 **return** $Data(CellID_q)$;

The data cache is a common key-value in-memory storage where for a key-value pair its key is $CellID$ and its value references the corresponding POI data in a cell. This framework uses the cache by three basic operations: $Cache.load$ means that LA finds and reads the cache object of the cell $CellID$ in this cache; $Cache.save$ shows LA saves the cell $CellID$ in cache or refreshes it by a new one; $Cache.check$ is a Boolean function that return if the cell $CellID$ is cached.

The function $Parse$ outputs $CellID$ for this query q. In a session, a user submits a spatial query (such as kNN and range query) which the argument arg defines. Generally, in grid-based partition, query processing decomposes the original query into some basic sub-queries which download the content of a cell. Some existing schemes [19] also employ the similar processing. The majority of real-world LBS queries need to obtain one cell or its few neighbor ones. Therefore, we simplify that one query needs one cell and then its user location actually lies upon the cell.

$CR \longleftarrow GCC(CellID_q)$ means the location anonymization algorithm GCC that creates a cloaking region from an uncached CellID for accessing LBS.

4.2 Cache-Aware Location Anonymization Function

We devise an improved location anonymization algorithm to gain benefits form cache management. For an uncached query, by similarly adapting the quad-tree index and the proxy-base structure, NewCasper [7] as a typical cloaking algorithm works readily at the above query framework but is independent of data cache.

The Greedy Cached Cloaking Algorithm (GCC) showed in Algorithm 2 is heuristic. GCC combines recursively an uncached cell whose probability is maximum until the selective cell set as a cloaking region satisfies the custom privacy requirement which involves two parameters, the anonymity degree k and the minimum size of the cloaking region A. Besides, the capacity L of cache zone is limited and is much less than the size of POI databases. For a time snapshot, the function $Count$ outputs the number of user queries in a cell and N is the total number of user queries.

Algorithm 2. GCC: Greedy Cached Cloaking Algorithm

    Input: CellID,Set Minimum A,Cache Size L,Anonymity Degree k
    Output: The cloaking region CR whose CellIDs are ready for downloading.
1 $CR \longleftarrow \{CellID\}, r \longleftarrow Count(CellID)$;
2 **while** $r < k$ OR $|CR| < A$ **do**
3        $SelectiveCID \leftarrow \varnothing$;
4        $MaxCounter \leftarrow 0$;
5   **for** $cid \in CounterDB$ **do**
6   **if** $Cache.check(cid)=false$ AND $cid \notin CR$ **then**
7   **if** $Count(cid) > MaxCounter$ **then**
8                    $SelectiveCID \leftarrow cid$;
9                    $MaxCounter \leftarrow Count(cid)$;
10   $r \longleftarrow r + MaxCounter$;
11   $CR.add(SelectiveCID)$;
12 **return** CR

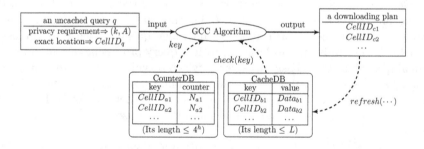

Fig. 2. Sketch of Greedy Cached Cloaking

As shown in Fig. 2, GCC needs two sub databases $CounterDB$ and $CacheDB$ which are maintained locally in the proxy in a trusted domain. Primary keys of both are CellID. $CounterDB$ is an array of counters which dynamically count up the number of sending a request in a cell in a timestamp for helping GCC to comparing the request probabilities of these cells. $CacheDB$ in the cache zone temporarily stores POI data. Generally, their length are less than the number 4^h of cells which a level-h full quad-tree partition creates. If the counter value of a cell is zero, the corresponding counter does not need to be stored in $CounterDB$. Note that, even for a fine-grained partition, using a large height of quad-tree which means the cell number of exponential growth, $CounterDB$

still is relatively small since it is created for one timestamp. By comparison, *CacheDB* is continuously refreshed with a series of downloading operations in all timestamp of proxies running in a customized size L.

4.3 Full Quad-Tree Based Partition on Geographic Coordinates

In rush hours of accessing real-world LBS, LA needs to handle frequent user queries in run-time and maintain efficiently the quad-tree data structure. An existing scheme NewCasper [7] employs a pyramid structure that actually is a full quad tree to achieve this. Our scheme can use a similar full quad tree. For simplifying its calculation and adapting data format of real geographic locations, a geographical coordinate (lng, lat), which means a couple of longitude lng and latitude lat, is proportionally mapped into a point (x, y) in the overlap grid plane using Eq. 1 where the left-bottom point (lng_0, lat_0) of a rectangle region that is mapped into the original point $(0,0)$ of a plane-coordinate system. Note that, since latitude and longitude of its bounds are aligned and the Earth surface is sphere, it actually is an approximate rectangle.

$$\begin{bmatrix} x \\ y \end{bmatrix} = \begin{bmatrix} \frac{2^l}{\Delta lng} & 0 \\ 0 & \frac{2^l}{\Delta lat} \end{bmatrix} (\begin{bmatrix} lng \\ lat \end{bmatrix} - \begin{bmatrix} lng_0 \\ lat_0 \end{bmatrix}) \tag{1}$$

In the full quad-tree partition, each node except leafs divides the current rectangle region into four smaller rectangles. In every level of the hierarchical partition, a location point must fall into one of four child rectangles. Therefore, in the height-h quad-tree, these level-h child rectangles whose ranges are narrowed down exponentially from top to bottom, can target a leaf that references a cell. The root is defined as the zero level and a leaf that is a cell is the unit rectangle. Clearly, the side length of a level-i rectangle is 2^{h-i}. In the partition computation, the relative position (divided by the unit 2^{h-i-1} that is the side length of a next-level rectangle) of a location point (x, y) in a level-i rectangle is computed recursively using Eq. 2 and for convenience all four possible results $(0,0),(0,1),(1,1),(1,0)$ can be encoded into four characters 0,1,2,3 respectively. By this way, we can employ a length-h string as CellID whose $h - 1$ prefixes label parent nodes of this leaf naturally.

$$L_0 = 2^h, R_0 = (0,0), L_i = 2^{h-i}, R_i = \lfloor \frac{(x,y) - L_{i-1} * R_{i-1}}{L_i} \rfloor \tag{2}$$

A Toy Example. In the level-2 quad-tree partition, its root rectangle's side length is $2^2 = 4$ units and it consists of 4^2 cells. We obtain a point $(1.2, 3.4)$ in the plane after normalizing a location on the earth surface. The CellID string can be computed through following steps: First, $R_1 = \lfloor ((1.2, 3.4) - 2^2 * (0, 0)) \div 2 \rfloor = (0, 1) \Rightarrow$ '1'; Second, $R_2 = \lfloor ((1.2, 3.4) - 2^1 * (0, 1)) \div 1 \rfloor = (1, 1) \Rightarrow$ '2'; At last, we obtain a string '12' as the CellID of the point $(1.2, 3.4)$.

5   Analysis and Evaluation

In the user privacy problem of LBS, an untrusted LBS provider or server as an Honest-But-Curious adversary responses requests of mobile users by following prescribed steps, and thus easily monitors and collects these requests where user locations are shared. An untrusted LBS intends to learn twofold sensitive information. First, run-time exact locations of users, which often refer to location privacy, can leak real-world identities or other sensitive personal information to an adversary that holds proper background knowledge. Second, user-location linkability means that an adversary can identify which user issues the targeted query and the query anonymity controls privacy leakage on it.

5.1   Experimental Configuration

The experimental dataset comes from a useful data generator MNTG [20] which can generate location data of mobile users with custom parameters. By MNTG, on the map of a real city, we acquire a dataset which involves movement of about 1000 users on 20 continuous timestamps (from 0 to 19). All user locations distribute over a rectangle region whose area is about 67 square kilometers (km^2).

Location points of user queries and POIs are indexed by a full quad tree whose height is 5 and whose leaf cells divide the rectangle region into $1024 = 4^5$ smaller rectangle regions each of which covers an area of about 0.065 km^2.

On the POI data cache module which relies on a common cache lib EhCache, time cost per operation is about 1 millisecond (ms) locally. In comparison, downloading data from a LBS database usually is more expensive. In our test environment one downloading task consumes about 5 ms within a LAN.

5.2   Comparison on Workload of Different Schemes

As shown in Fig. 3, the proposed scheme can improve performance of location anonymizer proxies by reducing their workload. We implement the following four

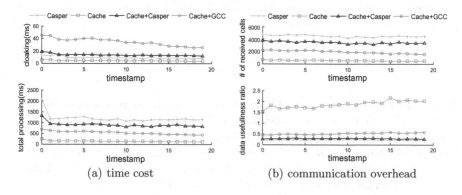

(a) time cost   (b) communication overhead

Fig. 3. Workload of location anonymizers and proxies

schemes using java in a development laptop for comparing their performance of in the same setting.

- The Cache scheme is a direct way of downloading POI data via a proxy and aims at improving communication efficiency between client-side and server-side. But it is noteworthy that this actually has a very limited protection for user privacy by removing user identities of queries and submitting a small cell instead of exact locations.
- Casper [7,21] is k-anonymous and blurs an exact location into a small region which is a set of cells and whose area must be larger than a threshold.
- The Cache + Casper scheme integrates Casper into the CachedCell framework (Algorithm 1) seamlessly.
- The Cache + GCC scheme represents the proposed location anonymization algorithm GCC (Algorithm 2) in $CachedCell$.

Figure 3(a) shows that GCC expends less time cost of whole processing than Casper via caching in middle servers and achieves location k-anonymity. The upper sub-figure is time cost of local processing which includes location cloaking algorithms and operations of accessing cache except downloading from a remote LBS server by each timestamp. GCC expends more time than Casper since frequently calling cache functions and finding eligible cells. But in the lower sub-figure GCC is more insensitive for workload of downloading and reduce time cost of whole processing. The Cache scheme has best efficiency without controlling user privacy.

Figure 3(b) demonstrates that GCC has a better usefulness ratio of bandwidth under location k-anonymity and can trade off data transmission efficiency and user privacy protection. We use the number of received cells to compute the amount of data transmitted under a simplification that each cell holds the near amount of POI data. Thus the data usefulness can be defined as $\frac{D_o}{D_i}$ where D_o is the number of demand cells which proxies output to clients and D_i is the number of downloaded cells that are inputted into proxies from remote LBS servers.

5.3   Comparison on Privacy Issues

Figure 4 displays performance of these three location anonymizers under different user privacy requirements where includes two arguments, k-anonymity and minimum size of cloaking region. Generally, a stronger privacy requirement leads to heavier workload of location anonymizers. We compute the average number of received cells of a user in a timestamp to assess quality of service in a proxy since location cloaking acquires additional POI data for hiding requested real location. This metric can be computed by $\frac{D_i}{D_o}$ that is the reciprocal of the aforesaid data usefulness ratio.

In the experiment of Fig. 4(a), the lower limit of cloaking region being set to 2, we change the k-anonymity value from 2 to 20. Cache + GCC has the minimum cost and cache feature can provide large cost savings for achieving k-anonymity.

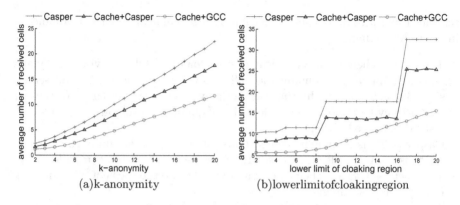

Fig. 4. Effect of user privacy requirement

For learning effect of the lower limit of cloaking region A, as shown in Fig. 4(a), A is changed from 2 to 20 and the k-anonymity value is constantly 10. Two curves of using Casper are of stair-step shape. Cache + GCC reduces cost for a larger threshold and trends to generate a small region that can satisfy user privacy requirement.

5.4 Effect of Cache Management

Cache Management involves two major faces replacement strategy and capacity. Figure 5 shows performance of the proposed scheme under different k-anonymity values and cache parameters.

According to Fig. 5(a), increasing the cache capacity can visibly improve efficiency of Cache + GCC along with k-anonymity values enlarged. The cache whose structure is a key-value in-memory storage can share temporarily requested POI data with other users.

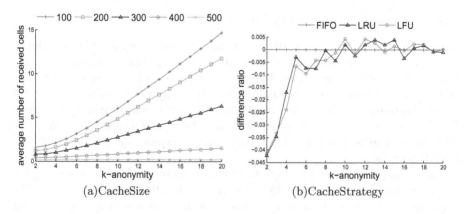

Fig. 5. Effect of cache management

However, Fig. 5(b) shows that different replacement policies affect this performance with small effects. In this figure, to compare three traditional replacement polices, FIFO, LRU (Least Recently Used) and LFU (Least Frequently Used), we compute the download difference rate as a benchmark that is denoted by $R(x) = \frac{D_i(x) - D_i(FIFO)}{D_i(FIFO)}$ where x is FIFO, LRU or LFU and $D_i(x)$ means the amount of downloaded cells via the replacement policy x.

6   Conclusion and Future Works

This paper aims at how to protect user privacy in query processing of LBS. In a proxy-based location anonymizer server, spatial cloaking algorithms generally need to retrieve a large amount of additional POI data from remote LBS servers. To alleviate this problem, we introduce the framework *CachedCell* which supports cache feature in location anonymizer proxies and a novel location cloaking algorithm *GCC*. We evaluate this solution by experiments to show its usability.

In future works, with recent location privacy metrics [5], we intend to design cache-aware location cloaking techniques which can support a stronger privacy assurance than location k-anonymity.

Acknowledgment. This work is partially supported by the HGJ National Significant Science and Technology Projects under Grant No. 2012ZX01039-004-009, Key Lab of Information Network Security, Ministry of Public Security under Grant No.C11606, the National Natural Science Foundation of China under Grant No. 61170263.

References

1. Liu, X., Zhao, H., Pan, M., Yue, H., Li, X., Fang, Y.: Traffic-aware multiple mix zone placement for protecting location privacy [22], pp. 972–980. IEEE (2012)
2. Sweeney, L.: K-anonymity: a model for protecting privacy. Int. J. Uncertainty Fuzziness Knowl. Based Syst. **10**(5), 557–570 (2002)
3. Gruteser, M., Grunwald, D.: Anonymous usage of location-based services through spatial and temporal cloaking. In: MobiSys, USENIX (2003)
4. Yang, D., Fang, X., Xue, G.: Truthful incentive mechanisms for k-anonymity location privacy. In: INFOCOM, pp. 2994–3002. IEEE (2013)
5. Andrés, M.E., Bordenabe, N.E., Chatzikokolakis, K., Palamidessi, C.: Geo-indistinguishability: differential privacy for location-based systems. In: Sadeghi, A.R., Gligor, V.D., Yung, M. (eds.) ACM Conference on Computer and Communications Security, pp. 901–914. IEEE (2013)
6. Shokri, R., Theodorakopoulos, G., Boudec, J.Y.L., Hubaux, J.P.: Quantifying location privacy. In: IEEE Symposium on Security and Privacy, pp. 247–262. IEEE Computer Society (2011)
7. Mokbel, M.F., Chow, C.Y., Aref, W.G.: The new casper: Query processing for location services without compromising privacy. In Dayal, U., Whang, K.Y., Lomet, D.B., Alonso, G., Lohman, G.M., Kersten, M.L., Cha, S.K., Kim, Y.K. (eds.) VLDB, pp. 763–774. ACM (2006)

8. Chow, C.Y., Mokbel, M.F., Liu, X.: A peer-to-peer spatial cloaking algorithm for anonymous location-based service. In de By, R.A., Nittel, S. (eds.) GIS, pp. 171–178. ACM (2006)

9. Gruteser, M., Liu, X.: Protecting privacy in continuous location-tracking applications. IEEE Secur. Priv. **2**(2), 28–34 (2004)

10. Xu, T., Cai, Y.: Location anonymity in continuous location-based services. In: Samet, H., Shahabi, C., Schneider, M. (eds.) GIS, vol. 39. ACM (2007)

11. Pan, X., Meng, X., Xu, J.: Distortion-based anonymity for continuous queries in location-based mobile services. In: Agrawal, D., Aref, W.G., Lu, C.T., Mokbel, M.F., Scheuermann, P., Shahabi, C., Wolfson, O. (eds.) GIS, pp. 256–265. ACM (2009)

12. Wang, Y., Wang, L., Fung, B.C.M.: Preserving privacy for location-based services with continuous queries. In: ICC, pp. 1–5. IEEE (2009)

13. Pingley, A., Zhang, N., Fu, X., Choi, H.A., Subramaniam, S., Zhao, W.: Protection of query privacy for continuous location based services. In: INFOCOM, pp. 1710–1718 (2011)

14. Lee, H., Oh, B.S., Kim, H.I., Chang, J.W.: Grid-based cloaking area creation scheme supporting continuous location-based services. In: Ossowski, S., Lecca, P. (eds.:) SAC, pp. 537–543. ACM (2012)

15. Wang, Y., He, L.P., Peng, J., ting Zhang, T., Li, H.Z.: Privacy preserving for continuous query in location based services. In: ICPADS, pp. 213–220. IEEE Computer Society (2012)

16. Ostrovsky, R., Skeith III, W.E.: A survey of single-database private information retrieval: techniques and applications. In: Okamoto, T., Wang, X. (eds.) PKC 2007. LNCS, vol. 4450, pp. 393–411. Springer, Heidelberg (2007)

17. Khoshgozaran, A., Shahabi, C., Shirani-Mehr, H.: Location privacy: going beyond K-anonymity, cloaking and anonymizers. Knowl. Inf. Syst. **26**(3), 435–465 (2011)

18. Papadopoulos, S., Bakiras, S., Papadias, D.: Nearest neighbor search with strong location privacy. PVLDB **3**(1), 619–629 (2010)

19. Galdames, P., Cai, Y.: Efficient processing of location-cloaked queries [22], pp. 2480-2488 (2012)

20. Mokbel, M.F., Alarabi, L., Bao, J., Eldawy, A., Magdy, A., Sarwat, M., Waytas, E., Yackel, S.: MNTG: an extensible web-based traffic generator. In: Nascimento, M.A., Sellis, T., Cheng, R., Sander, J., Zheng, Y., Kriegel, H.-P., Renz, M., Sengstock, C. (eds.) SSTD 2013. LNCS, vol. 8098, pp. 38–55. Springer, Heidelberg (2013)

21. Mokbel, M.F., Chow, C.Y., Aref, W.G.: The new casper: a privacy-aware location-based database server. In: Chirkova, R., Dogac, A., Özsu, M.T., Sellis, T.K. (eds.) ICDE, pp. 1499–1500. IEEE (2007)

22. Greenberg, A.G., Sohraby, K., (eds.): Proceedings of the IEEE INFOCOM 2012, Orlando, FL, USA, 25–30 March 2012

International Workshop on Human Aspects of Making Recommendations in Social Ubiquitous Networking Environments (HRSUNE 2014)

A Review of Factors Affecting Recommender Decisions in Social Networks for Educational Purposes

Estefanía Martín[1(✉)], Isidoro Hernán-Losada[1], and Pablo A. Haya[2]

[1] Dep. Lenguajes y Sistemas Informáticos,
Universidad Rey Juan Carlos, Madrid, Spain
{estefania.martin,isidoro.hernan}@urjc.es
[2] Instituto de Ingeniería del Conocimiento,
Universidad Autónoma de Madrid, Madrid, Spain
pablo.haya@iic.uam.es

Abstract. There are a multiple papers focused on recommendation techniques and algorithms. However, less attention has been dedicated to social factors that influence in the recommendation process. Depending on the context where the recommendation system is applied, it is necessary to understand human and social factors that affect the output of the recommendation algorithm. This study sheds light on design and decision making of recommender system. In particular, we conducted a survey where 126 students were asked to extract which are the main factors for improving suggestions when they are interacting in an Online Social Network (OSN) or in a specific domain such as an Educational Social Network (ESN). The results show that different factors have to be considered depending on the type of the network.

Keywords: Social networks · Recommender system · Social factors · e-learning

1 Introduction

The huge quantity of contents that users face daily and the time limits they have to search for and find products that fit their needs means that, sometimes, users cannot find what they are looking for, and the abundance of information overwhelms them. Recommender systems (RS) can help users filter and find useful content. A RS suggests items that could be relevant, performing an estimation of what could be useful for the users, and what they could like or be interested in maximizing the value of usefulness of content for the user [1].

RSs have also become an important part of Online Social Networks (OSNs). They are used to suggest news, friends and different types of content. Following a tendency called homophily in sociology [2], interests and connections could be propagated through the social network owing to contacts have similar interests probably. This information could be used by RSs in order to make suggestions.

Nowadays, RSs are used in e-learning [3, 4, 5]. Most of the areas where RSs are applied are movie suggestions followed by shopping RSs [3]. One example of

© Springer International Publishing Switzerland 2014
Y. Chen et al. (Eds.): WAIM 2014, LNCS 8597, pp. 261–271, 2014.
DOI: 10.1007/978-3-319-11538-2_24

video-based RS is Youtube. Its recommender makes use of similarities that exist between different videos based on their content or the average viewing time of users [6]. Thanks to the recommendation algorithm, users are driven to watch other videos in the platform [7]. Other video-based approaches use the tags associated to videos and the user profile for providing useful recommendations [8, 9]. Other factor to improve recommendations is the social similarity between videos, expressed as popularity distribution across social circles [10].

Education is a main domain of adaptive systems where RSs are classified. The most popular features to model the user as an individual are the user's knowledge, interests, learning goals, background, individual traits such as personality or learning styles, and user's context [11]. Educational recommender systems should be personalized by educational objectives, and not only by users' preferences [12]. However, the assessment of both learner knowledge and learning activities are more difficult than users' interests [13]. In the educational area, there is also growing interest in Educational Social Networks (ESNs). Two examples of RSs in ESNs were proposed by Dwivedi et al. [14] and Fazeli et al. [15]. The first approach combines two main factors in order to provide suggestions to learners: the learner's own preference on a particular type of resources and friends' suggestions. The second one provides a social recommender for teachers.

Any RS needs information to provide suggestions to users. In the case of social networks, the explicit information given by the user, the previous interactions and his network could be used in order to provide recommendations. However, what are the main factors to be considered in the recommendation mechanism? Do these factors change depending on the type of social network (e.g. friendship-based or educational)? This paper tries to bring light about these questions. It presents a the results of a survey where 126 students were involved in order to extract which are the main factors for improving suggestions when we are interacting in a friendship OSN or in a specific domain such as ESNs.

2 Research Questions

OSNs can be mainly oriented to two different approaches: interests or relationships. On the one hand, social networks based on users' interests provide suggestions matching the interests of users with information about contents or previous collective knowledge of other users. On the other hand, social networks based on relationships take into account the type of relation between the user and their neighbours and also the trust of other users. However, depending on the context where the recommendation system is applied, it is necessary to understand human and social factors that affect the output of the recommendation algorithm. These factors could be different and have to match user's goals. Against this background, some questions arise related to the factors to be considered in the recommendation process. Two main research questions arose:

- When users are interacting with social networks, what factors do they consider important to make suggestions?
- Do users behave in the same way when they are involved in an ESN? Should recommender algorithms consider the same factors as in current social networks?

The main goal of our study is to analyse the factors to be considered in general social networks, and if these factors change when users are members of a social educational network. Furthermore, we would like to show which are the most important factors are and which is their relative importance to make suggestions. We set up an anonymous questionnaire of 28 questions that was anonymously delivered for preventing privacy issues. 126 participants expressed their opinion related to the use of social networks and specifically, social networks in education. There were a mix of free text, scale factor and multiple-choice questions. The structure of this questionnaire was the following. The first set of questions was focused on personal features of the users including age, degree, year, work, age and sex. The second group asked about previous knowledge, interactions and task performed in current social networks such as Facebook, Youtube, Twitter, LinkedIn, etc. Next, questions were related about their opinions of recommendations provided by these OSN according to their usefulness, suitability and which aspects they considered more relevant to take into account in the recommendation algorithm. When, users answered the questions related to OSN, the same questions were presented regarding to a hypothetical ESN.

The results of the questionnaire are presented in the next section classifying the information in these four groups: general data, background in social networks, and main factors and recommendations in social networks (both general social networks and educational ones).

3   General Data Set

The number of participants in this survey was 126 people who are undergraduate students from Universidad Rey Juan Carlos. They have previous experience using online virtual learning environment such as WebCT or Moodle. Their background is diverse and they can be classified in two different groups according to their background. 28 students answered the questionnaire with engineering background (e.g. Computer Engineering, Telecommunications, Chemistry Engineering, etc.). Other 98 people were involved in this survey with a non-engineering background. Most of the students of this last group have an educational degree. All the degrees studies take 4 years to complete. 55 participants were in first year (44 %), 29 were in second year (23 %), 33 were in third year (26 %) and finally, 9 were in fourth year (7 %). The sex-based distribution was 102 women (81 %) and 24 men (19 %). The age of participants was distributed as following: 22 users were 18 years old, 31 participants are between 19–20 years, 28 students between 21 and 22, 16 users between 23 and 24, 17 participants were between 24 and 30, and finally, 12 students were over 30 years old.

The questionnaire was answered in a computer lab with 35 PCs. Participants were structured in 4 different groups. At the beginning of the session, the interviewer explained the goal of the questionnaire and next, users answered it.

3.1 Experience with Social Networks

Participants provided their background about what social networks known, and where they had an active profile. All users knew Facebook followed by Youtube (121), Twitter (121), Tuenti (120), Instagram (109), Google + (86), MySpace (77) and Badoo (76). Less known Social networks were LinkedIn (46), Hi5 (33), Tumblr (32), Flickr (31), Pinterest (14) and Foursquare (5). Furthermore, they are active users in the following OSNs: Facebook (106), Tuenti (99), Twitter (93), Youtube (71), Google+ (53) and Instagram (51). The number of participants registered in the other SNs is below 50.

Most of the participants (72 %) check the news in their social networks more than one time in the day, 13 % of users check once a day, and only 2 % (3 users) does not use OSNs in their daily live owing to privacy issues or this type of platforms do not interest to users. The rest (13 %) enters in SNs once a week or several times during the week. 93 % participants, who has an active profile in some OSN, use a mobile device such as a smartphone or tablet.

Most frequently activities performed by users registered in OSNs are: sending messages (88 %), reviewing the timeline and interactions of their friends/contacts (87 %), watching videos or listening music (84 %), and publishing contents such as pictures, news, videos (73 %), and learning (71 %). Some people use OSNs with other purposes such as professional purposes (47 %), on-line gaming (30 %) or meeting new people (29 %).

3.2 Recommendations in Online Social Networks (OSN)

The goal of the next group of questions was to know if users are aware of the recommendations included in OSNs. Most of the participants (95 %) know that OSNs provide personalized recommendations to users. However, only 51 % of users like these suggestions. Most valuable recommenders were Facebook (30 %), Youtube (30 %), Twitter (20 %) and Google + (7 %).

Regarding the main factors to be considered in order to suggestions fit to the users' interests, participants selected the characteristics they preferred from the following list: my contacts, the webpages visited in the OSN, time spent in the OSN, contents watched, comments written, comments watched by contacts, comments written by contacts, contents uploaded, contents shared, contents with my best ratings, trending topics, or websites visited outside of the OSN. These factors were rated on a five scale between 1 and 5. The main factors are related to personal features of the user (see Fig. 1) such as contents with my highest ratings ($M = 3.89$, $SD = 1.01$), contents watched ($M = 3.76$, $SD = 1.00$), contents shared ($M = 3.75$, $SD = 1.12$), webpages visited ($M = 3.69$, $SD = 0.62$), and contents uploaded ($M = 3.38$). The first social factors to be considered in the recommendation process are the trending topics ($M = 3.39$, $SD = 1.17$) and the network of contacts ($M = 3.15$, $SD = 1.01$). Participants always considered more important the contents watched than the comments provides by users of the SN. They also prioritize contacts' information over similar users' information. This fact shows that friendships are important in social networks and they prefer that recommender systems take into account the friendship network before doing suggestions based on similarities between users.

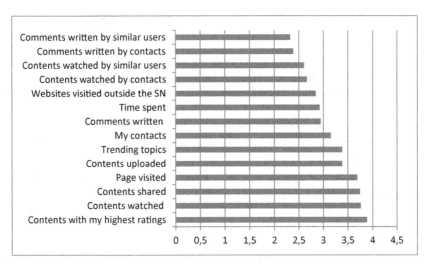

Fig. 1. Average of scores for the main factors to be considered in the recommendation process where users are interacting in online social networks

If we provide recommendation based on the type of content, it would be helpful to know what type of contents (text, images, documents, videos, audios, interactive elements or other types) users demand in OSNs. The results obtained to this multiple-choice question are presented in the Fig. 2. Videos and documents combining both texts and images were the most popular types of contents for recommending purposes (24 % and 21 %, respectively). Furthermore, it must be highlighted that 99 and 85 participants prioritize the recommendation of videos and documents as the perfect type of materials for using in OSNs. Lowest scores are obtained by texts and audios. Additionally, they proposed to include other type of content in the suggested contents such as news or events.

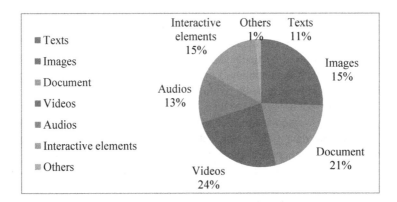

Fig. 2. Types of contents to be recommended in a general online social network

3.3 Recommendations in a Educational Social Network (ESN)

Once the main factors of general OSNs were obtained, the last part of the questionnaire focus on answering if these factors vary in an educational context. We asked for putting themselves in the shoes of someone using a ESN where it is possible to upload, to share and to comment different learning materials (documents, images or videos, among others).

Before asking for specific questions, we would like to know which characteristics should have an educational content to be interesting for participants. The answers are grouped and ordered by relevance according to the users:

- The content must be aligned with their personal learning interests. The learning material uploaded and shared in the platform must strengthen the study and stimulate to students both at the personal and the professional level. Real studies and experiences help to transfer the knowledge in educational environments. Furthermore, creative material or complementary approaches from different points of view can help to learning process. This material can be complemented with news about the personal interests of each user. It is important to consider the novelty of the content in order to provide suggestions.
- The published material must be reliable or be uploaded from a trust source, such as institutional entities (e.g. governments, universities, journals, etc.), websites which you are subscribed or a trust user (e.g. renowned professor, expert in the area, followee or classmates).
- Related to the appearance of content, students prefer visual formats such as videos or images. These formats promote students pay attention to these materials. Textual resources without visual information are boring for them. They point to the combination of textual and visual formats where textual information is brief and concise.

Next, they were asked for aspects to be considered by a recommender algorithm in a social educational network such as type of contents to be suggested or personal features to be considered.

Regarding to types of contents to be recommended in this educational environment, Fig. 3 presents the results obtained. This is a multiple choice option where students selected the types that they considered more appropriated. The most popular formats are videos (91 %) followed by documents that combine text and images (84 %), and interactive elements such as animations or interactive videos (67 %). Texts, images and audios obtain 58 %, 55 % and 44 % votes. Furthermore, participants provided us feedback about other possible types to be considered such as quizzes, electronic books, research papers, blogs, external links, events, games, workshops or conferences.

If we compare this data to the results obtained in the previous section about general OSNs, we can observe there are more votes per type of content. Although videos, documents and interactive elements continue to be the most popular types of contents, there is also a small change in the users' preferences: images are less popular in an educational context than text information. Finally, new types of elements to be considered recommended appear in this context owing to the own characteristics of the domain.

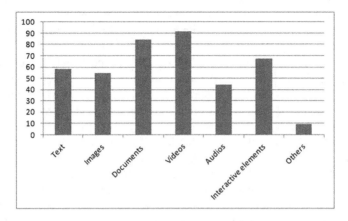

Fig. 3. Types of contents to be recommended in a social educational network

The personal features that recommender systems could be taken into account in the recommendation process inside a social educational network are presented in Fig. 4. The average for each factor is represented in a Likert scale between 1 and 5. The results obtained in this question are compared to the obtained in general OSNs.

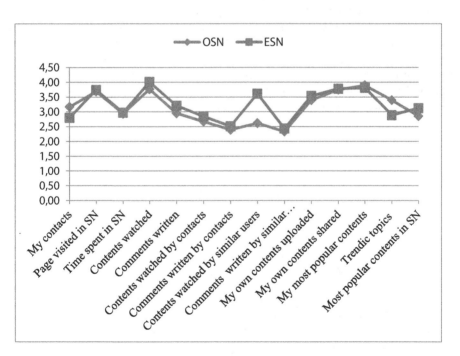

Fig. 4. Comparative of factors to be considered by the recommendation process in social networks and social educational networks

Most popular factors are related to the own user such as contents watched in the ESN (M = 4.01, SD = 0.94), contents rated with high scores (M = 3.81, SD = 1.06), contents shared with other users (M = 3.78, SD = 0.94), page visited in the ESN (M = 3.74, SD = 1.04), contents uploaded (M = 3.53, SD = 0.99) and comments written (M = 3.21, SD = 0.94). Apart from the own personal features of the user, the reputation of the source of the suggestions has high important in ESNs. Participants consider relevant the contents watched and the comments written both classmates (M = 3.88, SD = 1.06) and workmates (M = 3.98, SD = 0.95) interacting in the ESN. Comments of classmates and workmates are also important for users but less than contents watched. The last factor considered relevant is the content watched by users with similar features in ESN (M = 3.61, SD = 1.03). As it can be seen, the priorities of users change owing to the specific domain of the social network. The higher differences are in trending topics (M = 2.88, SD = 1.26) and contacts in the network (M = 2.79, SD = 1.20). The means differences are 0.51 and 0.37 respectively. In both cases, the relevance of trending topics and contacts in the network for participants is more important in OSNs than ESNs.

Finally, we asked about how to solve the problem when a new user is registered in a recommender system based on the relationships between the user and the rest of the roles in the ESN (see Fig. 5). On the one hand, if we analyzed the most valuable recommendations based on the roles in this type of platforms, the most important suggestions will provide when the teacher is involved suggesting relevant contents (M = 4.50, SD = 0.62) or commenting contents (M = 4.34, SD = 0.73), and the partners of the same workgroup interact with the platform or the classmates (M = 4.09, SD = 0.72). Friendships and suggestions of other students enrolled in the same studies but not in the same classroom should not be considered. Furthermore, as they mentioned before, reputation is also a valuable feature in ESNs because although they prefer the recommendation of their classroom circle (teachers, workmates and classmates), they suggest to take in mind the recommendation of trustworthy person in the ESN as a teacher of other university or an expert in the area.

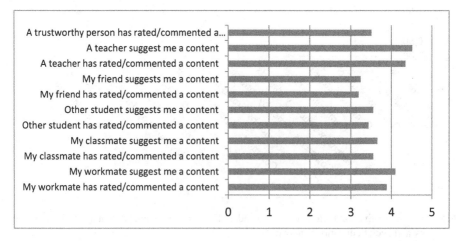

Fig. 5. Comparative of factors to be considered by solving the cold-start problem

Therefore, the same factors and with the same relative importance are proposed to solve the problem when a new user appears in the platform. They prefer that the recommender system will use information of the closed circle, their classroom. Suggestions of teachers, workgroup partners and classmates are valuable for them as well as trustworthy people in the ESN.

4   Discussion

The goal of our work was to understand human and social factors of making suggestions and sharing resources in social networks as an emerging platforms people use in their daily life. Currently, people are joined to social networks with different purposes and they check the last news every day mostly using a mobile device. Only few people have not use these platforms owing to privacy issues in most cases.

People are aware that most of OSNs provide custom suggestions in order to improve the user experience with this platform. However, users are not satisfied with these recommendations. A possible reason of this fact could be that recommender systems are more focused on a general solution and they have not taken into account the specific domain of the OSN and the main factors demanded by users. Furthermore, the advertising suggestions included in these platforms disturb users frequently.

We are focusing on knowing the main factors to be considered during the recommendation process and if any differences exist depending on the type of social network either a general purpose OSN or an ESN. The results presented in the previous section show that there are differences of understanding recommendations depending on the social networks' domain. The study presented in this paper focus on the main factors to take in mind in the recommendation process and the type of the most suitable contents to be presented.

Related to the factors to be considered in the recommendation process in an OSN, they change depending on if the goal of the network is educational or it is a general purposes network based on relationship. As it could be expected, the importance of the relationships varies in the educational context. In both types of social networks is important to take in mind the personal interests and features of each user to provide recommendations. However, when the user is in an educational environment, he prefers that recommender algorithms combine their suggestions with information about his classmates, partners of the same workgroup and similar users. Suggestions based on the friendship in social educational environments seem not be relevant in ESN. However this factor is really important in general OSNs where users demand the recommendation of contents based on friendships and they consider less important the information of similar users. These differences could be because of the users' goals change depending on the network. If you are in a relationship social network you prefer to keep track what contents your contact have watched or commented. However, if you are a user of an ESN, your goal is focused on learning and you want the suggestions fit better to your personal features. Other difference was that the importance of trending topics decreases in an educational context. Finally, people always considered more important the contents watched than the comments provides by users. The time when content is published must be taken into account in both types.

Regarding ESNs, the main factors pointed by students are: their personal learning goals and interests, reliable and novel learning material and visual presentation of the material. ESNs are networks driven by the personal learning interests of users. Thereby, a key aspect for an interesting learning material is the alignment with the user's personal learning goals and future career. Furthermore, they are interesting on novel, creative and different learning material to know the last news about their interests and understanding the topics from approaches with different points of view. The inclusion of real experiences and studies improves the transfer of knowledge from the theoretical point of view to the real world. One of the main concerns of students is the trust in the published material. Currently, a lot of information is available about any topic in Internet. However, the great amounts of materials provoke users are confused and disoriented for obtaining top quality learning resources. Recommender systems could help in this goal using a reputation mechanism where users of the ESN have associated a score. Aspects related to the user's expertise area or number of followers could also be taken into account for selecting the best learning resources and suggesting them to other users.

The presentation of material is also important for users. Most information in our daily lives is provided by visual channels such as videos or pictures. OSN are not an exception. People demand that the appearance will be attractive and the best types of materials to suggest are videos, documents and interactive elements. They are the most popular types of contents both in OSN and ESN. There is a small change between images and texts. Images are more popular in a relationship social network than in educational environments. Additionally, new types could be suggested in both networks such as news or events. However, owing to the specific domain of ESNs, new types of materials arise related to educational area such as quizzes, electronic books, research papers, workshops or conferences.

5   Conclusions and Future Work

This paper has presented the results of a survey where 126 undergraduate were involved. Its goal is to understand the main social factors for providing suggestions in OSNs. These social factors are different depending on the type of social network and they must be taken into account by recommendation algorithms to giving suggestions to users. The users' interests change if they are enrolled in a relationship-oriented social network or in an educational-oriented social network. Therefore, the features considered in the recommendation process should be different in each case.

The conclusions obtained in this study will be applied in the recommendation mechanism inside ClipIt in the future, an ESN that promotes the learning of STEM concept thanks to sharing resources and discussion between peers.

Acknowledgment. The research leading to these results has received funding from the European Community's Seventh Framework Programme (FP7/2007-2013) under grant agreement n° 317964 JUXTALEARN.

References

1. Adomavicius, G., Tuzhilin, A.: Toward the next generation of recommender systems: a survey of the state-of-the-art and possible extensions. IEEE Trans. Knowl. Data Eng. **17**(6), 734–739 (2005). doi:10.1109/TKDE.2005.99
2. McPherson, M., Smith-Lovin, L., Cook, J.M.: Birds of a feather: Homophily in social networks. Ann. Rev. Sociol. **27**, 415–444 (2001)
3. Bobadilla, J., Ortega, F., Hernando, A., Gutiérrez, A.: Recommender Systems Survey. Knowl.-Based Syst. **46**, 109–132 (2013). doi:10.1016/j.knosys.2013.03.012
4. Park, D.H., Kim, H.K., Choi, I.Y., Kim, J.K.: A literature review and classification of recommender systems research. Expert Syst. Appl. **39**(11), 10059–10072 (2012)
5. Santos, O.C., Boticario, J.G.: Educational recommender systems and technologies: practices and challenges. IGI Global, pp. 1–362 (2012). doi:10.4018/978-1-61350-489-5
6. Davidson, J., Liebald, B., Liu, J., Nandy, P., Van Vleet, T., Gargi, U., Gupta, S., He, Y., Lambert, M., Livingston, B., Sampath, D.: The YouTube video recommendation system. In: Proceedings of the Fourth ACM Conference on Recommender Systems, 26–30 September, Barcelona, Spain (2010). doi:10.1145/1864708.1864770
7. Zhou, R., Khemmarat, S., Gao, L.: The impact of Youtube recommendation system on video views. In Proceedings of the 10th ACM SIGCOMM Conference on Internet Measurement (IMC '10), pp. 404–410. ACM, New York (2010). doi:10.1145/1879141.1879193
8. Park, J., Lee, S.J., Lee, S.J., Kim, K., Chung, B.S., Lee, Y.K.: Online video recommendation through Tag-cloud aggregation. IEEE Multimedia **18**(1), 78–87 (2011). doi:10.1109/MMUL.2010.6
9. Bertini, M., Del Bimbo, A., Ferracani, A., Gelli, F., Maddaluno, D., Pezzatini, D.: Socially-aware video recommendation using users' profiles and crowdsourced annotations. In: Proceedings of the 2nd International Workshop on Socially-Aware Multimedia (SAM '13), pp. 13–18. ACM, New York (2013). doi:10.1145/2509916.2509924
10. Ma, X., Wang, H., Li, H., Liu, J., Jiang, H.: Exploring sharing patters for video recommendation on Youtube-like social media. Multimedia Systems, 1–17 (2013). doi:10.1007/s00530-013-0309-1
11. Brusilovsky, P., Millán, E.: User models for adaptive hypermedia and adaptive educational systems. In: Brusilovsky, P., Kobsa, A., Nejdl, W. (eds.) Adaptive Web 2007. LNCS, vol. 4321, pp. 3–53. Springer, Heidelberg (2007)
12. Santos, O.C., Boticario, J.G.: Modeling recommendations for the educational domain. Proc. Computer Sci. **1**, 2793–2800 (2010). doi:10.1016/j.procs.2010.08.004
13. Drachsler, H., Hummel, H., Koper, R.: Identifying the goal, user model and conditions of recommender systems for formal and informal learning. J. Digit. Inf. North Am. 10, January 2009. http://journals.tdl.org/jodi/index.php/jodi/article/view/442. Last access: 05 February 2014
14. Dwivedi, P., Bharadwaj, K.K.: e-learning recommender system for learners in online social networks through association retrieval. In: Proceedings of the CUBE International Information Technology Conference, pp. 676–681. ACM, New York (2012). doi:10.1145/2381716.2381846
15. Fazeli, S., Drachsler, H., Brouns, F., Sloep, P.: A Trust-based social recommender for teachers. In: Proceedings of 2nd Workshop on Recommender Systems for Technology Enhanced Learning (RecSysTEL 2012) in conjunction with the 7th European Conference on Technology Enhanced Learning (EC-TEL 2012), Saarbrücken, Germany, 18–19 September 2012 (2012)

Exploring Arduino for Building Educational Context-Aware Recommender Systems that Deliver Affective Recommendations in Social Ubiquitous Networking Environments

Olga C. Santos[✉] and Jesus G. Boticario

aDeNu Research Group, Artificial Intelligence Department,
Computer Science School, UNED, C/Juan del Rosal 16, 28040 Madrid, Spain
{ocsantos,jgb}@dia.uned.es

Abstract. One of the most challenging context features to detect when making recommendations in educational scenarios is the learner's affective state. Usually, this feature is explicitly gathered from the learner herself through questionnaires or self-reports. In this paper, we analyze if affective recommendations can be produced with a low cost approach using the open source electronics prototyping platform Arduino together with corresponding sensors and actuators. TORMES methodology (which combines user centered design methods and data mining techniques) can support the recommendations elicitation process by identifying new recommendation opportunities in these emerging social ubiquitous networking scenarios.

Keywords: Contextual recommender systems · Educational recommender systems · Ubiquitous computing · Affective computing · Arduino · Sensors · Actuators · ISO 9241-210

1 Introduction

Recommender systems in educational scenarios can be used to support learning [1]. Their capability of observing learners' behavior and making personalized suggestions accordingly reaches its full potential when context is taken into account [2]. Furthermore, recommender systems are nowadays integrated in applications available in the cloud, mainly following a centralized approach and produced at the recommendation server [1]. Therefore, they can integrate information sources coming from different components.

Context-aware ubiquitous learning is a computer supported learning paradigm for identifying learners' surrounding context and social situation to provide integrated, interoperable, pervasive and seamless learning experiences [3]. Several studies have reported on the benefits of context-aware ubiquitous learning, for instance in the promotion of learning motivation [4] and the improvement of learning effectiveness [5]. In this way, the learning system is able to sense the real-world situation of the learners, interact with them and provide them with adaptive support accordingly [6].

© Springer International Publishing Switzerland 2014
Y. Chen et al. (Eds.): WAIM 2014, LNCS 8597, pp. 272–286, 2014.
DOI: 10.1007/978-3-319-11538-2_25

As discussed in [2], contextual information can be captured by sensors. However, from the review done to 11 context-aware educational recommender systems and reported in Sect. 2, it results that the automated gathering of contextual information when dealing with learners' affective states is still a challenge. This challenge is to be tackled due to the interplay between the cognitive aspects of learning, the external behavior showed by the learner and affect [7]. In fact, there are benefits in creating sensor based personalized affective feedback for each individual student in a classroom environment, and this can be done with minimally invasive sensors [8].

In this situation, there is a need to understand the human and social factors of making recommendations in Social Ubiquitous Networking Environments (SUNE). By SUNE we refer to those environments that support social interactions and are interconnected through different networks (e.g., Internet and wireless local networks) and provide advanced computing features, where computing is made to appear everywhere and anywhere using any device, in any location, and in any format, involving sensors, microprocessors, as well as new input/output user interfaces. There is a widely used open source physical computing platform called Arduino[1], which is characterized by its flexible and simple to use hardware as well as a relatively easy to use C like programming language. This prototyping platform seems appropriate for building complex electronics at low cost, such as those that can support SUNE.

Thus, in this paper, we explore the potential of Arduino for delivering contextualized recommendations in e-learning SUNE. First, we analyze the contextual features considered in technology enhanced learning scenarios and how educational recommender systems deal with them. Next, we review the usage of Arduino to deal with contextual features, with emphasis on those that are the most challenging (i.e., affective states). After that, we propose TORMES methodology to identify context-aware affective recommendation opportunities in SUNE. Following, we discuss some open issues identified, and end with some conclusions and the outline of future work.

2 Context in Educational Recommender Systems

As learning environments are increasing their diversity and richness, the incorporation of contextual information about the learner in the recommendation process is attracting major attention [9]. After discussing several approaches to categorize the context, Verber et al. have proposed a context framework that considers 8 categories (i.e., computing, location, time, physical conditions, activity, resource, user, and social relations) for classifying context-aware recommender systems in technology enhanced learning scenarios [2]. In order to understand how recommender systems deal with context in educational scenarios, we have analyzed those found in our review of the literature. Table 1 compiles the 11 educational recommender systems analyzed. For each of them, the analysis has identified the application scope, the context considered, the recommendation approach followed and the evaluation carried out (if any). First, they are chronologically ordered, and then by system name (when available).

---

[1] http://arduino.cc/

Table 1. Main features of reviewed context-aware educational recommender systems

System	Scope	Context	Recommendation approach	Evaluation
CALS Yau and Joy, 2007 [10]	activities	Retrieve contextual information from the scheduled events database (time available for learning and type of location) and two sensors (GPS for location detection and a microphone for noise detection). Other types of sensors to detect other environmental attributes are mentioned (light levels and temperature).	Identifies the learning context which the learner is currently situated in, and then recommends appropriate learning activities for them based on the learning context using if-then rules.	n/a
SPA Gonzalez et al., 2007 [11]	course	Emotional features from tests and interactions (which theoretically can include physiological signals obtained through wearable computing).	Generate emotional arguments with predictive models to suggest enrolling a course (by activating or inhibiting excitatory attributes to be shown in the message).	Proposed predictive models were evaluated with data extracted from users registered in a website.
Affective e-learning model Shen et al., 2009 [12]	learning materials	Information collected from physiological sensors (skin conductance, blood volume pressure and electroencephalograph activity, heart rate) tagged with self-reports from the learner.	Emotion-aware recommendation rules implemented on a intelligent content recommender.	Physiological data from 1 subject (several weeks, close-to-real world setting). Recommendations evaluated by analyzing the times the user changed content recommended.

(Continued)

Table 1. (*Continued*)

System	Scope	Context	Recommendation approach	Evaluation
PL–CR2 Luo et al., 2010 [13]	course	The context sensor module collect dynamic activities data from learners, such as connecting mode, bandwidth and resource access response time (download time).	A resource is recommended using a hybrid recommendation algorithm that consider if the resource's type matches learner's connecting mode and the resource access response time satisfies learner's lowest requirement.	Simulations to evaluate recommendation algorithm performance.
Semantic Learning Space + ELAA Yu et al., 2010 [14]	learning content	Aggregates context information gathered from physical sensors (camera, microphone, RFID, GPS, pressure sensors) and virtual sensors (calendar service, organization workflow, etc.) aimed to identify the learning schedule and task.	Recommended content defined by the personal context such as prior knowledge, goals, user's situation context (i.e., subject studied for a long time, recommend another subject) and the presentation by the infrastructure context (device, network) are managed in terms of ontologies.	14 participants used the system and answered a questionnaire to measure the user acceptance based on the content provisioned, response time, and user interface.
n/a Wang and Wu, 2011 [15]	resources	RFID is used to sense the location of the learning objects in the actual environment.	Once the system detects learning targets in the actual environment, the learning recommendation module automatically generates a recommendation list that matches the learner requirements.	30 participants in a control *vs.* experimental group analysis to verify with a t-test the effect on learning effectiveness (i.e., if provided learning support helped learner to complete task quicker and more accurate)

(*Continued*)

Table 1. (*Continued*)

System	Scope	Context	Recommendation approach	Evaluation
SCROLL Li et al., 2012 [16]	learning content	Collects learner's context information from: (i) activity (motion and actual action, such as listening to music, surfing Internet), (ii) device status (battery, Internet connection, ring tone), (iii) environmental data (location, time, temperature, weather).	When physical learning objects are detected nearby, recommend location-dependent quizzes or information shared by other learners on that object. Additionally, if the learner is in her preferred learning context, the system will show messages to encourage her to study.	11 students used the system to support their learning. Engagement, system usability, user satisfaction and stimulation to learn were measured.
A3 Boff and Reategui, 2013 [17]	peers, contents, exercise	Questionnaires and GUI controls to let the learner indicate her affective states.	Recommendation provided in natural language through a virtual character. The type of language and the emotional appearance of the virtual character depend on the learner's affective state. Peer recommendation takes into account the mood state of the peer to be recommended.	Performance of recommended items' descriptions in terms of processing time and accuracy
Auto-Tutor D'mello and Graesser, 2013 [18]	response given by a pedagogical agent	Detects the emotions (boredom, flow/engagement, confusion, frustration) of a learner by monitoring conversational cues, gross body language, and facial features.	Synthesizes affective responses through animated facial expressions and modulated speech to address the presence of negative emotions. Production rules informed by theories on emotion and learning as well as recommendations made by pedagogical experts.	Performance of classification techniques when detecting learners' affective states using experimental simulations.

(*Continued*)

Table 1. (*Continued*)

System	Scope	Context	Recommendation approach	Evaluation
BISPA Kaklauskas et al., 2013 [19]	learning materials	Learners control from a GUI the gathering of the physiological measures (skin temperature, skin humidity, skin conductance, touch intensity, heart rate, blood pressure) computed with external equipment every 45 minutes.	Recommendations to improve learning productivity are selected from a database of 5000 recommendations.	Data from 1 student was used to compare her learning productivity and interest in learning with her physiological parameters.
LRAR Leony et al., 2013 [20]	peers, activities and resources	Learners select affective state from a static list. Future work is to develop plug-ins to include sensors to gather the affective state (galvanic skin response, face gesture recognition).	The recommendation engine is developed on top of Apache Mahout machine learning engine and the algorithm to calculate the recommendations is left for the implementation of each case.	n/a (plans focus on evaluating performance in terms of the response time of a recommendation request to the server).

The application **scopes** identified (i.e., what is recommended) are *peers* [17, 20], *courses* [11, 13], *learning items* (referred as activities, materials, contents, resources and exercises) [10, 12, 14–17, 19, 20] and *responses* provided by pedagogical agents to respond to learners' affective and cognitive states [18].

Regarding the **context**, reported systems involve several dimensions of the afore-mentioned contextual framework [2], as follows: (1) *computing* (dealing with net-works, hardware and software characteristics) [13, 14, 16], (2) *location* (information to place the learner, either at a high grain such as classroom, home or outdoor, or at a lower grain by specifying geometric coordinates) [10, 14–16], (3) *time* (during which contextual information is known or relevant, either at a higher grain such as week, month, etc., or at a lower grain by given the time span) [10, 14, 16], (4) *physical conditions* (environmental conditions where the system or user is situated, commonly including measures for heat, light and sound) [10, 16], (5) *activity* (i.e., the task, objectives or actions of the learner) [10, 14, 16], (6) *resource* (used for learning) [14, 15], (7) the *affective information* within the user dimension [11, 12, 17–20], and (8) *social relations* (describing associations, connections or affiliations between two or more persons) [none].

Recommendation **approaches** followed are *rules* [10, 12, 16, 18, 19], predictive *models* [11], recommendation *algorithms* [13, 15, 17, 20] and *ontologies* [14].

As for the **evaluation**, users were involved in the evaluation of less than half of the systems analyzed [12, 14–16, 19]. Only in one of them (i.e., [15]) the effect of rec-ommendations provided on learning effectiveness was statistically evaluated.

From this analysis of context-aware recommender systems in technology enhanced learning scenarios follows that most recommendations focus on learning items, affective information is the most common contextual feature considered while no system dealing with social relations contextual features has been found, recommen-dation approaches are mainly rules or recommendation algorithms, and the evaluation of the learning effectiveness with statistical significance is hardly carried out.

Having this in mind, in the next section we analyze if affective recommendations can be produced in educational context-aware recommender systems with a low cost approach using the open source electronics prototyping platform Arduino together with corresponding sensors and actuators.

3   Arduino Support for Contextualized Recommendations

Arduino is a tool for making computers sense and control the physical world, taking inputs from a variety of sensors, and can affect its surroundings by controlling lights, motors, and other physical outputs (i.e., actuators) [21]. Arduino projects can be stand-alone, or they can communicate either with software running on a computer via a serial communication or with an external component using various wireless mechanisms (Wi-Fi, Bluetooth, GSM, GPRS, Xbee, etc.).

We have researched in the literature the usage of Arduino to produce context-aware recommender systems that deal with related affective information. This implies two perspectives: (i) gathering context information, and (ii) delivering context-based feedback.

Regarding context information gathering, we found 10 works that report the usage of Arduino to gather physiological measures. Although in most cases the data collected was not used to detect the affective state of the user, the same data can be used (as reported in the corresponding works in Table 1) with these other purpose. Thus, in Table 2 we compile the works found in the literature that use Arduino for gathering physiological data and identify the system goal, the type of Arduino board, and the sensors used. In addition, it is specified if sensors have been developed by the authors or an available component was used, stating in the former case the manufacturer and/or model. All works have been published in 2013, which suggests that this is an emerging research line with a trend in the appearance of more research and thus, an opportunity to be caught up by educational context-aware recommender systems. Works reviewed have been ordered alphabetically with the first authors' last name.

Works compiled in Table 2 take measures through physiological sensors that can be used to detect affective states [33]. Additionally, some of them also consider physical measures, such as movement with the accelerometer [23, 27, 32]. All the systems analyzed (except [26]) include the validation of the system developed, showing that the deployment carried out properly detects the physiological signals aimed to be recorded with the Arduino infrastructure.

Regarding the delivery of context-based feedback, [33] suggests audiovisual adaptation to reflect or evoke certain affective state in the user by making use of either music or light. We have found 2 works in this respect. In [22], Arduino was used to indicate with color lights (specifically, red-green-blue light emitting diodes - RGB LEDs) the popularity of the recommended item (in this case, a bottle of wine). In [23], the system provides an actuator, specifically a LED to give visual feedback to the user through changes in the light conditions.

Although works reported are not specific for the educational domain, the usage of Arduino in educational context-aware recommendations seems of interest to manage contextual information about the learner. This information can be of value to deliver affective recommendations in SUNE that take into account learners' affective state.

4 Affective Contextual Recommendations in SUNE

From previous sections, it seems that integrating Arduino low cost infrastructure for sensing the learning context is of potential interest in SUNE. In particular, it seems feasible to build the physiological infrastructure for affective contextual information gathering using Arduino boards and available physiological sensors, such as the e-Health platform used in [29]. However, new recommendation opportunities in SUNE that go beyond those already identified in technology enhanced learning settings have to be identified. These are to take into account the possibilities of existing sensors and actuators that can be used to deliver affective educational recommendations.

The works reported in Sect. 2 already provide some ideas for educational context-aware affective recommendations. In particular, in [12] the following emotion-aware recommendation rules are used: (1) if learner is engaged, deliver contents according to subject's learning progress and objective; (2) if learner is confused, deliver examples or case studies in line with current knowledge; (3) if learner is bored, deliver video/music

Table 2. Arduino works that deal with physiological signals. Physiological sensors acronyms: **ECG:** Electrocardiography; **EDA:** Electrodermal Activity, **EEG:** Electroencephalography, **EMG:** Electromiography; **GSR:** Galvanic skin response; **HR:** Heart Rate.

System	Goal	Board	Sensors
BITalino Guerreiro, 2013 [23]	Multimodal biosignal acquisition for rapid prototyping of end-user applications in physiological computing.	Arduino Uno	ECG (developed), EMG (developed), EDA (developed), accelerometer (Analog Devices), phototransistor (TEMT6000).
Kishore et al., 2013 [24]	Recognize human activities, such as if the user is sitting or standing.	Arduino	Pulse Sensor[a] to evaluate HR variability.
Koga et al., 2013 [25]	Capture the transition of mental conditions.	Arduino Mega ADK	EMG (Oisaka Electronic), EEG (Mindflex headset by Mattei), HR (Polar).
Lotlikar et al., 2013 [26]	Help people measure vital signs at home and keep their record.	Arduino Duemilanove	Pulse, skin temperature, blood pressure (manufacturer not specified).
Lung et al., 2013 [27]	Person monitoring to extract behavioral patterns for assistance.	Arduino ChipKit Max32	HR (chest belt BM-CS5R) and accelerometer (EZ Chronos watch).
Mansor et al., 2013 [28]	Remote health monitoring system.	Arduino Uno	Body temperature (LM35); HR also considered but not reported.
Orha and Oniga, 2013 [29]	Assessment of health status for disease prevention. Emotion determined from GSR data.	Arduino Uno	Body temperature, blood pressure, respiratory rate, ECG and GSR (all from the e-Health platform[b])
Sinha et al., 2013 [30]	Polygraph tests for lie detection.	Arduino Uno	GSR and HR (developed)
Rahim, 2013 [31]	Check the stress level on the user.	Arduino Uno	Temperature sensor (LM 35), GSR (developed) and pulse sensor (same as in [24]).
Roy et al., 2013 [32]	Wearable electronic rescue system that detects abnormal condition of heart as well as sudden accidental fall.	Arduino Mega	ECG (developed) and accelerometer (ADXL 345).

[a] http://pulsesensor.myshopify.com/

[b] http://www.cooking-hacks.com/documentation/tutorials/ehealth-biometric-sensor-platform-arduino-raspberry-pi-medical

according to subject's preference to ease the tension; and (4) if learner is hopeful, deliver music according to subject's preference to enhance meditation. In [19], the following examples have been identified: (1) learning the most difficult subjects is recommended in the morning, leaving the easier ones for the evening, (2) limit contacts with people who cause negative stress, (3) for students who get up in the morning in bad mood get recommendations on ways to improve their learning efficiency, which are different than those given when she is studying in the evening and in good mood. In [20] some use cases are suggested: (1) recommend peer in good mood with adequate level of knowledge and who felt the same sensations when carrying out the activity in which the current learner is blocked, (2) if the learner is confused, recommend to read related exploratory material, (3) recommend a resource accessed by similar learners (in terms of having the same affective state when accessing other resources). Finally, when pedagogical agents are used to deliver recommendations, the user's affective state can be considered when choosing the type of language for the character to talk at a given moment [17]. In addition, in the pedagogical agent presented in [18], rules are also dynamically adapted to each individual learner, such as if the current emotion is classified as boredom and the previous emotion is frustration then the learner can be told "Maybe this topic is getting old. I'll help you finish so we can try something new".

However, the above sample rules compiled from the literature do not consider the particularities of SUNE, which aim to support social interactions and are interconnected through different networks where computing is made to appear everywhere and anywhere using any device, from any location and with any format. In addition, we already commented in Sect. 2 that no recommender system dealing with social relations contextual features was found in the review of the state of the art.

In order to identify recommendation opportunities in e-learning SUNE that properly take advantage of available contextual information, TORMES methodology can be used. TORMES [34] follows the standard ISO 9241-210 and is based on: (1) user centered design methods to gather educators' tacit experiences in supporting learners emotionally in e-learning scenarios, and (2) data mining techniques to detect learners' emotional states during interactions with the system. Regarding data mining for emotions' detection, it follows a multimodal detection approach as described elsewhere [35, 36].

To apply TORMES, we will contact some educators with experience in social ubiquitous learning. In this way, we expect to elicit relevant SUNE scenarios where educational context-aware recommendation opportunities are identified and managed in terms of emotion-aware recommendation rules. The required infrastructure to deploy these scenarios can take advantage of current progress in the field (e.g., cloud support for the recommendation logic that automatically collects contextual information from sensors). To cope with interoperability, a semantic modeling of the recommendations elicited is to be done [37], which considers available emotional focused specifications, such as W3C Emotion ML [38].

Contextual features such as sound, light, environment temperature and so on that can also be managed with Arduino through corresponding sensors and actuators will be specially taken into account as they can potentially enrich the way recommendations are delivered. To facilitate the elicitation process regarding the recommendation possibilities offered to support ubiquitous learning, we are considering the use of the board

Arduino Esplora,[2] which already provides a number of built-in, ready-to-use set of on board sensors and actuators for interaction. In particular, it has onboard sound and light outputs, and several input sensors, including a temperature sensor, an accelerometer, a microphone, and a light sensor. For testing other sensors and actuators, the Groove system[3] (which plugs into an Arduino) might also be an interesting option.

5 Discussion

Human factors such as users' affective states matter during the recommendation process in SUNE, for instance when rating a resource to feed content-based or collaborative filtering recommendation approaches. Therefore, there is a need to automatically gather this information with a low cost infrastructure so educational context-aware recommender systems can take that information into account during the recommendation process. However, although this is a relevant feature for building educational context-aware recommender systems, the state of the art compiled in Table 1 shows that existing recommenders do not have that capability. It is also relevant to note that during 2013 several works have been published showing how Arduino open source electronics prototyping platform was successfully used to measure physiological signals (although with different goals than those pursued by educational context-aware recommender systems, mainly dealing with medical and health related projects[4]). Therefore, this recent trend can also be followed for affective states detection with Arduino based approaches.

However, several open issues exist. For instance, although sensors have been shown to accurately measure physiological aspects, the translation from physiological aspects to affective states is all but straightforward. Thus, it has to be further researched how accurate are these sensors in capturing affective states and how accurate do they need to be to fit the recommendation delivery process. Moreover, even if these sensors are very accurate, it has to be investigated how much can these measurements contribute to the recommendation process and check if recommendations delivered get perceptibly better by including affect in the prediction process. Finally, even if the predictions get better by including affective states, it is also needed to evaluate if this has an impact on actual learning and these potentially better recommendations indeed translate into observably improved learning outcomes. In fact, evaluating adaptive systems (such as recommenders) is very challenging, even more if they are deployed in educational scenarios. This situation is also reflected in our analysis of the state of the arte of educational context-aware recommender systems, as only 1 system from the 11 analyzed has attempted to verify an effect on the learning effectiveness. This requires developing an appropriate user-centric evaluation framework for educational recommender systems in SUNE, which should be informed by available approaches to evaluate generic recommender systems [39, 40] as they can help to determine the

---

[2] http://arduino.cc/en/Guide/ArduinoEsplora

[3] http://www.seeedstudio.com/wiki/GROVE_System

[4] http://medicarduino.net/

behavioral and attitudinal improvements, as well as the potential mediating variables that could explain these effects.

Another issue that requires further analysis deals with the implementation, that is, how the different technological parts involved might work in conjunction to cope with the scenarios obtained with TORMES. In particular, TORMES application will serve to show the kind of learning scenarios (formal, non-formal, informal) that are best supported, the sensors that are to be used to capture the affective state, how the recommendations are to be provided through appropriate actuators and how they should look like, as well as the most appropriate context for Arduino based affect detection. Privacy issues are also to be considered.

For a real world deployment, wearable electronics (which are included as part of the users' cloths or complements such as watches, bracelets or glasses) are to be used. In this sense, there is already a wearable version of Arduino microcontroller called Lylipad.[5] LilyPad is a set of sewable electronic pieces designed to build soft interactive textiles and it can also sense information about the environment and act on it.

6   Conclusions and Future Work

In this paper we have explored the potential of Arduino for producing contextualized recommendations in e-learning SUNE and discussed if an Arduino prototyping approach can help to identify the human factors involved in the recommendation process, both in the input (i.e., detecting contextual information that reflects learners' affective states) and the output (i.e., delivering the affective recommendation) within the context in which the learner is placed.

Addressing affective issues is of relevance for e-learning SUNE as emotional intelligence is "a type of social intelligence that involves the ability to monitor one's own and others' emotions, to discriminate among them, and to use the information to guide one's thinking and actions" [41]. Additionally, social-affective information seems to be of value for promoting the communication and collaboration among learners [17].

Acknowledgements. Authors would like to thank Spanish Ministry of Economy and Competence for funding MAMIPEC project (TIN2011-29221-C03-01).

References

1. Drachsler, H., Verbert, K., Santos, O.C., Manouselis, N.: Panorama of Recommender Systems to Support Learning, 2nd edition (under review)
2. Verbert, K., Manouselis, N., Xavier, O., Wolpers, M., Drachsler, H., Bosnic, I., Duval, E.: Context-aware recommender systems for learning: a survey and future challenges. IEEE Trans. Learn. Technol. **5**(4), 318–335 (2012)

---

[5] http://lilypadarduino.org

3. Yang, S.J.H., Okamoto, T., Tseng, S.-S.: Context-aware an d ubiquitous learning (guest editorial). Educ. Technol. Soc. **11**(2), 1–2 (2008)
4. Chu, H.C., Hwang, G.J., Huang, S.X., Wu, T.T.: A knowledge engineering approach to developing e-libraries for mobile learning. Electron. Libr. **26**(3), 303–317 (2008)
5. El-Bishouty, M.M., Ogata, H., Yano, Y.: PERKAM: personalized knowledge awareness map for computer supported ubiquitous learning. Educ. Technol. Soc. **10**(3), 122–134 (2007)
6. Hwang, G.J.: Paradigm shifts in e-learning: from web-based learning to context-aware ubiquitous learning. In: Huang, R., Spector, J.M. (eds.) Reshaping Learning. New Frontiers of Educational Research, pp. 253–271. Springer, Heidelberg (2013)
7. Blanchard, E.G., Volfson, B., Hong, Y.J., Lajoie, S.P.: Affective artificial intelligence in education: from detection to adaptation. AIED **2009**, 81–88 (2009)
8. Cooper, D.G., Arroyo, I., Woolf, B.P., Muldner, K., Burleson, W., Christopherson, R.: Sensors model student self concept in the classroom. In: Houben, G.-J., McCalla, G., Pianesi, F., Zancanaro, M. (eds.) UMAP 2009. LNCS, vol. 5535, pp. 30–41. Springer, Heidelberg (2009)
9. Manouselis, N., Drachsler, H., Verbert, K., Duval, E.: Recommender Systems for Learning. Springer Briefs in Electrical and Computer Engineering. Springer, New York (2013)
10. Yau, J., Joy, M.: A Context-aware and adaptive learning schedule framework for supporting learners' daily routines. In: Second International Conference on Systems, (INCOS, 2007), pp. 31–36 (2007)
11. Gonzalez, G., De la Rosa, J.L., Montaner, M., Delfin, S.: Embedding emotional context in recommender systems. In: IEEE 23rd International Conference on Data Engineering Workshop, pp. 845–852 (2007)
12. Shen, L., Wang, M., Shen, R.: Affective e-learning: using emotional data to improve learning in pervasive learning environment. Educ. Technol. Soc. **12**(2), 176–189 (2009)
13. Luo, J., Dong, F., Cao, J., Song, A.: A context-aware personalized resource recommendation for pervasive learning. Cluster Comput. **13**(2), 213–239 (2010)
14. Yu, Z., Zhou, X., Shu, L.: Towards a semantic infrastructure for context-aware e-learning. Multimedia Tools Appl. **47**(1), 71–86 (2010)
15. Wang, S.L., Wu, C.Y.: Application of context-aware and personalized recommendation to implement an adaptive ubiquitous learning system. Expert Syst. Appl. **38**(9), 10831–10838 (2011)
16. Li, M., Ogata, H., Hou, B., Uosaki, N., Mouri, K.: Context-aware and personalization method in ubiquitous learning log system. Educ. Technol. Soc. **16**(3), 362–373 (2013)
17. Boff, E., Reategui, E.: Mining social-affective data to recommend student tutors. In: Pavón, J., Duque-Méndez, N.D., Fuentes-Fernández, R. (eds.) IBERAMIA 2012. LNCS, vol. 7637, pp. 672–681. Springer, Heidelberg (2012)
18. D'mello, S., Graesser, A.: AutoTutor and affective autotutor: learning by talking with cognitively and emotionally intelligent computers that talk back. ACM Trans. Interact. Intell. Syst. **2**(4), 23:2–23:39 (2013). (Article 23)
19. Kaklauskas, A., Zavadskas, E.K., Seniut, M., Stankevic, V., Raistenskis, J., Simkevičius, C., Stankevic, T., Matuliauskaite, A., Bartkiene, L., Zemeckyte, L., Paliskiene, R., Cerkauskiene, R., Gribniak, V.: Recommender system to analyze student's academic performance. Expert Syst. Appl. **40**(15), 6150–6165 (2013)
20. Leony, D., Gélvez, H.A.P., Merino, P.J.M., Pardo, A., Kloos, C.D.: A generic architecture for emotion-based recommender systems in cloud learning environments. J. Univers. Comput. Sci. **19**(14), 2075–2092 (2013)
21. Banzi, M.: Getting Started with Arduino. O'Reilly Media, Sebastopol (2009)

22. Garcia-Perate, G., Dalton, N., Conroy-Dalton, R., Wilson, D.: Ambient recommendations in the pop-up shop. In: Proceedings of the 2013 ACM International Joint Conference on Pervasive and Ubiquitous Computing (UbiComp '13), pp. 773–776 (2013)
23. Guerreiro, J.A.: Biosignal embedded system for physiological computing. Instituto Superior de Engenharia de Lisboa. Master Thesis (2013)
24. Kishore, P., Saraf, S.S., Onkari, S.M.: Human activities – their classification, recognition and ensemble of classifiers. Int. J. Comput. Appl. **76**(14), 6–11 (2013)
25. Koga, K., Nakayamal, I., Kobayashi, J.: Portable biological signal measurement system for biofeedback and experiment for functional assessment. In: 13th International Conference on Control, Automation and Systems (ICCAS 2013), pp. 412–416 (2013)
26. Lotlikar, S., Dolas, K., Rane, A., Paradkar, D.: Smart phone based e-health. In: International Conference on Computer Science and Engineering (CSE), pp. 12–15 (2013)
27. Lung, C., Oniga, S., Buchman, A., Tisan, S.: Wireless data acquisition system for IoT applications. Carpathian J. Electron. Comput. Eng. **6**(1), 64–67 (2013)
28. Mansor, H., Helmy, M., Shukor, A., Meskam, S.S., Rusli, N.Q.A.M., Zamery, N.S.: Body temperature measurement for remote health monitoring system. In: Proceedings of the IEEE International Conference on Smart Instrumentation, Measurement and Applications (ICSIMA), pp. 1–5 (2013)
29. Orha, I., Oniga, S.: Automated system for evaluating health status. In: IEEE 19th International Symposium for Design and Technology in Electronic Packaging (SIITME), pp. 219–222 (2013)
30. Sinha, A., Pavithra, M., Sutharshan, K.R., Subashini, M.: A MATLAB based on-line polygraph test using galvanic skin resistance and heart. Aust. J. Basic Appl. Sci. **7**(11), 153–157 (2013)
31. Rahim, S.N.Z.B.: Stress detector. University of Technology of Malaysia, Master Thesis (2013)
32. Roy, J.K., Deb, B., Chakraborty, D., Mahanta, S., Banik, N.: The wearable eletronic rescue system for home alone elderly- labview & arduino evaluation. IOSR J. Electron. Commun. Eng. **8**(6), 50–55 (2013)
33. Novak, D., Mihelj, M., Munih, M.: A survey of methods for data fusion and system adaptation using autonomic nervous system responses in physiological computing. Interact. Comput. **24**, 154–172 (2012)
34. Santos, O.C., Saneiro, M., Salmeron-Majadas, S., Boticario, J.G.: A methodological approach to eliciting affective educational recommendations. In: Proceedings of the 14th IEEE International Conference on Advanced Learning Technologies (ICALT '14), pp. 529–533 (2014). doi:10.1109/ICALT.2014.234
35. Saneiro, M., Santos, O.C., Salmeron-Majadas, S., Boticario, J.G.: Towards emotion detection from facial expressions and body movements to enrich multimodal approaches. Recent Adv. Inf. Technol. **2014**, 14 (2014). (Article ID 484873)
36. Salmeron-Majadas, S., Santos, O.C., Boticario, J.G.: Affective state detection in educational systems through mining multimodal data sources. In: D'Mello, S.K., Calvo, R.A., Olney, A. (eds.) 6th International Conference on Educational Data Mining. pp. 348–349. International Educational Data Mining Society, Memphis (2013)
37. Santos, O.C., Boticario, J.G., Manjarres, A.: An approach for an affective educational recommendation model. In: Manouselis, N., Drachsler, H., Verbert, K., Santos, O.C. (eds.) Recommender Systems for Technology Enhanced Learning: Research Trends & Applications, pp. 123–143. Springer, New York (2014)
38. Schröder, M. (ed.): Emotion Markup Language (EmotionML) 1.0, W3C Candidate Recommendation, Cambridge, Mass, USA (2012)

39. Pu, P., Chen, L., Hu, R.: Evaluating recommender systems from the user's perspective: survey of the state of the art. User Model. User-Adap. Inter. **22**, 317–355 (2012)
40. Knijnenburg, B.P., Willemsen, M.C., Gantner, Z., Soncu, H., Newell, C.: Explaining the user experience of recommender systems. User Model User-Adapt. Inter. **22**, 441–504 (2012)
41. Mayer, J.D., Salovey, P.: The intelligence of emotional intelligence. Intelligence **17**, 433–442 (1993)

The Ethics of a Recommendation System

Pinata Winoto[(✉)] and Tiffany Tang

Department of Computer Science, Kean University, Wenzhou, China
{pwinoto,yatang}@kean.edu

Abstract. In this paper, we extend the current research in the recommendation system community by showing that users' did attach ethical utility to items. In an experiment (N = 111) that manipulated several moral factors regarding the potentially harmful contents in movies, books and games, users were asked to evaluate the appropriateness of recommending these items to teenagers and dating couples (adults). Results confirmed with previous studies that gender plays a key role in making moral judgment especially regarding the ethical appropriateness of an item. The study further identifies degrees of aversion regarding the appeal of these elements in media for ethical recommendations. Based on the study, we propose a user-initiated ethical recommender system to help users pick up morally appropriate items during the post-recommendation process. We believe that the ethical appropriateness of items perceived by end users could predict the trust and credibility of the system.

Keywords: Recommendation system · Moral factor · Empirical study · Gender

1 Introduction

A household of 4 including parents Alex and Mary, 15 years old Chloe, and 12 years old John have subscribed an online movie rental service, for example Netflix, for over 2 years. Occasionally, the family receives recommendations from its highly successful and profitable personalized recommendation service (known as *Cinematch* in Netflix) based on the family's rental histories. Alex and Mary both enjoy war and action movies the most (for example, they both gave high ratings to movies such as *Schinlder's List*, the *Bourne* franchises); therefore two movies, *The Kite Runner* and *Mission Impossible 4: Ghost Protocol* are among the recommended items. However, both these movies should not be recommended without warnings to this account, as they are not appropriate for both John and Chloe (two children in this household). From the system's perspective, both movies will be favored by this user (a collective account): they are algorithmically but not ethically appropriate. (Both movies are listed in the IMDB as 'PG-13'; the first contains child rape which is especially not appropriate for young children). So, should the recommender system (RS) make the suggestions or not? This question highlights the ethics of the RS which has largely left ignored in the RS research community.

An RS has become an integral part of many modern applications. It is capable of providing users with a list of recommended items they might like, or suggest to users an ordered list of items based on the system's guess on how much users might prefer

© Springer International Publishing Switzerland 2014
Y. Chen et al. (Eds.): WAIM 2014, LNCS 8597, pp. 287–298, 2014.
DOI: 10.1007/978-3-319-11538-2_26

the items. Therefore, a RS can facilitate users' decision-making especially when choosing from a large collection of items is beyond users' cognitive capabilities [1]. A wide variety of recommendation algorithms has been studied extensively in the research community [2, 3].

The opening exemplifies the specific issue we intend to address in this paper: should a RS be made ethically? As Rodriguez and Watkins [4] put it "humans seek books and movies to stimulate their cognitive faculties,…" Therefore, technological artifacts can influence/alter human moral behavior [5]. This is especially true in applications with persuasive technology [5], such as that the RS directly or indirectly persuades users to consume certain items and/or insidiously built user interfaces (UIs) that are meant to influence users' consumption attitudes or behaviors (persuasive user interface). Moreover, personal moral values have been linked to media enjoyment [6]. The way how moral values and media influence each other is said to be reciprocal: in the long-term, users' moral values might be shifted after repeated exposure to media use. While in the short-term, the conflicts between their moral values and those being communicated to them tend to affect users' appraisal of the media (in our example, the ratings users give to the movie) [7]. Hence, it is imperative to employ an ethical mediator to influence media users, which motivates our study. In particular, we propose a framework for a two-layered ethical recommender system (eRS) to generate ethically appropriate items based on content analysis of candidate items (including books, movies and games) and a set of ethical rules. The ethical rules would be similar to those proposed in [8], but ours are not religion-related. As an initial probe into the role of ethics in making recommendations, we conducted a user study to investigate whether or not users differ on ethical recommendations, especially if the target users are children or adults. We identify a set of potentially harmful elements in movies, books and games, and ask users' responses during the recommendation process. Results confirmed with previous studies in that gender plays a key role in making moral recommendation. The study further identifies degrees of aversion regarding the appeal of these elements in media for ethical recommendations.

The structure of this paper is as follows. In Sect. 2, previous works will be presented. We will then outline the general architecture of an ethical recommender system in Sect. 3. A user study where we attempt to obtain how users are looking at the ethical and non-ethical aspects of a RS appears in Sect. 4. Lessons learned from our user study will be presented in Sect. 5.

2 Related Work

2.1 Morality and Information Technology

In the business ethics research community, Kohlberg's [9] cognitive moral development (CMD) theory has been widely adopted. The theory focuses primarily on the cognitive process of making moral judgments. It is stated that morality is developmental and moral rules can be developed gradually during the growth of each individual and his/her interactions with the social environment, including computer software [11]. Moreover, Veebeek [5] indicates that technology should be designed to mediate users' moral behaviors so as to force them to reflect on their moral actions. In the context of a RS, the user interacts with the system through the user interface: providing ratings/comments to the movies/music pieces/books the system suggested, browsing and exploring each item of the recommended list in order to make rental/ purchase decisions. The users' exposure to these items will also shape their actions and reflections toward the world. These two processes are said to standing at the intersection of how the RS acts as a mediating agent between the user and the world [4]. And the de-abstractions of moral values/rules should be understood and translated in the context where the algorithm(s) are being used [5]. As such, the recommendation algorithms are said to have moral implications [10], a view held by another body of research called value sensitive design (VSD) and values in design (VID) [11]. They pointed out that computer technology reflects the values of its designers; such values include human well-being, human dignity, justice, welfare, privacy, and human rights. Due to the varied nature of the domain in which technological artifacts would reside, how such values play out differ and change considerably over time [11]. For example, an algorithm can be sophisticated and intelligent enough to pin-point a specific person from the huge amounts of public data, which in turn calls for significant improvements of algorithms to secure user privacy.

In the mass communications and psychology community, the media effect refers to the many ways individuals may be affected by both news and entertainment media - including movies, television, books, magazines, newspapers, websites, video games, and music [12]. Media effects function through a two-way process: the medium shapes the users who, in turn, influence the originator(s) (which can be a web service, the software system, or the movie/music/book itself etc.). The two-way process, from another perspective is a means for technology be moralized to help shape users' moral values.

2.2   Media Effect on Human Behavior and Moral Judgment

Previous empirical evidences have consistently suggested that watching violent television/movies scenes and playing violent video games may lead to real-life aggressive behaviors and attitudes such as desensitization to violence, to name a few [13–15]. Longitudinal studies further revealed that watching violent television/movies potentially induces stronger aggressive behaviors, especially for children at a later age.

Younger children exposed to profanity shows an increase in real-world profanity use, aggressive attitudes and behaviors [17]. Some researchers regard profanity as verbal aggression which is different from physical aggression in that verbal aggression may be more easily imitated than physical aggression and repeated exposure to the former can lead to the latter [17]. The portrayal of sexual behavior affects children and adults negatively. It is especially damaging to children since early exposure have a major impact on their life, and could shape their values and attitudes when they grow up [18]. Exposure to pornography also negatively influences family values.

A number of studies have demonstrated a gender difference regarding the negative effects of the media. For example, [15] uncovered stronger correlations between exposure to violence and aggression for boys than for girls. One recent study, however, obtained mixed results: a significant increase in aggression was identified by only one of the measures [14]. As for the age differences regarding the negative effects of profanity use, results are mixed. Coyne et al. [17] claimed that among younger children, swearing seems to be more normal; while adult males are more likely to become verbally aggressive (use more profanity).

Recently, personal moral values have been linked to media enjoyment [6]. According to Zillmann's *Moral Sanction Theory*, during media consumption users tend to scrutinize the moral values reflected in character intentions, behaviors and outcomes based on their own personal moral values [6], and if there are conflicts between the two, users will further assess both and be influenced morally; meanwhile, the extent to which they are willing to adjust the moral conflicts reflect their affection towards the characters and the media. This process is said to be reciprocal: In the long-term, users' personal values might be shifted after repeated media exposure; while in the short-term, the conflicts between their personal moral values and those being communicated to them tend to affect users' appraisal of the media (in our example, the ratings users give to the movie), as argued in [7]). The process is in line with research on media effect by Bryant and Thompson [12] and more broadly, with the way how technology could act as mediator between users and their moral actions [5].

It is evident that an individual's moral value is one of the predictors to media appraisal. Meanwhile, due to the impact of media use on an individual's moral development, especially young immature users - it is imperative to employ an ethical mediator to influence media users (Fig. 1).

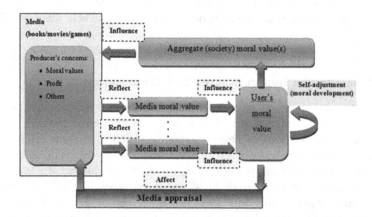

Fig. 1. The reciprocal affect among media use, moral judgment and media appraisal

2.3   The Ethics of a Recommender System: A Comparison of Previous Works and Ours

In the RS community, to our knowledge, there are only two similar studies. Reference [4] first proposed that a computational system should enable individuals to fulfill their moral responsibilities. They proposed a framework for an ethical recommender system

(Eudaemonic Systems) which is built upon a large-scale network of data and resources that are pertinent to reflect the moral value of an individual. The linked web of data and resources (called Linked Data Cloud) extracts information from such publicly available data sets as "Amazon.com's RDF book mashup, MusicBrainz.org's metadata archive, and the Internet Movie Database's (IMDB) collection of movie facts etc. Although these data, when strategically put, can help inform the recommendation mechanism during the process, it is extremely challenging to extract any personalized information to identify a user across these different web service providers and therefore establish a link between these resources and the user in an effort to build the user's profile for personalized recommendations. Moreover, because the service providers have been adopting different mechanisms for their metadata, unless there is a unified decoding framework on top of the Cloud, the data obtained are scattered, and unidentifiable. As such, the ethical RS can only make non-personalized suggestions since not enough information can be extracted to build a user profile [4] and the computational cost remains to be one of the most challenging issues.

Souali et al. [8] proposed a framework on which an automatic ethical-based RS is built. In addition to the core components in a typical RS, the ethical RS holds an ethical database which acts as a pre-recommendation filter to categorize ethically appropriate candidate items. These data include country-sensitive habits and customs, which indicate that the notion of ethics is more culture-based. Thus, to a certain degree, the main role of their ethical RS is not to mediate users' moral behaviors; instead, to provide recommendations before building a complete user profile and to facilitate that process [8].

Recognizing the impact of media use on an individual's moral development, especially young immature users, it is imperative to employ an ethical mediator to influence media users, which motivates our study.

3   The Architecture of the Two-Layered Ethical Recommender System (eRS)

The first layer is capable of matching a target user's preferences against an item database and other users with similar interests (neighbors of the target user) and making suggestions; the second layer is an ethical filter picking up ethically appropriate items based on a given set of ethical rules and content analysis of candidate items. We agreed with [4] that most of the typical RSs have limited performance in that the recommendations come out from one domain only (e.g., movie or music or book). Hence our proposed RS is built upon the *user ratings × item* matrix where candidate items come from three domains: movie, music and books. This type of RS (multi-domain RS) yields more satisfactory performances than the typical RS [19].

The ethical filter shall be imposed in the post-recommendation process implemented through a user interface where end users (parents) choose to enforce the control level limiting movie presentations. Figure 2 visualizes the process of the filter which consists of an ethical analyzer (algorithmic part) and an ethical rule database.

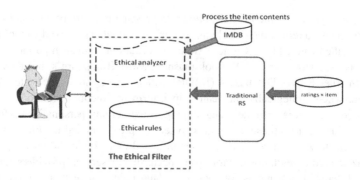

Fig. 2. The ethical filter implemented in the post-recommendation process

The ethical analyzer is capable of extracting and examining movie contents to determine the ethical appropriateness of a given movie. The contents relating to the ethical considerations of a movie can be found under the section 'Motion Picture Rating MPAA' on the IMDB site. For example, the ethics-related contents associated with the movie *The Kite Runner* is '*strong thematic material including the rape of a child*'. Due to the sensitivity of this, the *e*RS should suggest this movie to the Household with great care.

As an initial probe into the role of ethics in making recommendations, we conducted a user study to investigate whether or not users differ on ethical recommendations, especially if the target users are children or adults. We identify a set of potentially harmful elements in movies, books and games, and ask users' responses during the recommendation process. In the next section, we will report the study and provide discussions.

4   The User Study: Analysis and Discussions

Similar to other studies in the field [20–22], we administrated a questionnaire on a student population to obtain their views on the ethical appropriateness of making recommendations on books, movies and games containing various moral factors such as violence, profanity, nudity/sexual contents, and drug/alcohol abuse.

4.1   Study Goals and Data Collection

The present study aims at examining the following three issues:

- whether or not subjects' attitudes differ when the target audiences/users are children and adults, if yes, how far the differences would be;
- their degree of aversion regarding the potentially harmful factors in media when making ethical recommendations, such as which factor would be considered as the most important one; and
- the role of gender, religion, and other subjects' characteristics on ethical recommendations.

The first goal is mainly for the data validation purpose. The second and third issues substantiate the first in an attempt to associate the extent to which subjects make ethical recommendations when considering the potentially harmful elements in media use. A short questionnaire is distributed to each undergraduate student enrolled in five classes where the authors are the instructors. All participants filled the questionnaire voluntarily in class. All questions and instructions are written in Korean (their native language). Figure 3 partially shows the questions. Students are required not to put their name or identity on the questionnaire (for anonymity).

Q1 to Q9 aim at obtaining basic information on recommenders ranging from their age to religion, while Q10 to Q13 instruct subjects to specify the degree of willingness they will make recommendations on items containing potentially harmful elements (moral factors) in three domains: books, movies and games. Specifically, subjects were asked to indicate their recommendation attitude using a 7-point Liker-type scale on each question (Q10-1 to Q13-5): a higher rating indicates a stronger reluctance to recommend items containing the corresponding moral factor. The 7-point scale has also been used in many previous studies such as [17, 20, 21].

Q11 Specify which type of movies do you think are not good for young dating couples (18 – 25 years old)?	Q11-1 Movies showing many violent scenes (fighting among people)
	Q11-2 Movies contain coarse language (yelling, cursing, or other bad words)
	Q11-3 Movies with many nudity/sexual content
	Q11-4 Movies showing drug/alcohol abuse
	Q11-5 Movies showing domestic violence (fighting between husband & wife)
Q12 Which types of computer or console games (Xbox, PS3, Wii, or others) do you think are not good for teenagers (12 – 17 years old)?	Q12-1 Games with violence among people (fighting/war games)
	Q12-2 Games with violence of fighting monsters
	Q12-3 Game with horror/thriller content
	Q12-4 Games with many nudity/sexual content
	Q12-5 Betting games with real money (online casino games, such as poker, blackjack, roulette, etc.)
	Q12-6 Betting games without real money
	Q12-7 Games contain coarse language (yelling, cursing, or using bad words)

Fig. 3. Partial questions used for data collection

4.2 Key Findings

Since some subjects may not fill the questionnaire truthfully, we performed data cleaning before data analysis, removing 14 of the original 125 questionnaires. We assumed that if a subject filled the questionnaire seriously, some variations should be shown on her/his answers. To further check the validity of subject feedback, we calculate the Cronbach's α on the ratings given to Q10 to Q13. The resulting Standardized Cronbach's α using Spearman correlation is 0.898, and is 0.845 when using Kendall correlation,[1] or both are higher than 0.7 (the minimum value suggested in social/psychological study), hence they are acceptable for further analysis.

---

[1] Spearman/Kendall correlation is a rank-order version of the Pearson correlation coefficient; it is chosen due to the nature of our data—ordinal [23].

We obtained genera statistics among 111 feedbacks: female/male ratio is 23/88; 90 subjects are older than 21 years, 21 are between 18 and 21 (inclusive), and none younger than 18; 64 subjects identified themselves without a religion, 20 are Buddhist, 24 are Catholic/Christian, and 3 are others; 82 have driver license while 29 do not; 13 subjects have teenager siblings while 98 do not; 36 subjects smoke and/or drink alcohol frequently, while 75 do not.

How the Differences on the Target Audiences Affecting the Ethical Recommendations. We first examine whether or not subjects' attitudes over each moral factor differ when the target audiences are children or adults. Since our data used rank-order measurement instead of numeric [23], we applied the rank-order version of the t-test called Wilcoxon Signed Rank (pair) test on the ratings subjects given to Q10 to Q13. The basic statistic for the test is the sum rank (W) which picks the minimum of the sum of the positive rank (W^+) and the sum of the negative rank (W^-). Figure 4 shows the total sum ranks ($S = W^+ + W^-$) after comparing ratings given to Q10 ("Movies to teenagers") and Q12 ("Movies to young dating couples (adults)"). A higher rank means a higher disapproval. And a positive value S for a given moral factor means it is ranked higher in data 1 (Q10) compared to that in data 2 (Q11), or it is more disapproved for teenagers. The table also shows the corresponding Z-value and the measure of effect size r.

		Q11 (Data 2: Movies for dating couples)		
	Moral factors	S	Z	r
Q10 (Data 1: Movies for teenagers)	violent scenes (vio.)	2343***	−5.319	−.505
	coarse language (c.lang.)	2606***	−5.811	−.552
	nudity/sexual content (nu./sex)	3523***	−7.455	−.708
	drug/alcohol abuse (drug/alc.)	2627***	−5.963	−.566
	domestic violence (dom./v)	1290***	−4.029	−.383

*$p \le 0.05$, **$p \le 0.01$, ***$p \le 0.001$

Fig. 4. The effects of ethnical recommendations on target audiences

In general, we observe that subjects attached a higher degree of responsibility when recommending movies to teenagers than to more mature audiences (young couples), in which the strongest difference is for movies containing nudity/sexual content (with the highest S = 3523, Z = −7.455, r = −.708), the second strongest difference is for movies containing drug/alcohol abuse (S = 2627, Z = −5.963, r = −.566). The results are statistically significant with p < 0.001 (the corresponding Z and r value are shown next to S respectively.

Degree of Aversion Over Various Moral Factors. If we compare the ethical recommendation within a context, say whether or not a certain movie is suitable to suggest to a teenager (Q10), we should obtain subjects' degree of aversion over various moral factors that affect their ethical recommendation for teenagers. For example, whether the factor "Nudity/sexual content" should be considered more seriously than another factor

"Coarse language" in their decision-making. Figure 5 shows the values of S after pair-wise comparison between two factors, with negative values highlighted in shading. Note that a higher rating indicates stronger reluctance on recommending the items. As shown here, "Pet/animal abuse" (the last row) is rated higher than other moral factors, that is, the values of S in this row are all positive or the number of positive signs, or Σ (+) = 5, is the largest with three S are statistically significant (Σ (*) = 3). Here, we define Σ (*) as the sum of significant indicators, in which +1 is assigned to a positive significant S and −1 to a negative significant, as shown in Fig. 5.

| | Factors | S (after comparison to other factors in Q10 (Data 2)) | | | | | | Σ(+) | Σ(*) |
		Vio.	C/lang.	Nu./sex	Drug/alc.	Dom./v	Pet/a		
	Vio.	–	−881***	−1170***	−1340***	−1309***	−1805***	0	−5
	C/lang.	881***	–	−133	−382	−453	−870**	1	0
Data 1	Nu./sex	1170***	133	–	−250	−386	−779**	2	0
(Q10)	Drug/alc.	1340***	382	250	–	−79	−467	3	1
	Dom./v	1309***	453	386	79	–	−328	4	1
	Pet/a	1805***	870**	779**	467	328	–	5	3

*$p \leq 0.05$, **$p \leq 0.01$, ***$p \leq 0.001$

Fig. 5. The comparison of moral factors on movie recommendation to teenagers

Results indicate that "Domestic violence" is the most morally worrisome factor to be considered for dating young couples. Later in the next sub section, we will show that this is gender-dependent. In the games category, "Betting with real money" is the most morally worrisome factor and "Horror/ thriller" and "Fighting monsters" are the least. In the domain of books, "Good guy eventually win" books are the least prohibitive categories, aligning with our intuition.

Gender Difference on Ethical Recommendation. Figure 6 visualizes the test results which painted a more complex and mixed picture of violent effects on the ethical recommendation by gender difference.

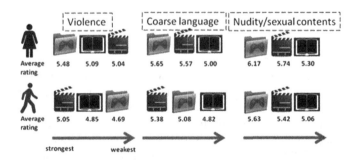

Fig. 6. Order of aversion for ethical recommendation among the three domains

Interestingly, while both Male and Female groups agree on the ordering of Books, they disagree whether Movies or Games should be put first in "Violence" and in "Nudity/Sexual content". Female subjects showed a greater sensitivity to violence in games than in movies, partially because they were far less exposed to games than males [14–16], which meant they were less desensitized to violence. Male subjects, being the largest consumer group in the video game industry, are less sensitive to violence in games. The results are supported by previous empirical studies suggesting males are more aggressive in general than females [14], and the grueling evidence that repeated and excessive exposure to violent video games can cause players to become less sensitive to real-life violence (desensitization) [16].

Not surprisingly, results indicate that subjects were the least prohibitive when making recommendations on books, possibly due to the non-interactive depiction of violence and sexual contents in books. In fact, [24] argued that harmful elements portrayed in a fictionalized fashion such as books posed less damage to children than those in a more interactive fashion such as movies and those involving users to assume an active role such as games [13]. These seem to be consistent with our finding that female subjects are extremely cautious when making movie and game recommendations.

5 General Discussion and Concluding Remarks

Our findings revealed that the moral factors associated with candidate items are predictive to subjects' ethical recommendation. In this study, the moral factors refer to the potentially harmful elements in the candidate item; they vary on such features including presentation type (interactive and vivid vs. fictionalized), intensity, and how the user will consume the item (active playing vs. passive watching). Behavioral scientists and psychologists have linked these features to the degree of negative effects and physiological arousal [13, 16, 17, 24]. Our present study is consistent with these findings. For example, subjects relaxed their ethical considerations when making recommendation on books; female subjects set the highest alert when recommending games depicting nudity and sexual content compared to books and movies; regarding "coarse language" use in books, movies, and games, subjects demonstrated the greatest reluctance on movies due to the accompanied affective display that may amplify verbal abuse, and showed the least reluctance on books.

The ethical analyzer is capable of extracting and examining movie contents to determine the ethical appropriateness of a given movie. It can be regarded as the technological artifact that mediates users' moral behaviors and attitudes. The main reason of implementing the content analyzer instead of relying on the rating given by the corresponding industry standard is due to the overwhelming criticism over the lack of controls in the entertainment industry. In particular, [18] have raised serious concerns on the unreliability of the rating systems including the Motion Picture Association of America (MPAA) (on movies) and the Entertainment Software Rating Board (ESRB) (on computer and video games) and the inadequacy of warning systems regarding profanity use in games and TV programs [17].

For example, in IMDB *The Kite Runner* is rated PG-13 in US (not suitable for children under 13 years old without parents' supervision), but it is rated 18A in British Columbia (Canada), or it might be deemed as harmful to teenagers older than 13 as well. Therefore, we argue that the content analyzer should extract the information pieces and either process it algorithmically (technology intervention) or present them as compound features in the UI to allow parents to determine (user control). In the latter implementation, the RS becomes a critique-like one, also known as Conversational RS [2] which allows users to specify their preferences gradually through multiple interactions with the system. However, the cost of evaluating and comparing diverse choice alternatives cannot be ignored [1], which prevents these semi-automated conversational recommendation approaches from gaining widespread popularity in commercial systems [2]. We do not intend to further elaborate on this issue in this paper, since it is beyond the scope to be addressed in this study.

Regardless of approach, the ultimate goal is to make the RS be morally responsible for the potentially harmful consequences it might incur, that is, moralizing the RS.

From software engineering's perspective, the inclusion of such analyzer is subject to the explicit requests from the client, as suggested in the previous section. Therefore, we believe the ultimate responsibility rests on the hands of parents who should guide their children.

References

1. Shugan, S.M.: The cost of thinking. J. Consum. Res. **7**(2), 99–111 (1980)
2. Winoto, P., Tang, T.Y.: The role of user mood in movie recommendations. Expert Syst. Appl. **37**(8), 6086–6092 (2010)
3. Adomavicius, G., Mobasher, B., Ricci, F., Tuzhilin, A.: Context-aware recommender systems. AI Mag. **32**(3), 67–80 (2011)
4. Rodriguez, M.A., Watkins, J.H.: Faith in the algorithm, part 2: computational eudaemonics. In: Velásquez, J.D., Ríos, S.A., Howlett, R.J., Jain, L.C. (eds.) KES 2009, Part II. LNCS, vol. 5712, pp. 813–820. Springer, Heidelberg (2009)
5. Verbeek, P.-P.: Moralizing Technology: Understanding and Designing the Morality of Things. University of Chicago Press, Chicago (2011)
6. Zillmann, D.: Basal morality in drama appreciation. In: Bondebjerg, I. (ed.) Moving Images, Culture, and the Mind. University of Luton, Luton (2000)
7. Tamborini, R.: Moral intuition and media entertainment. J. Media Psychol. **23**(1), 39–45 (2011)
8. Souali, K., Afia, A.E., Faizi, R.: An automatic ethical-based recommender system for e-commerce. In: Proceedings of the ICMCS'11, pp. 1–4 (2011)
9. Kohlberg, L.: Stage and sequence: the cognitive-developmental approach to socialization. In: Goslinfed, D.A. (ed.) Handbook of and Socialization Theory and Research. Rand McNally, Chicago (1969)
10. Nissenbaum, H.: How computer systems embody values. Computer **34**(3), 118–120 (2001)
11. Friedman, B., Kahn Jr., P.H.: Human values, ethics, and design. In: Sears, A., Jacko, J.A. (eds.) The Human–Computer Interaction Handbook: Fundamentals, Evolving Technologies, and Emerging Applications, 2nd edn, pp. 1241–1266. Lawrence Erlbaum Associates, New York (2008)

12. Bryant, J., Thompson, S.: Fundamentals of Media Effects, 1st edn. McGraw-Hill Higher Education, New York (2002)
13. Anderson, C., Bushman, B.: Effects of violent video games on aggressive behavior, aggressive cognition, aggressive affect, physiological arousal, and prosocial behavior: a meta-analytic review of the scientific literature. Psychol. Sci. **12**(5), 353–359 (2001)
14. Bartholow, B.D., Anderson, C.A.: Effects of violent video games on aggressive behavior: potential sex differences. J. Exp. Soc. Psychol. **38**, 283–290 (2002)
15. Huesmann, L.R., Taylor, L.D.: The role of media violence in violent behavior. Annu. Rev. Public Health **27**, 393–415 (2006)
16. Carnagey, N.L., Anderson, C.A., Bushman, B.J.: The effect of video game violence on physiological desensitization to real-life violence. J. Exp. Soc. Psychol. **43**, 489–496 (2007)
17. Coyne, S.M., Stockdale, L.A., Nelson, D.A., Fraser, A.: Profanity in media associated with attitudes and behavior regarding profanity use and aggression. Pediatrics **128**(5), 867–872 (2011)
18. Gentile, D.A., Maier, J.A., Hasson, M.R., de Bonetti, B.L.: Parents' evaluation of media ratings a decade after television ratings were introduced. Pediatrics **128**, 2010–3026 (2011)
19. Winoto, P., Tang, T.Y.: If you like the Devil Wears Prada the book, will you also enjoy the Devil Wears Prada the movie? A study of cross-domain recommendations. New Gener. Comput. **26**(3), 209–225 (2008)
20. Khazanchi, D.: Unethical behavior in information systems: the gender factor. J. Bus. Ethics **15**, 741–749 (1995)
21. Kreie, J., Cronan, T.: How men and women view ethics. Commun. ACM **41**(9), 70–76 (1998)
22. Magee, R.G., Kalyanaraman, S.: The perceived moral qualities of web sites: implications for persuasion processes in human–computer interaction. Ethics and Info. Tech. **12**, 109–125 (2010)
23. Motulsky, H.J.: Intuitive Biostatistics. Oxford University Press, Oxford (2010)
24. Atkin, C.: Effects of realistic TV violence vs. fictional violence on aggression. Journalism Quarterly **60**(4), 615–621 (1983)

The Negative Effects of 'Too Accurate' Recommendation Results: A Pilot Study of the Effects of Search Engine Overuse on Chinese Students

Lotus Xinhe Zhou[1(⊠)] and Tiffany Y. Tang[2]

[1] Department of English, Wenzhou Kean University, Wenzhou, China
zhouxin@kean.edu
[2] Department of Computer Science, Wenzhou Kean University,
Wenzhou, China
yatang@kean.edu

Abstract. Googling, sharing, recommending and assembling knowledge pieces have become ubiquitous in social learning environments. However, there are a number of well-documented research efforts indicating the disadvantages of over-relying on the internet especially during students' academic studies. While the majority of these research focuses on the overuse of social media, we are interested in the effects of the overuse of the search engine since googling information been synonym with problem-solving for students. And searched results have long been regarded as recommendation. In this paper, we describe our pilot study to examine the extended effects of search engine overuse as well as the perceived information quality. Mixed results were obtained, although most of the preliminary results are mostly consistent with our speculations in the over-use of search engine on students' academic studies.

1 Introduction

Googling, sharing, recommending and assembling knowledge pieces have become ubiquitous in social learning environments (among them, [1, 27]). In higher education and for scholars, googling information has been synonym with problem-solving and research (to name a few, [9, 11, 14, 23]). Hence, there has been an increasing interest on how to improve the research results which have been regarded as more group-wise recommendations [30, 33]. There are a number of well-documented research efforts arguing against the over-relying on the internet especially during students' academic studies, relatively few have been known for the various effects of search engine over-use by young Chinese students which have accounted for a relatively high percentage among young Chinese net-citizens [4, 6, 12]: while students become more expert in using the Internet to search things they don't know about, enjoying the seemingly efficiency and proficiency of the acquisition of whatever kind of information they get online; Sutherland-Smith [28] revealed that students tend to be '*shallow, random and often passive*' in the way they use the internet to search for information (p.781). Students affected by the efficiency of the Internet gradually change their study mode

© Springer International Publishing Switzerland 2014
Y. Chen et al. (Eds.): WAIM 2014, LNCS 8597, pp. 299–306, 2014.
DOI: 10.1007/978-3-319-11538-2_27

from using their mind to think to online-surfing [20]: googling becomes to be the synonymous with research, an essential step before students sharing information. Controversy about the Internet making people stupid have never been neglected by people for its crucial effects on the development of human beings' offspring. With the popularity of the Internet increasing overwhelmingly in China, and the widespread of search engines dominating the Internet, a great number of Chinese students acclimatize themselves to such cushy way of obtaining information.

Noticing this tendency and doing more research and analytical work are required to determine the effects of the over-relying of search engines on young educated Chinese students. Researches done to test the effects of browsers on students around the world, but few research was done targeted on Chinese students, and examine some of the issues related to the overuse of the Internet for pedagogical development, which motivates our research. In particular, in this paper, we describe our pilot study attempting to answer the questions like: in general, what kinds of effects can browsing too much have among educated students in China? How do they perceive internet information quality? Are there any hidden relationships between perceived information qualities with the extent of internet use?

The organization of this paper starts with the brief abstract and introduction of current web use situation among Chinese students, then continues with the summary of related works, introducing some of the scientific research conducted that have connections with our research work. Then we described the protocol of our study and experimental discussions. After that we propose the solutions based on the study result and provide suggestions. We concluded our paper in the end.

2 Related Work

While most studies focus on the human factors on information and knowledge sharing and recommendation, one area was largely left unknown; that is, the over-use of internet, especially search engine among students, as searched results have been perceived as recommended elements, though not too personalized due to the technological difficulties of following users' every internet footprints [30, 32, 33]. Many previous studies have established the negative effects of Internet use on students' academic performance and social performances (among them, [2, 12, 13, 15, 31]). For example, Anderson [2] revealed that over-use of Internet can not only affect students' academic work, but also their social skills. Too much Internet use can lead to Internet addiction [31], and Internet addiction has been very common among college students in Western countries and China [12, 13, 25]. Richtel [24] pointed out that too much time spent on online technology can lead to damaging of the ability of sustaining attention, but studies addressed upon what exactly are the effects of online browsers on the performance of Chinese students are sparse, whether Chinese educated students fit in all those descriptions still remain unknown, which motivates our study here.

Nicholas Carr [5] famously argues that the advantages of having immediate access to such an incredibly rich store of information are many but they come at a price. While providing us with fruitful information, search engines also tend to encourage skimming on the surface of knowledge and rather than dive to its depth. Moreover, the advent of the Internet makes brains with extraordinary memory become ordinary, and hinders the interpersonal intellectual communications among human brains [21, 22]. Carr [5] pointed out that only printed texts allow humans' brains to remember and interpret the visual and auditory stimuli. By using search engines too often, we gain the ability of searching information on an enormous database provided but lose the ability of dealing with complex information [19]. And Carr also argues that as people use "intellectual technologies" they inevitably begin to take on the qualities of them. Such a claim manifests that Chinese students browsing the Internet too hard can not only cause their brains to decrease thinking but transfer them into an intellectual robot, featuring all the mechanical characteristic a smart computer possesses. Furthermore, as Richtel [24] depicts people who are in a state of being constantly connected and distracted by the floating information online. The attractions from social networks, the feeling of being permanently connected to the outside world render some of the students who lack self-control fail to concentrate on the work they're undertaking, making them feel impossible to sustain their attention on what they're currently up to.

Sparrow et al. [26] argue that the Internet together with search engines provides easy access of information, thus rendering people to be more prone to seek help from computers based on their recall of the place where the information is stored instead of the information itself. Carroll [7] raised his concern that teachers currently have the difficulty to use dominant modal of technology efficiently in their classroom, in other words, teachers are still on the way seeking the optimum pedagogical method to enable students to learn fully within the technological environment, echoed in [19] as well. Thus, at the present stage, teachers may adopt teaching approaches that do not correspond with the needs of students. For example, current resources available for teachers to teach students knowledge with is online search engines, whose creditability is not so ideal because of teachers' personal preference, free editing style of online encyclopedia and the influence of different interest group [17]. Students may get misled by using search engines too much, their opinions can easily reshaped by some tendentious opinions supported by a particular group. Therefore we can infer that the use of search engines as an immature pedagogy can influence students in a negative way.

In the next section, we will present our pilot study as well as discuss some preliminary results.

3 Our Pilot Study and Preliminary Results-Up

3.1 Study Setup

A questionnaire was distributed to undergraduate students in the spring of 2014. A total of 20 questions were administrated. The questions were designed to obtain general information regarding their internet use, with a focus on the use of search engine in their high-school and university life; user perception over information quality was also obtained. The administration of questionnaires is common in the research community ranging from behavioral economics, social science, education to human computer interaction (to name a few, [3, 10, 16, 18, 29]).

Subjects and study procedure

Fifty undergraduate students aged averagely twenty participated in the research. Forty-five valid answers were obtained. Participants were required to complete a brief questionnaire confirming their Internet experiences, the difference of their Internet using between high school and university, and then access the avail (velocity and effectiveness) of search engines on a five-point scales anchored at "very slow or very ineffective" (1) and "very quick or very effective" (5). The analysis part is based on the questionnaire completed by all forty-five students.

User study analysis

Reliability test

Cronbach's alpha for all the data collected from the questionnaires is approximately 0.53. Although it is lower than 0.7, but given the sample size of our pilot study, it suggests the reliability of our data [8].

Preliminary analysis and discussions

The change of online help-seeking behaviors

We performed a chi-squire test on both female and male students to examine the relation between their help-seeking behaviors in high school and university. The relations between these variables were significant for both male and female students, $p < .01$, and there is no significant difference between them. In other words, all students' help-seeking behaivors changed, more of them chose to seek help from online resources (55 % and 50 % for female and male students respectively) rather than classmates and textbooks.

The information–seeking behaviors: Quick fix, fast surfing, broad scanning and deep diving

We investigated whether students look for quick answers, broad scanning or deeper diving in order to examine whether or not online help seeking through searching can help students understand the require generic information skills as well as the knowledge of the discipline. Unfortunately, a higher percentage of both female and male students attempt to look for quick answers instead of assembling knowledge pieces: 71 % and 64 % respectively. Our finding is consistent with previous research that students tend to be 'shallow, random and often passive' in the way they use the internet to search for information [28].

The overuse of googling on memory and attention

A two-sample t-Test was performed to establish relationships between the (over)use of search engine and memory and attention between female and male students. Results indicate that male students ($M = 3.30$) are more prone to the memory diminishing than female students ($M = 2.96$), $t (124) = -2.07, p < .05$.

Some general perception of search engines and task performances

Most students thinks the fastest way to seek help from is to go online, while over half of them admit that the most efficient way to seek help from is to ask their teachers and classmates for help. Apparently, students' choices reveal how they use the Internet in daily life. Generally the most efficient way doesn't have what the students ask for, the speed is what they want, and therefore searching through the browsers is their first choice for problem-solving (Figs. 1 and 2).

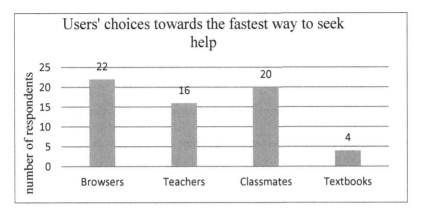

Fig. 1. User perception over the search engine Baidu: the time saved

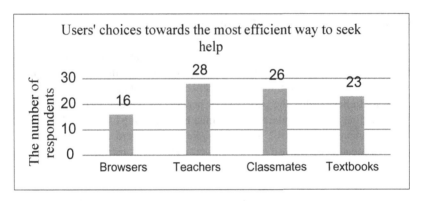

Fig. 2. User perception over the search engine Baidu: the efficiency

While browsing online, most students find it hard to read and take notes online, for it generates tiredness caused by the radiation and the fixed posture they had when reading. And scrolling materials make them uncomfortable to read. Besides, the links and the advertisements hidden among the materials will easily distract their attention. The problems students have when browsing online can have a negative effect onto their understanding upon the information they find (Fig. 3).

They are certainly unaccustomed to such way of learning and reading, therefore the action of browsing online may degenerate their abilities to analyze and think. Most students agree that they are haunted by tons of information online daily, and it adds pressure on their task of searching for the right information, forcing them to take a glimpse into single piece of information instead of probe deeply into the information they get.

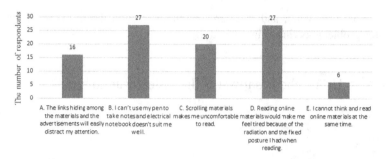

Fig. 3. User perceptions of the drawbacks of online materials

Regarding the different Internet-use behaviors between female and male students, the recommender systems can assign different tasks. For example, female students are more prone to remember online information than male students, so they can be suggested more with memory task to play their strengths, while male students can be improved with intensive training on memory, or be assigned with tasks that call for less memory ability.

4 General Discussions and Concluding Remarks

Jackson et al. [15] revealed that that there's a positive correlation between Internet use and academic performance. The more the students used the Internet, the higher their performance. And it also fits in the relation between Internet use and reading skills. The initial results indicated that the pedagogical benefits of the over-relying of internet is small yet the time average students spent online is astonishing high. Another interesting finding is students with above average reading skills were neither negatively nor positively influenced by the use of the Internet, which demonstrates that the use of the Internet doesn't necessarily benefits those who have already had higher reading abilities, the use of the Internet wasn't limited in reading to enlarge one's horizon.

The study is limited in that it only conducted a pre-questionnaire on the use of search engine, although there are few research on this topic. In the future, we intend to include goal-oriented tasks in order to objectively observe the behaviors of users and the co-correlations among users' internet use, academic performances and task nature, since we believe that the helps users seek from 'baidu-ing' are closely associated with the nature of the tasks.

Acknowledgements. We would like to thank all participants in this study. The financial support from Wenzhou Kean University Academic Affairs Office is greatly acknowledged.

References

1. Allsop, D.T., Bassett, B.R., Hoskins, J.A.: Word-of-mouth research: principles and applications. J. Advertising Res. **47**(4), 388–411 (2007)

2. Anderson, K.J.: Internet use among college students: an exploratory study. J. Am. Coll. Health **50**(1), 21–26 (2001)
3. Arnab, S., Brown, K., Clarke, S., Dunwell, I., Lim, T., Suttie, N., Louchart, S., Hendrix, M., de Freitas, S.: The development Approach of a pedagogically-driven serious game to support relationship and sex education (RSE) within a classroom setting. Comput. Educ. **69**, 15–30 (2013)
4. Bristow, M.: China's Internet 'Spin Doctors'. 12/16/2008
5. Carr, N.: Is Google Making Us Stupid? The Atlantic.com. July 2008
6. CNNIC. The 29th statistical survey report on the internet development in China (2012)
7. Carroll, J.: From encyclopaedias to search engines: technological change and its impact on literacy learning. Literacy Learn. Middle Years **19**(2), 27–34 (2011)
8. Cronbach, L.: Coefficient alpha and the internal structure of tests. Psychometrika **16**, 297–334 (1951)
9. Crane, B.: Google in the classroom – more than just research. Inf. Searcher **17**(3), 16–24 (2007)
10. Ellison, N.B., Steinfield, C., Lampe, C.: The benefits of facebook friends: social capital and college students' use of online social network sites. J. Comput.-Mediat. Commun. **12**, 1143–1168 (2007)
11. Fry, J.: Researchers migrate to search engines. Libr. Inf. Update **5**(9), 5 (2006)
12. Huang, R.L., Lu, Z., Liu, J.J., You, Y.M., Pan, Z.Q., Wei, Z., He, Q., Wang, Z.Z.: Features and predictors of problematic internet use in Chinese college students. Behav. Inf. Technol. **28**(5), 485–490 (2009)
13. Huang, H., Leung, L.: Instant messaging addiction among teenagers in China: shyness, alienation, and academic performance decrement. Cyberpsychol. Behav. **12**(6), 675–679 (2009)
14. Jamali, H.R., Asadi, S.: Google and the scholar: the role of Google in scientists' information-seeking behavior. Online Inf. Rev. **34**(2), 282–294 (2010)
15. Jackson, L.A., Eye, A.V., Witt, E.A., Yong, Z., Fitzgerald, H.E.: A longitudinal study of the effects of Internet use and videogame playing on academic performance and the roles of gender, race and income in these relationships. Comput. Hum. Behav. **27**(1), 228–239 (2011)
16. Kamenica, E.: Behavioral economics and psychology of incentives. Ann. Rev. Econ. **4**, 427–452 (2012)
17. Leaver, T.: Wikipedia what's in it for teachers? Screen Educ. **53**, 53–60 (2009)
18. Pawar, U.S., Pal, J., Gupta, R., Toyama, K.: Multiple mice for retention tasks in disadvantaged schools. In: Proceeding of the 2007 Conference on Human Factors in Computing Systems, ACM CHI'2007 (2007)
19. McDowell, L.: Electronic information resources in undergraduate education: an exploratory study of opportunities for student learning and independence. Brit. J. Educ. Technol. **33**(3), 255–266 (2002)
20. Mostafa, J.: Seeking better Web searches. Sci. Am. **292**(2), 67–73 (2005)
21. Prasad, T.: The 'Google effect:' may be good, may be bad. The Hindu. 15 May 2012
22. Prensky, M.: Digital Natives, Digital Immigrants. On the Horizon, NCB University Press, vol. 9(5) (2001)
23. Rieger, O.Y.: Search engine use behavior of students and faculty: user perceptions and implications for future research. First Monday, vol. 14, 12–7 December 2009
24. Richtel, M.: Attached to Technology and Paying a Price. New York Times (2010)
25. Siomos, K.E., Dafouli, E.D., Braimiotis, D.A., Mouzas, O.D., Angelopoulos, N.V.: Internet addiction among Greek adolescent students. Cyberpsychol. Behav. **11**, 653–657 (2008)

26. Sparrow, B., Liu, J., Wegner, D.M.: Google effects on memory: cognitive consequences of having information at our fingertips. Science **2011**(333), 776–778 (2011)
27. Su, A.Y.S., Yang, S.J.H., Hwang, W.-Y., Zhang, J.: A Web 2.0-based collaborative annotation system for enhancing knowledge sharing in collaborative learning environments. Comput. Educ. **5592**, 752–766 (2010)
28. Sutherland-Smith, W.: Weaving the literacy Web: changes in reading from page to screen. Reading Teacher **55**(7), 662–669 (2002)
29. Tao, Z.: Understanding online community user participation: a social influence perspective. Internet Res. **21**(1), 67–81 (2011)
30. Taneja, N., Chaudhary, R.: Query recommendation for optimizing the search engine results. Int. J. Comput. Appl. **50**(13), 20–27 (2012)
31. Wu, H.R., Zhu, K.J.: Path analysis on related factors causing Internet addiction disorder in college students. Chin. J. Public Health **20**(2004), 1363 (2004)
32. Ni, X., Yan, H., Chen, S., Liu, Z.: Factors influencing Internet addiction in a sample of freshmen university students in China. Cyberpsychol. Behav. **12**(3), 327–330 (2009)
33. Zahera, H.M., El-Hady, G.F., Abd El-Wahed, W.F.: Query recommendation for improving search engine results. Int. J. Inf. Retrieval Res. **11**, 45–52 (2011)

Supporting Growers with Recommendations in RedVides: Some Human Aspects Involved

Olga C. Santos[✉], Sergio Salmeron-Majadas, and Jesus G. Boticario

aDeNu Research Group, Artificial Intelligence Department,
Computer Science School, UNED, C/Juan del Rosal, 16, 28040 Madrid, Spain
{ocsantos,jgb}@dia.uned.es, ssalmeron@bec.uned.es

Abstract. This paper discusses some human aspects that are to be considered when designing recommendations for RedVides, a cloud based networking environment that collects the status of the crop with sensors and can take decisions through corresponding actuators. The goal behind is to support growers in decision making processes, which can be benefited from collaborations among growers and with other stakeholders.

Keywords: Agriculture recommender systems · Sensors · Actuators · Cloud-based environments · Ubiquitous computing

1 Introduction

Agricultural production has traditionally required a follow-up process that relies mainly on information collected by growers themselves, either from their direct involvement in a manual data collection or by asking personnel involved in crop maintenance [1]. To support these agricultural processes with information and communication technologies, Agricultural Information Systems (AIS) have been developed. Through distributed sources of information, AIS provide real-time data monitoring, support on recommendation and decision-making and farm simulation, amongst others [2]. A more specific type of AIS are the so called Farm Management Information Systems (FMIS). FMIS collect, process, store and disseminate data providing information which is needed to carry out farming operating functions [3]. These systems also handle information which is needed to support correct decisions and real-time management [4]. They can integrate precision farming [5], which uses sensors to avoid manual data collection, and can also be coupled with some farming equipment (e.g., actuators) in order to provide automatic execution of decisions [6]. Both AIS and FMIS relay on service oriented architectural approaches (e.g., [2, 4, 7]). In this context, cloud computing has been considered as an enabler for providing scalable computation, software, data access, and storage services [6].

In this context, besides growers themselves, there are other stakeholders that can support and/or benefit from decision making support, such as providers of products to be used in the field (e.g. seeds, pesticides) and external information sources (e.g., meteorological forecast agencies and phytosanitary surveillance organizations).

© Springer International Publishing Switzerland 2014
Y. Chen et al. (Eds.): WAIM 2014, LNCS 8597, pp. 307–314, 2014.
DOI: 10.1007/978-3-319-11538-2_28

These aforementioned issues have been considered in the project RedVides (TSI - 020100-2011-29).[1] The resulting RedVides system manages real-time information to support decision making process in crops management. For this, it combines several kinds of input: (1) data collected by independent and autonomous sensor units, (2) information provided by the grower on both field tasks and plants' evolution, and (3) external information obtained from weather forecast agencies and phytosanitary surveillance organizations.

In this paper, we discuss the potential of integrating a recommender system into RedVides to support existing social interactions in this ubiquitous networking environment, and discuss the human aspects of the recommendations involved. Recommender systems are applicable here since they have been acknowledged for years as valuable systems to guide users when there is information overload [8], such as the one experienced by growers [4]. In fact, by monitoring, analyzing and understanding the behaviour of growers, their demographics, opinions, preferences and history, the effectiveness of produced recommendations may be significantly improved [9].

The structure of the paper is the following. First, we present some background on recommender systems in the agricultural domain. Then, we discuss our ideas for considering the human aspects in agricultural recommender systems. The paper ends with some conclusions and suggestions for future work.

2 Background

Several attempts have been made to develop decision support systems for optimizing agriculture operations [10]. Due to the aforementioned potential value of recommenders to deal with the information overload experienced by growers, here we review the usage of recommender systems in the agricultural domain. Works found in the literature show that recommender systems can be used in different ways.

First works were focused on supporting agricultural e-commerce applications, following a similar approach than those used to recommend songs, movies or books, which mainly focused on the target object to be recommended taking into account quantitative opinions (i.e., ratings) from the users. For instance, in [11] a recommender system for searching and selecting wines that matches buyers' preferences was developed, which can also be used by sellers to attract new buyers and promote wines, as marketing methods for wine are more advance and sophisticated than those for other agricultural products [12].

Following the same approach, which is based on collaborative filtering, agricultural web portals have also integrated recommender systems to help users finding information of interest. For instance, the portal of the Embrapa Information Agency aims to organize, process, store and disseminate agricultural technological information [13]. Here, the recommender system provides (a) detailed information of production stages, with indications of reference material and different sources of reliable statistics; (b) information that favours efficient production considering technical, environmental,

---

[1] http://www.cubenube.com/index.php?pag=redvides&lang=en

social and economic aspects; (c) support to public and private stakeholders in planning and decision making, and (d) transfer of current knowledge with a friendly language for the sugarcane industry. In the same vein, Organic.Edunet is a web portal to facilitate access, usage and exploitation of digital content related to organic agriculture and agroecology. This portal focuses on organic and sustainable education and provides recommendations on relevant resources for agriculture educators [14].

Other works, while keeping the same idea, have addressed the production process. Specifically, RecOrgSeed [9] extracts knowledge about growers and seeds using data mining and semantic annotations to recommend organic seeds that the grower can buy. In particular, it takes into account growers' requirements and needs, product ratings and matching criteria with other users that demonstrate similar behaviour in relation to the acquisition of the seeds. This is of particular importance considering our approach since growers are also supported in collaboration and communication in terms of virtual communities.

Other systems focus on diseases control. CuSSRS [15] recommend when to spray the crop with a given phytosanitary product, since product under-spray leads to inadequate disease control and over-spray leads to environmental concerns and potential phytotoxicity in the fruit. To tackle this issue the grower is provided with advanced notice of the need to re-spray based on fruit grow, and rain and wind weathering.

In [16], a semantic rule-based event-driven service-oriented architecture is proposed to deliver personalized recommendation driven by real-time events and user preferences that take into account the user context, location and requirements. It considers experts and information sources as service providers, and agro-professionals as service consumers. From the perspective of the users, growers subscribe to events related to climate, agriculture extension, government schemes, agroindustry and agromarket that meet their decision-making requirements. Recommendations suggest appropriate cultivation practice for each user that are based on rules triggered, such as "if temperature increases, then rice price goes down", which corresponds to a given climatic condition event.

A recommender system for sustainable agricultural techniques has also been developed using semantic rules [17]. ARIDUS recommends sustainable alternatives for cultivation with lower impact on sustainability dimensions. The recommender mechanism points out combinations of crops that can help farmers to planning more profitable intercropping (i.e., a well-known sustainable technique where crops with different root systems are used in order to reach different layers of the soil profile and recycling the nutrients, promoting positive effects on its biological, chemical and physical attributes), by respecting the sustainability principles.

Recommender systems have also been proposed to support policy planning for agricultural research. For instance, regarding funding, a recommender system has been developed to map a project proposal profile with the strategic goals to support funding decision making, evaluating the effectiveness of ongoing funding allocation and suggesting some actions based on the policy guideline [18].

Last but not least, there have been other systems that also take into account agricultural products in the recommendation process. In particular, a preference-based recommendation procedure to suggest vegetables to eat [19].

According to this review, the diversity of usages of recommender systems in the agricultural domain can be classified into (1) websites for selling (or eating) products (e.g., wine, vegetables), (2) portals with contents about agriculture (e.g., Embrapa Information Agency, Organic.Edunet), (3) support for agricultural process (e.g., seeds selection, pest control, cultivation practice, policy planning). The latter can be of relevance for RedVides approach.

As for the recommender systems focused on supporting the agricultural process (i.e., [9, 15–18]), human aspects have not been explicitly considered in their design. However, there are some related issues that can be identified in them, which refer to user information, such as requirements, profiles, needs, context and preferences. This information is to be used to recommended appropriate items (such as seeds, products, information) and actions to take (such as planning, spray) when specific conditions take place regarding control and timing (dealing with "when to do") and in most of the cases guide the deployment of certain cultivation practice.

3 Human Aspects to Be Considered When Recommending in RedVides

Redvides is a cloud based networking environment aimed to support growers in decision making processes by collecting the status of the crop with sensors and taking decisions through corresponding actuators. It relays on a ubiquitous wireless networked infrastructure deployed in the crop, which communicate to the processing unit that is available in the cloud platform. Besides growers themselves, there are other stakeholders that can support and/or benefit from the decision making process, such as providers of products to be used in the field and external information sources that, among others, provide meteorological forecasts or indications on how and when to fumigate the crop.

In this context, there is a need to support the growers' needs as well as the social interactions that can take place in RedVides ubiquitous networking environment among growers and with other stakeholders. As in afore-analyzed works, it is desirable to extend RedVides with the infrastructure to host virtual communities [9] where stakeholders involved are supported by appropriate recommendations that take into account the user context, location and requirements [16] along the different use cases involved in the agricultural process, which have been compiled elsewhere [6].

Defining appropriate recommendations in these use cases requires an elicitation process that involves the different stakeholders in identifying available recommendation opportunities. An initial approach to address this process is to consider the TORMES methodology [20]. This methodology has been proposed to support the elicitation process in the educational domain, but we argue that it can be revised to support the elicitation process in other domains, such as the agricultural one. TORMES follows the user-centred design process as defined by the standard ISO 9241-210 [21] in an iterative manner, and thus, covers the following activities: (1) Understanding and specifying the context of use: identifying the people who will use the system, what they will use it for, and under what conditions they will use it; (2) Specifying the user requirements: identifying any requirements or user goals that must be met for the system to be successful, considering the variety of different viewpoints and individuality; (3) Producing design solutions to meet user requirements, which can be made in stages to encourage creativity, from an initial rough concept to a final full-blown design; and (4) Carrying out user-based evaluation of the design against the requirements.

In TORMES, recommendations' descriptions are semantically modelled in terms of 7 dimensions [22]. Adapting their meaning to the agricultural context, these dimensions can be defined as follows: (i) *What* is to be recommended, that is, the action to be done on an agricultural object (e.g., buying certain seeds, spraying some phytosanitary product); (ii) *How and Where* the recommendation is delivered to the grower be defining the communication channel and presentation mode (e.g., an SMS sent to the grower's mobile phone); (iii) *When and to Who* the recommendation is produced, which depends on defining the runtime information that deals with the grower and crop context that trigger the recommendation (e.g., climatological conditions that require the grower to spray some chemicals to prevent the appearance of some disease, joint purchase of seeds among growers in the same sustainable cropping cycle to reduce the cost); (iv) *Why* a recommendation has been produced, that is the rationale that provides an explanation to the user (e.g., to comply with a certain regulation); and (v) *Which* are the recommendation features that additional semantic information that characterise the recommendations themselves (e.g. relevance, category). At this point and in a tentative way, recommendations that deal with human aspects can be classified as follows: (i) health benefits, by using the minimum amount of phytosanitary product (or any) and thus reducing the amount of chemistry received when eating the product, (ii) environmental care, selecting appropriate kind of crops to grow each seasons (e.g. appropriate sustainable techniques to be applied in the crop), and (iii) interactions among other growers and other stakeholders, which can refer to very different issues, such as selecting the seeds to buy (e.g., those that are more appropriate for the conditions of the soil, buying them altogether to reduce costs), sharing data about plant disease and/or pest appearance and evolution, synchronizing crops in intercropping to assure that although crops from nearby growers are planted in the same season, they are harvested at different times, so when a crop is harvested, a new one from a different family substitutes the previous one.

4 Conclusions and Future Work

Despite the potential benefits of recommender systems to deal with the information overload experienced by growers, up to the moment, we have not found many systems in the literature that report their usage in the agricultural domain, and in any case, very little is reported from the human-system viewpoint, including social issues involved. Recommender systems presented in Sect. 2 consists of websites for selling products, portals with contents about agriculture, and systems providing support for agricultural process. The latter deal with seed selection, pest control, cultivation practice and policy planning, and thus, is of relevance for RedVides approach.

In Sect. 3, after drawing those functional and structural issues that are to be considered in the recommendations' elicitation process as defined by TORMES in order to support users in the decision making process in agriculture, we have pointed out a tentative classification of the recommendations from a human perspective, including social related issues, that can be taken into account when modelling recommendation opportunities in the agricultural domain. In particular, those categories that have been identified are health benefits, environmental care, and interactions among other growers and stakeholders.

However, there is a need to go further on the identification of the human aspects involved in the recommendation process in this domain. Hence, in order to identify the real-time information needs for decision making process in crops management that take into account the human aspects involved explicitly, we have proposed to take into account our own experience on eliciting user-centred recommendations in the educational domain [20]. The purpose here is to refine the TORMES methodology in order to meet a deep involvement of growers in the elicitation of recommendation process. To this end we are considering in RedVides those needs that can be extracted from the use cases identified in [6]. As discussed elsewhere, the adoption of a user centred design approach when developing decision support systems can facilitate the uptake of the system by farmers [23].

Acknowledgements. Authors would like to thank all our colleagues from Cubenube (coordinator), Zignux and Encore companies who have been involved in RedVides project. This work has been partially funded with the National Plan of Scientific Research Development and Technological Innovation 2008-2011 in the call Avanza 2 by the Spanish Ministry of Industry, Energy and Tourism under grant no. TSI - 020100-2011-29, including support from the European Regional Development Fund (ERDF).

References

1. Peets, S., Mouazen, A.M., Blackburn, K., Kuang, B., Wiebensohn, J.: Methods and procedures for automatic collection and management of data acquired from on-the-go sensors with application to on-the-go soil sensors. Comput. Electron. Agric. **81**, 104–112 (2012). http://dx.doi.org/10.1016/j.compag.2011.11.011
2. Arroqui, M., Mateos, C., Machado, C., Zunino, A.: RESTful Web Services improve the efficiency of data transfer of a whole-farm simulator accessed by Android smartphones. Comput. Electron. Agric. **87**, 14–18 (2012). http://dx.doi.org/10.1016/j.compag.2012.05.016
3. Salami, P., Ahmadi, H.: Review of farm management information systems (FMIS). New York Sci. J. **3**(5), 87–95 (2010)
4. Sørensen, C.G., Fountas, S., Nash, E., Pesonen, L., Bochtis, D., Pedersen, S.M., Basso, B., Blackmore, S.B.: Conceptual model of a future farm management information system. Comput. Electron. Agric. **72**(1), 37–47 (2010). http://dx.doi.org/10.1016/j.compag.2010.02.003
5. Lehmann, R.J., Reiche, R., Schiefer, G.: Future internet and the agri-food sector: state-of-the-art in literature and research. Comput. Electron. Agric. **89**, 158–174 (2012). http://dx.doi.org/10.1016/j.compag.2012.09.005

6. Kaloxylos, A., Eigenmann, R., Teye, F., Politopoulou, Z., Wolfert, S., Shrank, C., Dillinger, M., Lampropoulou, I., Antoniou, E., Pesonen, L., Nicole, H., Thomas, F., Alonistioti, N., Kormentzas, G.: Farm management systems and the Future Internet era. Comput. Electron. Agric. **89**, 130–144 (2012). http://dx.doi.org/10.1016/j.compag.2012.09.002

7. Murakami, E., Saraiva, A.M., Ribeiro Jr., L.C.M., Cugnasca, C.E., Hirakawa, A.R., Correa, P.L.P.: An infrastructure for the development of distributed service-oriented information systems for precision agriculture. Comput. Electron. Agric. **58**, 37–48 (2007)

8. Burke, R.: Hybrid recommender systems: survey and experiments. User-Model. User-Adap. Inter. **12**, 331–370 (2002)

9. Markellos, K., Markellou, P., Liopa-Tsakalidi, A., Staurianoudaki, M.: Personalised web services for agricultural domain: a case study for recommending organic seeds to farmers and growers. Int. J. Electron. Democracy **1**(2), 170–187 (2009)

10. Pontikakos, C.M., Tsiligiridis, T.A., Yialouris, C.P., Kontodimas, D.C.: Pest management control of olive fruit fly (Bactrocera oleae) based on a location-aware agro-environmental system. Comput. Electron. Agric. **87,** 39–50 (2012). http://dx.doi.org/10.1016/j.compag.2012.05.001

11. Manouselis, N., Costopoulou, C., Sideridis, A.: Introducing recommender systems in agricultural e-commerce applications. In: Proceedings of the International Conference on Information Systems in Sustainable Agriculture, Agroenvironment and Food Technology (HAICTA 2006), pp. 98–106 (2006)

12. Baourakis, G., Matsatsinis, N.F., Siskos, Y.: Agricultural product development using multidimensional and multicriteria analyses: the case of wine. Eur. J. Oper. Res. **94**(2), 321–334 (1996)

13. de Barros, F.M.M., de Medeiros Oliveira, S.R., de Oliveira, L.H.M.: Development and validation of a recommender system for technologial information on sugarcane. Bragantia **72** (4), 287–395 (2013)

14. Manouselis, N., Kyrgiazos, G., Stoitsis, G.: Revisiting the multi-criteria recommender system of a learning portal. In: Proceedings of the 2nd Workshop on Recommender Systems in Technology Enhanced Learning 2012. CEUR, vol. 896, pp. 35–48 (2012)

15. Albrigo, L., Beck, H., Timmer, L., Stover, E.: Development and testing of a recommendation system to schedule copper sprays for citrus disease control. J. ASTM Int. (JAI) **2**(9), 1–12 (2005)

16. Laliwala, Z., Sorathia, V., Chaudhary, S.: Semantic and rule based event-driven services-oriented agricultural recommendation system. In: Proceedings of the 26th IEEE Conference on Distributed Computing Systems Workshop (ICDSW'06), pp. 24–30 (2006)

17. Pereira, D.H.G., Dantas, C.F.F., Ribeiro, C.M.F.A.: A pragmatic approach for sustainable development based on semantic web services. In: Proceedings of the 14th International Conference on Information Integration and Web-based Applications & Services (IIWAS '12), pp. 82–90. ACM, New York (2012)

18. Buranarach, M., Porkaew, P., Supnithi, T.: A decision support system development to support rice research policy planning using an ontology-based framework. In: Proceedings of the Joint International Symposium on Natural Language Processing and Agricultural Ontology Service 2011 (SNLP-AOS 2011) (2011)

19. Liao, Y.: Green product retrieval and recommendations system. In: I. Management Association (ed.) Green Technologies: Concepts, Methodologies, Tools and Applications, pp. 848–868. Information Science Reference, Hershey (2011). doi:10.4018/978-1-60960-472-1.ch416

20. Santos, O.C., Boticario, J.G.: User-centred design and educational data mining support during the recommendations elicitation process in social online learning environments. Expert Syst. (in press). doi:10.1111/exsy.12041

21. ISO Ergonomics of human-system interaction - Part 210: Human-centred design for interactive systems. ISO 9241-210:2010 (2010)
22. Santos, O.C., Boticario, J.G., Manjarrés-Riesco, A.: An approach for an Effective Educational Recommendation Model. In: Manouselis, N., Drachsler, H., Verbert, K., Santos, O.C. (eds.) Recommender Systems for Technology Enhanced Learning: Research Trends & Applications, pp. 123–143. Springer, New York (2014)
23. Parker, C.G., Campion, S.: Improving the uptake of decision support systems in agriculture. In: First European Conference for Information Technology in Agriculture (EFITA), pp. 129–134 (1997)

Geographical Information in a Multi-domain Recommender System

Tiffany Y. Tang[1(✉)], Pinata Winoto[1], and Robert Ziqin Ye[2]

[1] Department of Computer Science, Kean University, Wenzhou, China
{yatang, pwinoto}@kean.edu
[2] Institute of Electronic and Information Engineering,
Central South Forestry University of Technology, Changsha, China
yezqin@gmail.com

Abstract. Multi-domain recommendation is more challenging than that in the traditional single-domain one. In our previous study on the cross-domain recommendation, we have uncovered the tradeoff between the accuracy and the coverage of the recommendation. Later, we have also reported our findings on uncovering the association between user's interests of items across domains that are related to each other to a certain degree using another dataset collected from users with different demographic information. In this paper, we further discuss the comparison between our previous two experimental results and some practical implications in the design of recommendation systems.

Keywords: Recommender system · Collaborative filtering · Cross domain · Geographical

1 Introduction

In the past, the majority of recommender systems (RSs) is for single domain only, such as for books, movies, *or* CDs. However, recent e-commerce sites have adopted cross-domain recommendation, commonly known as the cross-sell in marketing. For example, a user who browsed LCD monitors may be recommended with some graphic cards; or, a user who browsed lipsticks may be recommended with a skincare product.

As such, a cross-domain RS has a much greater potential in systems which consist of a wide-range of items, motivating a series of our study to uncover two different yet related issues: whether or not cross-domain interest exist, and whether or not a user's interest in items in certain categories can be used to predict the likeness of related items in other categories [15]. Since the data in this study was obtained from a set of homogeneous users, who mostly are students in Hong Kong, we later extended our study into another different geographical location: mainland China. Although Hong Kong is geographically close to mainland China, its culture was largely influenced by Britain due to its being a British colony, and thus Hong Kong has a richer western heritage and has developed subculture from mainland China [1, 5]. In the recommender system research community, to the best of our knowledge, little work has investigated the geographical issue which motivates the second part of our study. In particular, we examined whether there are any taste differences between subjects in these two regions;

© Springer International Publishing Switzerland 2014
Y. Chen et al. (Eds.): WAIM 2014, LNCS 8597, pp. 315–321, 2014.
DOI: 10.1007/978-3-319-11538-2_29

whether traditional recommendation approach (mainly collaborative filtering techniques, or CF) can still be used to blend the two groups and treat them as one big pool of data and on which make recommendations.

Hence, it is appealing to conduct empirical study to substantiate our understanding on the cross-domain RS through similar process but on different dataset. That is, we examined whether users' geographical location might have an effect on their preferences. Before we continue with the discussion on our work, we will provide a brief summary on cross-domain recommendation. Table 1 summarizes the pros and cons of the common recommendation approaches.

Table 1. A brief summary on various recommendation systems

	Single-domain RS			Multiple-domain RS
	Content-based	CF	Knowledge-based	Cross-domain
Main feature	Based on the analysis of items given users' past preferences	Based on 'like-minded' users	Based on preference-based feedbacks	A combination of content-based and CF techniques
Pros	− No need to have data on other users − Can deal with *'special'*-taste users − Can deal with *new/unpopular* items	− Can exploit the preferences of other users − Is application-independent − Can recommend *serendipity* items	− Offer better performance benefits − Users join in the process of recommendations— transparency	− Can diversify the recommended items − Offer the benefits of both the content-based and CF approaches − Can overcome data sparsity
Cons	− Is 'over-specialized' − Cannot exploit the preferences of other users − Rely on the representation of content features	− Cannot deal with *'special'* tastes − Cannot deal with *cold-start* problem − Cannot deal well with user/item matrix sparsity	− Cannot deal with items with complex compound features − Difficult to encourage users to contribute their time to participate in the processes	Rely on the similarity between domain items

Since the single domain RS has been dominating in both research and industry, substantiate understandings have gained especially with respect to quality of the system performance mainly in terms of the recommendation accuracy. However, in some commercial systems, it is obvious that recommendations made in a single domain cannot benefit both end users and the system. As such, cross-domain recommendation can be adopted to remedy the situation. In particular, cross-domain RS support scenarios that can help inform the system to pick items across related domains, thus, help selecting novel items that a user has not tried before. One notable issue in the cross-domain recommendation is that the underlying mechanism is in essence similar to that in the content-based filtering system in that it considers the features of the item. Though, in the single domain content-based filtering system, recommendations are made within one domain; while in the cross-domain RS, recommendations are generated across domains.

The rest of the paper is organized as follows. In the next section, we will discuss some related work to motivate the necessity and value of the cross-domain RS. Further, we will focus on a follow-up study in which we intend to substantiate our previous study in [13, 15], and examine whether users' geographical location might have an effect on their preferences. Then, we discuss the implication of our study and conclude this paper by a discussion of outstanding issues and our future work.

2 Related Work

Tsang et al. [14] studied the shopping behaviors in a Mainland Chinese city and Hong Kong malls, and revealed that Mainland Chinese shoppers' showed significant differences from Hong Kong shoppers in shopping motives, processes and outcomes. For example, Hong Kong shoppers usually shop with multiple intents; while Mainland Chinese shoppers turn out to make more planned purchases. Another recent study by Ramasamy et al. [10] also observed significant differences between consumers between Hong Kong shoppers and their Mainland Chinese counterparts.

All these differences, either subtle or significant, can fundamentally be traced to the subcultures characterized by class, age, location, gender and some other entities that might differentiate a micro-culture (e.g. mainland Chinese) from the macro-culture (e.g. Chinese) [1, 2].

Realizing the 'over-specialization' of recommendations, Ziegler et al. [16] attempted to diversify the returned items to include topical diversity among them. Li et al. [7] proposed a hybrid CF to diversify recommended items based on related keywords; for instance, a user may claim his/her interest in both 'Football' and 'English', but s/he may have only provided ratings on 'Football'-related items. Hence, the RS can help him/her by finding relevant items in English football, for example. Notice here that there is a major difference between them and ours: the items under scrutiny are not so closely related, while in our study, we are more interested in items that are related.

Cross-selling has been extensively studied in the marking and consumer behavior community [6]. Cross selling can include products or services belonging to different domains with an aim to build tighter customer relationships. Ratneshwar et al. [11] found out that cross-domain consideration is indeed goal-driven from the consumer psychology's perspective, that is, cross-domain buying patterns appear when a single product category could not meet all salient goals.

3 Study Set-up and Results

3.1 A Brief Summary of the Methodology Used in Our Previous Study

Our first study investigated the inter-domain relationships in user-rating matrix which grounded the cross-domain RS [15]. In the study we used ratings from 133 college students in Hong Kong (hereafter HKs) over 522 items. Results suggest that recommendation accuracy is affected by the closeness between the domains: the closer, the better. Most importantly, adding more domains may not increase the accuracy of RS. Later, we tried to verify our experiment by repeating it with 200 mainland Chinese students (hereafter MCs) [13].

We used similar data collection procedures for both HKs and MCs: in the first stage, each user is required to name five items s/he likes, among which 3 must come from different categories. After that, we compiled all the individual items and organized them into 11 categories (actors, actresses, books, games, directors, TV series, movies, musicians, singers, songs, and CDs). After data cleaning, there are 378 items. The numbers of items in each category are tabulated in Fig. 1. Then, in the second stage, users were asked to provide ratings (on a Likert scale of 0 to 5, from no rating to like it very much respectively) on all these items, which resulted in a total of 75600 ratings.

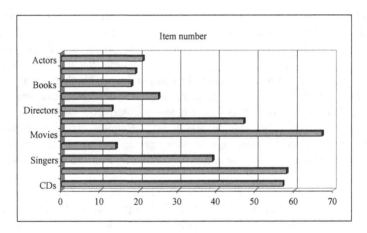

Fig. 1. The number of different items in our study

3.2 A Brief Summary of Our Previous Experimental Results

For more detail results, readers may refer to [13]. In the first part, we use Spearman correlation to compare the cross-domain relationship between MCs and HKs, in which mixed results are obtained. The results revealed some consistent pattern between MCs and HKs. For example, MCs who like an actor/actress also like him/her as a singer (correlation $r = 0.782$), and they are enthusiastic about his/her TV series/movies too ($r = 0.488$, $r = 0.407$, respectively, $p < 0.0001$). The three correlations in HKs are 0.844, 0.415, and 0.601 respectively.

Further analysis showed consistent association between items of the same genre in different domains. For example, subjects who show positive feedbacks over action/thriller TV Series, also show similar interests on action/thriller Books ($r = 0.9$ for MCs and $r = 0.975$ for HKs). And the results show strong correlation between Game and Movie, Game and Book, and Movie and Book in this genre too. We believe that cross-domain recommendation for {Movie, Game and Books} is sensitive to the genre shared among them in both MCs and HKs.

After grouping items into Western and Eastern ones, we performed t-test on a number of them. At the group level (MCs and HKs), we did not find many differences regarding ratings on Action, Thriller, Suspense-themed Game, TV Series and Movie ($p > 0.05$). However, the two groups differ significantly in their preferences over: (i) Romance, Drama and Comedy-themed TV Series, Movie and Book ($p < 0.05$); and (ii) Book and Game ($p < 0.005$). The results are consistent with our hypothesis that due to the geographical (and cultural) background, HKs were profoundly influenced by Western cultures; while MCs received less Western cultural influences. In fact, only 10 % of the items proposed by MCs are Western items, while this percentage is greater than 50 % for HKs.

Our further computational analysis on the predictive accuracy of collaborative filtering suggests the disadvantage of employing cross-culture recommendation for HKs, that is, it is better for them to find neighbors in the same group instead of from MCs (the true positive reduce from 66.8 % in < HKs, HKs > to 57.9 % in < HKs, MCs >), while their false negative do not change significantly. However, the results are not symmetric for MCs, that is, they are indifferent toward recommendation from HKs or MCs, as the true positive is 55.5 % and 54.8 % respectively. One possible explanation regarding the asymmetric property is that MCs have more diverse preferences than that in HKs. If this is the reason, then we may conclude that cross-location recommendation is acceptable when each group of users has diverse background, but unacceptable between those with more homogeneous yet different background.

4 Implication of Our Study

4.1 Cross-Domain RS: The Benefits

It is common that a wide variety of items, such as movies, books, music CDs, toys etc., exist in all commercial systems, such as ThisNext.com, Amazon.com. Therefore, one of the most significant potentials of cross-domain recommendation is its ability to exploit users' versatile interest on items in different domains to make serendipity and novel recommendations, and overcoming the 'overspecialization' that traditional RS suffers from [4, 8]. Another benefit it brings is to start up the system when new products are introduced. In particular, it is known that not until enough users have provided enough ratings in the database will the RS be able to make predictions on user ratings, a notoriously known phenomenon referred to as the cold-start problem [12], hence, when a new item is added, the system can only rely on the more complex content-based or hybrid approach to warm up the system in the traditional single domain RS. But in the cross-domain RS, the item can be matched to another item (the target user liked) from a different domain; as such, recommendation can be made.

Our previous analysis indicates that the key to the success of cross-domain recommendation is the choice of closely related domains for computation. However, the issue of how close is close enough remains unknown. In other words, in our study, we adopt the more natural and obvious relationships exhibited between domains such as Movies and TV series, Songs, CDs and Singers etc. There might have relationships between some seem-to-be-distant domains or cultures which require more experiments to uncover them. These experiments would certainly need more solid theoretical analysis between pairs of domains or among user consumption behaviors, and their associated feature space. Surprisingly, our current approach, although simple and direct, is able to generate satisfactory results. It is our hope that our effort illuminates one path through the cross-culture multi-domain RS design space where a wide range of items can be considered and can trigger more interesting and rigorous research in this area.

4.2 Human Factors in RSs

Earlier research on RS was largely focused on algorithmic design of the recommendation mechanisms: how to generate the most accurate recommendations through mostly objective evaluations such as precision, recall, Mean Absolute Error (MAE), etc. While these studies provide a solid understanding on how to make RS works, they largely ignore the various 'human factors' surrounding a RS. Recently, there have been a proliferation of research on the social aspects of a RS (thus the term 'human-recommender interaction' [9]), such as the usability of the system and transparency of the recommendations made [8, 9], how the recommendation results should be presented [3] mood and RS [15]. Our efforts in this study attempt to explore the cultural issues that might play a key role in the design of RS from software usability engineering's perspective.

Results from both studies lead us to believe that due to geographical diversity, user preferences over Western and Eastern items (both entertaining item and entertainer) are significantly different. In fact, in our second study, only 10 % of the items are Western items, while in the first study, this percentage is much high over 50 %. Although this finding is direct and natural, to our knowledge, no research has been done aiming at an understanding of whether cultural difference plays a key role in user preferences. The results are valuable for the algorithmic design and implementation of the cross-domain RS targeting at users from different countries.

5 Future Work

In order to further study the effect of geographical location toward recommendation accuracy, we plan to replicate a similar experiment using subjects from different locations within mainland China. For instance, we plan to compare students from the western part of China and those in the east coast. Although they are belong to MCs, they are different in term of their openness toward western culture since the east coasts of China have been opened almost two decades earlier than the west; hence, it will be interesting to study whether this geographical factor may affect the accuracy of recommendation, especially on various Western/Eastern items.

References

1. Chang, L.C.: Subcultural influence on Chinese negotiation styles. J. Glob. Bus. Manage. **2**(3), 189–195 (2006)
2. Chang, L.C.: Sub-cultural business negotiation: a Taiwanese and Japanese-Chinese case study. Afr. J. Bus. Manage. **5**(2), 389–393 (2011)
3. Herlocker, J., Konstan, J., Riedl, J.: Explaining collaborative filtering recommendations. In: Proceedings of the 2000 Conference on Computer Supported Cooperative Work (CSCW'2000), Philadelphia, PA, USA, pp. 241–250 (2000)
4. Herlocker, J., Konstan, J., Terveen, L., Riedl, J.: Evaluating collaborative filtering recommender systems. ACM Trans. Inf. Syst. **22**(1), 5–53 (2004)
5. Jozsa, L., Insch, A., Krisjanous, J., Fam, K.: Beliefs about advertising in China: empirical evidence from Hong Kong and Shanghai consumers. J. Consum. Mark. **27**(7), 594–603 (2010)
6. Lariviere, B., Van den Poel, D.: Investigating the role of product features in preventing customer churn, by using survival analysis and choice modeling: the case of financial services. Expert Syst. Appl. **27**, 277–285 (2004)
7. Li, Y., Liu, L., Li, X.: A hybrid collaborative filtering method for multiple-interests and multiple-content recommendation in E-Commerce. Expert Syst. Appl. **28**, 67–77 (2005)
8. McNee, S., Riedl, J., Konstan, J.A.: Being accurate is not enough: how accuracy metrics have hurt recommender systems. In: The Extended Abstracts of the 2006 ACM Conference on Human Factors in Computing Systems (CHI 2006), Montreal, Canada, pp. 1097–1101 (2006a)
9. McNee, S., Riedl, J., Konstan, J.A.: Making recommendations better: an analytic model for human-recommender interaction. In: The Extended Abstracts of the 2006 ACM Conference on Human Factors in Computing Systems (CHI 2006), Montreal, Canada, pp. 1103–1108 (2006b)
10. Ramasamy, B., Au, A., Yeung, M.: Managing Chinese consumers' value profiles: a comparison between Shanghai and Hong Kong. Cross Cult. Manage. Int. J. **17**(3), 257–267 (2010)
11. Ratneshwar, S., Pechmann, C., Shocker, A.D.: Goal-derived categories and the antecedents of across-category consideration. J. Consum. Res. **23**(3), 240–250 (1996)
12. Schein, A., Popescul, A., Ungar, L.H., Pennock, D.: Methods and metrics for cold-start recommendations. In: Proceedings of the 25th Annual International ACM SIGIR Conference on Research and Development in Information Retrieval (SIGIR'02) (2002)
13. Tang, T.Y., Winoto, P., Ye, R.Z.: Analysis of a multi-domain recommender system. In: Proceedings of the 3rd International Conference on Data Mining and Intelligent Information Technology Applications (ICMIA2011), Macao, pp. 471–476 (2011)
14. Tsang, A.S.L., Zhuang, G., Li, F., Zhou, N.: A comparison of shopping behavior in Xi'an and Hong Kong Malls: Utilitarian versus Non-Utilitarian Shoppers. J. Int. Consum. Mark. **16**(1), 29–46 (2003)
15. Winoto, P., Tang, T.: If you like the Devil wears Prada the book, will you also enjoy the Devil wears Prada the movie? a study of cross-domain recommendations. New Gener. Comput. **26**(3), 209–225 (2008)
16. Ziegler, C.-N., McNee, S., Konstan, J., Lausen, G.: Improving recommendation lists through topic diversification. In: Proceedings of the 14th International World Wide Web Conference (WWW '05), pp. 22–32 (2005)

International Workshop on Big Data Systems and Services (BIDASYS 2014)

Hashtag Recommendation Based on User Tweet and Hashtag Classification on Twitter

Mina Jeon, Sanghoon Jun, and Eenjun Hwang[(⊠)]

School of Electrical Engineering, Korea University, Seoul, South Korea
{jmina,ysbhjun,ehwang04}@korea.ac.kr

Abstract. With the explosive popularity of various social network services (SNSs), an enormous number of user documents are generated and shared daily by users. Considering the volume of user documents, efficient methods for grouping or searching relevant user documents are required. In the case of Twitter, self-defined metadata called hashtags are attached to tweets for that purpose. However, due to the wide scope of hashtags, users are having difficulty in finding out appropriate hashtags for their tweets. In this paper, we propose a new hashtag recommendation scheme for user tweets based on user tweet analysis and hashtag classification. More specifically, we extract keywords from user tweets using TF-IDF and classify their hashtags into pre-defined classes using Naïve Bayes classifier. Next, we select a user interest class based on keywords of user tweets to reflect user interest. To recommend appropriate hashtags to users, we calculate the ranks of candidate hashtags by considering similar tweets, user interest and popularity of hashtags. To show the performance of our scheme, we developed an Android application named "TWITH" and evaluate its recommendation accuracy. Through various experiments, we show that our scheme is quite effective in the hashtag recommendation.

Keywords: Twitter · Hashtag · User interest · Naïve Bayes classifier · Classification · Ranking · Recommendation · Android

1 Introduction

As a very popular online social networking service, Twitter is currently used by millions of users and organizations to quickly share and discover information. Users can access Twitter through the web or mobile devices and publish a message called tweet of up to 140 characters that can be sent or read by anyone. Each message can have replies from other users, which could lead to a real-time conversation around some hot topic or interesting content.

Furthermore, for grouping and searching of certain topics, users can utilize self-defined hashtags starting with a hash symbol (#) as a prefix to a word or a multi-word phrase without whitespace. A hashtag is a simple and convenient tool for users to categorize their tweets to represent some specific event or topic. In other words, users can attach appropriate hashtags to their tweet to join a relevant conversation about a specific topic on Twitter. The usage of hashtag in tweet is very easy and simple. People use hashtags for diverse purposes: for example, to classify a product (e.g. *#iphone5s*),

© Springer International Publishing Switzerland 2014
Y. Chen et al. (Eds.): WAIM 2014, LNCS 8597, pp. 325–336, 2014.
DOI: 10.1007/978-3-319-11538-2_30

to represent information or event (e.g. *#worldcup2014, #superbowl*), or to express an emotion (e.g. *#happy, #sad*), etc. If users talk about same topic using a specific hashtag, their tweets will appear in the same stream. In this way, hashtags can help to identify messages on a specific topic or event and facilitate conversation among users.

However, due to the wide range of topics discussed on Twitter, it is difficult for users to find out appropriate hashtags for their tweets. Besides, some users include too many hashtags in their tweet. For instance, in the tweet "*#This #is a #tweet #with #lots #of #hashtags,*" its hashtags do not deliver any clear purpose. Such tweets tend to confuse other users and are usually considered as a spam. Accordingly, hashtags in a tweet should be relevant to the topic and reflect user interest. To help users utilize hashtags appropriate to their tweets, it is necessary to understand the topics of tweets and recommended appropriate hashtags relevant to the topic to the users effectively and efficiently. Therefore, in this paper, we propose a hashtag recommendation scheme that helps users to utilize appropriate hashtags for their tweets by analyzing existing tweets and their hashtags. The main components of our proposed scheme are as follows.

- Extracting a set of keywords from each tweet
- Mapping the set of keywords into one of pre-defined classes
- Calculating the ranks of candidate hashtags
- Recommending the most appropriate hashtags to the user

Consequently, when a user completes a tweet with no hashtags, then our system can automatically recommend appropriate hashtags relevant to the tweet.

2 Related Work

So far, many studies have been done for tweet classification and hashtag recommendation. In this section, we first consider several works on tweet classification and then introduce works on hashtag recommendation.

2.1 Tweets Classification

On Twitter, it is important to classify a specific topic of tweet into general categories with high accuracy for better information retrieval or for easier understanding of topics. Therefore, a number of recent papers have introduced the classification of tweets on Twitter. Go et al. [1] introduced an approach for automatically classifying the sentiment of Twitter messages. These messages are classified as either positive or negative using tweets with emotions for distant supervised learning. They showed that machine learning algorithms (Naïve Bayes, Maximum Entropy and SVM) have high accuracy. Wang et al. [2] focused on hashtag-level sentiment classification, instead of presenting the sentiment polarity of each tweet relevant to the topic. They classified hashtags into three categories, a topic which is closely connected to a certain hashtag, sentiment hashtags which express subjective opinions and sentiment-topic hashtags which indicate a certain target word and the sentiment words. To capture the relationships among hashtags, they developed an undirected edge between two nodes if those particular hashtags appeared together in a single tweet. In addition, Lee et al. [3] classified trending topics into general categories such as sports, politics, technology, etc. by using two main supervised learning techniques. They employed a text classification technique called Naive Bayes and proposed a network-based approach to predict the category of a topic knowing the categories of its similar topics.

2.2 Hashtag Recommendation

Sometimes, people attempt to use a hashtag to categorize their tweets as broadcast media for certain topics or events such as elections. However, it might be difficult for them to select hashtags suitable for their tweets. To solve this problem, many recommendation schemes have been proposed for suggesting appropriate hashtags to the users. Zangerle et al. [4] suggested an approach for the recommendation of suitable hashtags to the user during the creation process. The recommender system retrieves a set of similar tweets using TF-IDF. Hashtags are extracted from the retrieved similar tweets and are ranked using their number of occurrences in the whole dataset, their number of occurrences in the retrieved dataset or similarity scores of tweets. Kywe et al. [5] proposed a personalized hashtag recommendation method based on collaborative filtering, which recommends hashtags found in the previous month's data. This method considers both a target user interest and tweet content. Given a user and a tweet, this method selects the top most similar users and top most similar tweets using TF-IDF. Hashtags are then selected from the most similar tweets and users assigned some ranking scores. Furthermore, Gordin et al. [6] proposed a method for unsupervised and content based hashtag recommendation for tweets. This method applied Naïve Bayes, Expectation-Maximization and Latent Dirichlet Allocation (LDA) to model the underlying topic assignment of language classified tweets. Even though a number of related studies have been conducted for the classification and recommendation, only a few works have addressed the problem of individual preferences for the hashtag recommendation.

3 Proposed Approach

In this section, we describe how to recommend appropriate hashtags to the user when he/she creates a tweet. The main steps for the recommendation include keyword extraction, keyword classification, hashtag ranking, and hashtag recommendation. The overall system architecture for hashtag recommendation is shown in Fig. 1.

Four main components for hashtag recommendation are as follows:

(1) Extracting a set of keywords from collected tweets using TF-IDF
(2) Classifying that set of keywords into one of pre-defined classes using Naïve Bayes Classifier
(3) Ranking candidate hashtags
(4) Recommending appropriate hashtags to the user

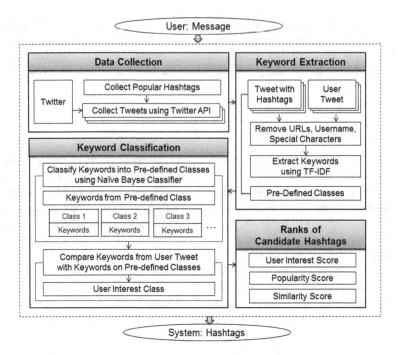

Fig. 1. The overall system architecture

3.1 Data Collection

To build a hashtag recommendation system, we first collected the top 500 most popular hashtags for the year 2013 from Statweestics [7]. By removing non-English and follow-activity hashtags, we have chosen 404 final hashtags with their ranking. Sample hashtag ranks are shown in Table 1. Hashtags are categorized by the preprocessing step into 16 classes defined in [8], which are *art & design, books, business, charity, entertainment, family, fashion, food & drink, funny, health, music, news, politics, science, sports,* and *technology*. Next, we collected 240,000 tweets (OTs) that contain at least one of those hashtags during one week (3-Apr-2014 to 10-Apr-2014) using Twitter Search API. For those tweets, we extract their keywords. For effective extraction of keywords, we remove words starting with the @ character (A username that can be used to send a message), URLs and special characters such as *%!? > $^&<{}.

Table 1. Sample hashtag ranking from Statweestics.com

Ranking	Hashtag
2	#android
18	#oomf
70	#fashion
123	#family
207	#football

3.2 Keyword Extraction

For more effective hashtag recommendation, the first step is to extract keywords from tweets. For that purpose, we use TF-IDF (Term Frequency-Inverse Document Frequency), which is one of the most common weighting methods. This method reflects the importance of each keyword in a tweet and can be defined by the following expression.

$$TFIDF_{t,d} = n_{t,d} \times \log \frac{|D|}{|\{d : t \in d\}|} \tag{1}$$

In the expression, $T = \{t_1, t_2, t_3, \ldots, t_n\}$ and $D = \{d_1, d_2, d_3, \ldots, d_n\}$ indicate the set of keywords t (term) and the set of tweets d (document), respectively. The result of keyword extraction is shown in Table 2.

3.3 Keywords Classification

Based on the extracted keywords of a tweet, we can decide the class type of the tweet. To do that, we first define the class type of each hashtag in the preprocessing step manually. As a result, each class contains a set of distinct hashtags. Also, by analyzing the keywords of each tweet and its hashtags, we can decide the relationship between keywords and hashtags. As a result, we can get a set of relevant keywords for each class type.

For more accurate hashtag recommendation, we calculate a user interest class by considering all the tweets created by a specific user. By analyzing all the user tweets, we can get a set of keywords and use a Naïve Bayes Classifier [9] to calculate the user interest class from those keywords [10], which is known to be simple and easy to implement. In some types of probability models, Naïve Bayes Classifiers can be trained efficiently in a supervised machine learning. Therefore, we employ this classifier to determine the class of keywords. In this paper, the probability of a tweet t being in class c is computed as follows:

$$C_{map} = \underset{c \in C}{\operatorname{argmax}} \ P(c) \prod_{i=1}^{n_t} P(k_i|c) \tag{2}$$

Here, P(c) is the probability of a document occurring in class c and $P(k_i|c)$ indicates the conditional probability of term k_i occurring in class c. $\{k_1, k_2, \ldots, k_{nt}\}$ are the keywords in t that are part of the vocabulary we use for classification and n_t is the number of such keywords in t. In some cases, keywords that do not exist in the training data might appear in the classification process. This can lead to zero probability and interrupt classification process. To prevent this problem, we use the Laplace smoothing (also called "add one smoothing") which simply adds 1 to the word count. Some of classified hashtags and their keywords are shown in Table 2.

Table 2. Classified hashtags and extracted keywords

Hashtags	Keywords	Class
#coffee, #food, #delicious, …	recipe, dinner, cook, yummi, snack, cake, …	Food & Drink
#nowplaying, #mtvhottest, #music,…	music, musicvideo, femaleartist, radio, …	Music
#oomf, #me, #lol, #cute,…	love, smile, awesome, want, friend,…	Funny

User interest classes play an important role in the personalized recommendation. Therefore, we perform the analysis of user tweets for recommendation based on the assumption that the user would frequently write tweets about the topic that are represented by the user interest class. Hence, by analyzing user tweets and associating their relationship with classes, we can decide the most relevant class, which is the user interest class. Keywords extraction from user tweets are already described in Sect. 3.2.

If the user has not published any tweets before, then we use timeline (following's tweets of the user). Table 3 shows user tweets, their keywords, and calculated user interest class.

Overall, by calculating user interest class of a specific user, our proposed scheme can achieve more effective personalized recommendation on Twitter.

Table 3. Extracted keywords and user interest class

User	User tweets	Keyword	Class
A	Just saying…… #soda #health #nutrition #diet	soda, health, nutrition, diet	Health
	Someone asked, "Do you have any tips for dieting?" My answer: "Step 1, Learn this photo, Step 2: Practice it daily."	tips, dieting, practice, daily	
B	Men's #streetstyle #menswear	men, streetstyle, menswear	Fashion
	Red and black together works! Agree? #fashion	red, black, fashion	

3.4   Ranks of Candidate Hashtags

To recommend hashtags for the user who creates a message more accurately, we further define the ranks of the candidate hashtags for recommendation. Here, candidate hashtags indicate those hashtags that are relevant to the keywords of the tweet. Subsequently, user interest class and the popularity of hashtags are applied. Consequently, top-n recommendations [11] can be provided to the user, where n indicates the number of recommended hashtags presented to the user. For detecting the most suitable top-n hashtags, the candidate hashtags have to be ranked. By detecting the similar keywords which belong to tweets and the most suitable class from the candidate hashtags, we proposed a ranking scheme that is composed of three steps. Their flow chart is shown in Fig. 2.

Fig. 2. Flow chart for ranking

(1) Calculating similar hashtags: This is based on the similarity between keywords of hashtags from collected tweets and keywords of user tweets. We detect the similar hashtags which belong to the keyword to candidate hashtags.

$$\text{Similiar Hashtag(SH)} = \frac{T_{keyword} \cdot H_{keyword}}{\left\|T_{keyword}\right\| \cdot \left\|H_{keyword}\right\|} \tag{3}$$

(2) Finding the class of each similar hashtag: This method is based on classification for user interest class as described in Sect. 3.3. It applies such classification as it analyzes published tweets by the user. If both SH class and UIH class are same, they are included in the candidate hashtags. If not, this step should be ignored and proceed to the next method (PH).

$$\text{User Interest Hashtag (UIH)} = \text{SH}_{class} \cap \underset{class \in c}{\arg\max} \, p(class) \prod_{i=1}^{n_t} p(keyword_i | class) \tag{4}$$

(3) Selecting the most popular hashtags: This method reflects the most popular hashtag ranking from Statweestics.com. In other words, it is based on the popularity of candidate hashtags. According to our collected hashtags, we consider those hashtag rankings as addressed in Sect. 3.1.

$$\text{Popular Hashtag (PH)} = \max\left(Hashtag_{popularity}\right) \tag{5}$$

The ranks of the candidate hashtags are important for hashtag recommendation because it can represent user interest and enable to find out more suitable hashtags to the user.

4 Experiments

For each experiment, we use three datasets as shown in Table 4. We use our collected 240,000 tweets (*OTs*) for training dataset as described in Sect. 3.1. Additionally, we collected tweets of 80 users which are 5 users for each class (*UTs*) for similarity between classes with a user interest. We also collected tweets containing 80 new hashtags (*NTs*) that did not appear in the classification process. This dataset is used for evaluating hashtag classification accuracy. Using these datasets, our experiments are described as follows.

Table 4. Description of our datasets

Dataset	Description	Number of data
OTs	Our collected Tweets for training	240,000 tweets
UTs	User Tweets	14,840 tweets from 5 users each class (total 80 users)
NTs	Tweets with New Hashtag	15,360 tweets

4.1 Hashtag Classification Accuracy

To evaluate the hashtag classification, we use *NTs* as a dataset in this experiment and measured how accurately the class of new hashtag is calculated. To do that, we first decide the class of a new hashtag manually and see whether the class from the classification is matched with it. The results are shown in Table 5. After modeling the classifier, it can classify new keywords into proper class by calculating the highest posterior probability. In the classification, a hashtag is classified into a class having the highest posterior probability. The figure shows that the precision is 62.9 % and recall is 71.5 %. Furthermore, we observed that a hashtag can belong to multiple classes. In that case, we decided to consider two classes for each hashtag by calculating two highest posterior probabilities. Precision and recall of such hashtag classification are shown in Fig. 3. In this experiment for the hashtag classification, recall is 77.1 %.

Table 5. The new hashtag classification result

New hashtag	Keyword	Expected Class	Class 1	Class 2
#superbowlxlviii	seahawks, watching, champions, broncos, ...	Sports	Sports	Entertainment
#worldcancerday	love,cancer, awareness, strong, ...	Health	Health	Family

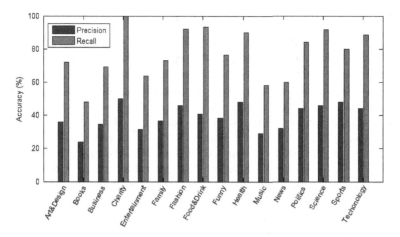

Fig. 3. Precision and recall for hashtag classification

4.2 Similarity Keywords of User Tweet

Users are likely to publish tweets to express their interest. By classifying keywords of user tweets, we can find a user interest class. Accordingly, we used *UTs* and extracted keywords from user tweets and classified them into pre-defined class. To evaluate the performance of the user interest classification, we compared the similarity between keywords of user tweet and a user interest class (i.e. pre-defined class). After that, we measured the similarity between the class based on user interest class and its pre-defined class by Cosine Similarity. The result is shown in Fig. 4. In this figure, *News* class has shown low similarity since *News* class usually covers a variety of topics such as sports, technology, business etc. By performing user interest classification, we achieved an average similarity of 0.67. *Sports* class gives the highest similarity score because tweets related to the sports usually have specific topics.

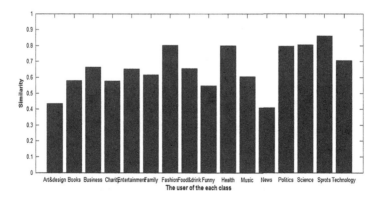

Fig. 4. Similarity between user interest class and its pre-defined class

4.3   Hashtag Ranking and Recommendation

This experiment is based on user tweets (*UTs*). In this experiment, we remove the existing hashtags from user tweets and investigate whether those hashtags are calculated by the recommendation or not. Messages without hashtags are used as the input data for this experiment. We carry out the experiment for ranking methods and top-n recommendation. We consider three ranking steps described in Sect. 3.4. For the evaluation of the recommendation, we considered three ranking methods respectively, Similar Hashtag (*SH*), User Interest Hashtag (*UIH*), Popular Hashtag (*PH*) method and their combination. We calculated precision and recall of the top-n recommendations with n = 1, n = 2, ..., n = 5 (the number of recommended hashtags). In this way, we experiment our three ranking methods with top-n hashtags. Precision and recall for top-n recommended hashtags can be seen in Fig. 5. As shown in this figure, the result of precision decreases with an increasing n and recall increases with an increasing n. In addition, our proposed ranking methods are suitable for hashtag recommendations rather than calculating a similarity of hashtag. We determined that hashtags can be recommended up to three, considering an average number of hashtags per message is one or two hashtags in [4]. Among three hashtags, a user can select hashtags via our user interface as described in Sect. 5.

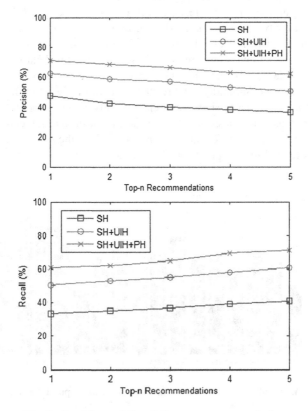

Fig. 5. Precision and Recall for top-n recommendations

5   Application

As shown in Fig. 6, we developed an application named "TWITH" (Twitter Hashtag) for our hashtag recommendation for Android. When a user writes a message without a hashtag, our system recommended and displayed three hashtags as shown on the left side of this figure. After that, a user can select the hashtags on the list and these hashtags are included in the message as shown on the right side of this figure. Then, a user can post a tweet with selected hashtags on Twitter.

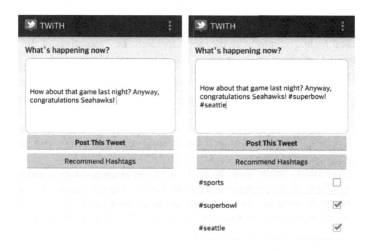

Fig. 6. Screenshot of user application for Android

6   Conclusion

Hashtag is a great tool for organizing information on Twitter. If a tweet is interesting to the users who follow a certain hashtag, then the tweet can be attached with the hashtag. In this paper, we proposed a new hashtag recommendation scheme for user tweets based on user tweet analysis and hashtag classification. To do that, we performed keyword extraction, keyword classification, and hashtag ranking. Especially, the ranks of candidate hashtags are calculated by considering similar tweets, user interest and popularity of hashtags. To show the performance of our scheme, we developed an Android application named "TWITH" and evaluated its recommendation accuracy. In conclusion, by recommending the most suitable hashtags up to three to the user our scheme achieved reasonable performance. This recommendation system can be helpful to the user which wants to know appropriate hashtags of their tweets.

Acknowledgements. This research was supported by Basic Science Research Program through the National Research Foundation of Korea(NRF) funded by the Ministry of Education (NRF-2013R1A1A2012627) and the MSIP(Ministry of Science, ICT and Future Planning), Korea, under the ITRC(Information Technology Research Center) support program (NIPA-2014-H0301-14-1001) supervised by the NIPA(National IT Industry Promotion Agency).

References

1. Go, A., Bhayani, R., Huang, L.: Twitter sentiment classification using distant supervision. CS224 N Project Report, Stanford, pp. 1–12 (2009)
2. Wang, X., Wei, F., Liu, X., Zhou, M., Zhang, M.: Topic sentiment analysis in twitter: a graph-based hashtag sentiment classification approach. In: Proceedings of the 20th ACM International Conference on Information and Knowledge Management, pp. 1031–1040. ACM, New York (2011)
3. Lee, K., Palsetia, D., Narayanan, R., Patwary, M.M.A., Agrawal, A., Choudhary, A.: Twitter trending topic classification. In: 2011 IEEE 11th International Conference on Data Mining Workshops (ICDMW), pp. 251–258 (2011)
4. Zangerle, E., Gassler, W., Specht, G.: Recommending#-tags in twitter. In: Proceedings of the Workshop on Semantic Adaptive Social Web (SASWeb 2011). CEUR Workshop Proceedings, pp. 67–78 (2011)
5. Kywe, S.M., Hoang, T.-A., Lim, E.-P., Zhu, F.: On recommending hashtags in twitter networks. In: Aberer, K., Flache, A., Jager, W., Liu, L., Tang, J., Guéret, C. (eds.) SocInfo 2012. LNCS, vol. 7710, pp. 337–350. Springer, Heidelberg (2012)
6. Godin, F., Slavkovikj, V., De Neve, W., Schrauwen, B., Van de Walle, R.: Using topic models for twitter hashtag recommendation. In: Proceedings of the 22nd International Conference on World Wide Web Companion, International World Wide Web Conferences Steering Committee, Republic and Canton of Geneva, Switzerland, pp. 593–596 (2013)
7. Statweestics. (http://statweestics.com)
8. Hong, L., Davison, B.D.: Empirical study of topic modeling in twitter. In: Proceedings of the First Workshop on Social Media Analytics, pp. 80–88. ACM, New York (2010)
9. McCallum, A., Nigam, K.: A comparison of event models for Naive Bayes text classification. Presented at the in AAAI-98 Workshop on Learning for Text Categorization (1998)
10. Jeon, M., Jun, S., Hwang, E.: Tweet-based hashtag classification scheme. In: Proceedings of the 2014 International Conference on Platform Technology and Service (PlatCon 2014), pp. 85–86 (2014)
11. Deshpande, M., Karypis, G.: Item-based top-N recommendation algorithms. ACM Trans. Inf. Syst. **22**, 143–177 (2004)

Business Ecosystem and Ecosystem of Big Data

Soonduck Yoo[1], Kwangdon Choi[1(⊠)], and Malrey Lee[2(⊠)]

[1] Hansei University, Gunpo-si, Republic of Korea
harry-66@hanmail.net, kdchoiyou@naver.com
[2] Chonbuk National University, Jeonju-si, Republic of Korea
mrlee@jbnu.ac.kr

Abstract. The purpose of this study is to explain the business ecosystem and ecosystem of Big Data: how does it work and to what extent does it influence the industry. This paper contextualizes Big Data in terms of previous studies, the current business ecosystem, and J.F Moore's business ecosystem theory. The business ecosystem of big data has three key areas: the core business, extended businesses and entire business ecosystem. The core business includes data holders, service providers, service users and infrastructure providers. This analysis characterizes the relevant aspects of the Big Data ecosystem and their relation to the broader industry.

Keywords: Big data · Ecosystem · Business ecosystem · Big data components · Industry

1 Introduction

In 2010, various organizations such as Gartner, noted the importance of big data, while recognizing the value of data analysis at the national level. Big data is a new force within the business ecosystem and serves as a platform for innovation and storing resource to all companies. Creating value from Big Data analysis and its related market is growing rapidly (Mckinsey 2011), (Gatner 2012a), (Gatner 2012b). This importance is reflected in aggressive investment plan by government to support industrial growth. The tendency to oversimplify Big Data as large-scale data or large data processing techniques itself, makes this research particularly needed (Korea Internet & Security Agency 2012), (National Information Society Agency 2012). Understanding Big Data ecosystem is required to analyze the Big Data market to promote related industries, such as smart phones and cloud computing market. The relationship within the Big Data ecosystem will be discussed in terms of value created using data.

This paper discusses the business ecosystem of Big Data and identifies main applications. The current characterization will yield insight into how Big Data industry will develop. To explain the ecosystem of big data, we need to identify the underlying biology and its relationship with businesses. We recognize the important contributions of natural science, the natural ecosystem in particular, to the study of Big Data and industrial ecosystems. One of best practices is the J.F Moore's business ecosystem. Based on the general mechanism of natural ecosystem, Moore defined business eco-system in terms of predator and prey theory. This paper explores the environment makeup of the business ecosystem of Big data and its components through Moore's

© Springer International Publishing Switzerland 2014
Y. Chen et al. (Eds.): WAIM 2014, LNCS 8597, pp. 337–348, 2014.
DOI: 10.1007/978-3-319-11538-2_31

definition of business ecosystem and the related theories. In order to derive the components of the ecosystem, we draw from theoretical background that analyzes several previous studies about Big Data ecosystem components. Finally, this research also finds characteristics of the components of Big Data ecosystem. These characteristics are used to breakdown the complex Big Data ecosystem. In addition, we organically analyze the business ecosystem based on the relationship between the actors of Big Data ecosystem.

2 Different Ecosystems

2.1 Ecosystem

The ecosystem concepts illuminate the wide range of industries, social, and economic impacts of Big Data. The following discussion of "ecosystem" is fundamentally based on biological communities that interact with the surrounding environment as presented by Roy Clapham (1992). A British ecologist and conservationist explained ecology as an 'approach to botany through the direct study of plants in their natural conditions' (Practical Plant Ecology 1923). He also pointed out that ecologists should be concerned with the structure of communities, or 'plant sociology' because plants exist in communities. This view became central to most British and American ecological theory.

Tansley coined the term, ecosystem, in (1935), although it may have been used earlier by his colleague Roy Clapham (1992). Tansley claimed that natural ecosystems are developed by openness, diversity and interaction through evolution. The energy circulation between producers and consumers through interaction is established through ecosystem theory. All living organisms act as independently existing objects that interact with each other. These interconnections form a network. Looking at the basic elements of the ecosystem, order is defined by the interaction, inter-dependence and cooperation between individuals. More specifically, these relations are formed through the principles of the food chain between organisms and energy circulation. It has the mutual coexistence and the evolutionary development through the co-evolution process. The actors of the ecosystem are defined into producers, consumers, and decomposers. They are connected to the food chain and the interaction between actors is maintained by the balance of the ecosystem, which evolves and adapts to the external environment.

Consumer is an organism that generally obtains food by feeding on other organisms or organic matter. By definition, consumers lack the ability to manufacture their own food from inorganic sources; a heterotrophy and the consumer are classified into three types such as primary consumer, secondary consumers and tertiary consumers. A producer, in a biological term, refers to the first trophic level in a simple food chain. These include plants and vegetables that utilize solar energy to generate starch through photosynthesis and they are the main source of energy that flows in the chain. A decomposer, in a biological term, refers to an organism that generates energy by breaking apart dead organic matter. A decomposer feeds on some of the released organic matter and excretes the rest as nutrients into soil and air, such as nitrogen and carbon dioxide. Decomposers play a crucial role in the ecosystem.

The basic principles of natural ecology are cross-applied to the economy or industry and the economic agents may be described in similar terms. This paper performs this type of analysis for Big Data to breakdown the complexities of business ecosystem. The following section introduces business ecosystem.

2.2 Business Ecosystem

Iansiti and Levien (2004) used business ecosystem as an analogy to describe and understand firm interactions. "We found that perhaps more than any other type of network, a natural ecosystem provides a powerful analogy for understanding a business network. Like business networks, natural ecosystems are characterized by a large number of loosely interconnected participants who depend on each other for their mutual effectiveness and survival." It should, however, be pointed out that there are differences between natural and business ecosystems. First of all, in business ecosystems the actors are intelligent and are capable of planning and seeing the future. Second, business ecosystems compete against other members. Third, business ecosystems are aiming at delivering innovations, whereas natural ecosystems are aiming at pure survival (Iansiti and Levien 2004, p. 39).

Power and Jerjian presented controversial view to the linear way of thinking. In their book "Ecosystem: Living the 12 Principles of Networked Business," they stated that one cannot manage single business on one's own; rather, one has to manage an entire ecosystem (Power and Jerjian 2001, p. 3). According to Power and Jerjian's work, there are four stakeholders to any enterprise, which should be taken into account: communities of shareholders, employees, businesses and customers (Power and Jerjian 2001, p. 18). The ecosystem standpoint is considered with the advantages of cooperation. In nature, different species help each other to produce wealth and prosperity to the community they belong to. Power and Jerjian produce one example about this phenomenon, namely a coral reef, where the structure for the whole community is created by coral polyps. In the same way, business ecosystem is often built on one single company, which is highly connected (Power and Jerjian 2001, p. 289).

Mirva Peltoniemi and Elisa Vuori (2005) considered a business ecosystem to be a dynamic structure, which consists of an interconnected population of organizations. These organizations can be small firms, large corporations, universities, research centers, public sector organizations, and other parties which influence the system. In different texts, business ecosystem is defined either consisting of several organizations or only one organization. In the latter, individual organization operate as an ecosystem in order to survive.

J.F Moore (1993) suggested the business ecosystem theory for the first time in "Predators and Prey: A New Ecology of Competition," which was published in Harvard Business Review. A Business Ecosystem as defined by Moore (1993) is: "An economic community supported by a foundation of interacting organizations and individuals the organisms of the business world. This economic community produces goods and services of value to customers, who are themselves members of the ecosystem. The member organizations also include suppliers, lead producers, competitors, and other stakeholders. Over time, the economic agents co-evolve their capabilities and roles, and they tend to align themselves with the directions set by one or more central companies.

Those companies holding leadership roles may change over time, but the function of ecosystem leader is valued by the community because it enables members to move toward shared visions to align their investments and to find mutually supportive roles."

Moore (1996) suggested the business ecosystem which had divided into three areas: core business, extended business, and the business ecosystem in his book, "The Death of Competitions." Looking at the components of the business ecosystem, Core Business is divided into the core contribution, a direct provider and a direct distribution channel (see Fig. 1). The extended business surrounds the core business and is the "suppliers of my suppliers." These surrounding operations support the core business, direct consumer and customer of my consumer. Typical contributors for a value network view of the business and its ecosystem are the co-creating new microeconomics and new wealth. A business ecosystem also includes the owners and other stakeholders of these primary species, as well as powerful species who may be relevant in a given situation, including government agencies and regulators, and associations and standard bodies representing customer or supplies. To one extent or another, an ecosystem includes direct competitors, along with companies that might be able to compete with any other important members of the community (Moore 2006).

Fig. 1. J.F Moore's typical business ecosystem

According to J Korhonen (2001), the industrial system operates through different principles of system development than the ecosystem. Industrial ecology can be a fruitful metaphor for facilitating the development of industrial systems toward the principles of system development of ecosystems.

In this study, we propose a model of the business ecosystem of Big Data based on Moore's model. According to the spheres of the business ecosystem, we identify the corresponding agents within the Big Data was based on the concept of the Moore.

3   The Ecosystem of Big Data

3.1   Previous Research on the Big Data Ecosystem

Leading researchers explain the definition of big data ecosystem. By analyzing 4 previous studies about Big Data ecosystem, the actors of Big Data ecosystem will be derived.

The first of the previous studies, #1 is "Future society forecast based on Big Data and Analysis Consulting of Industrial ecosystem" by Ryan & Co Analysis (2013). In this study, the actors of Big Data ecosystem were classified as User, Provider and Data. User is the Consumer of Big Data service and Provider is the producer of Big Data services. Data is the exchanged good in this classification. Data is sub-classified as private data and public data and Provider is sub-classified as service providers and solution providers. Users were classified as general users, governments, and institutions as professional groups. The difference of this study, relative to others is that data is presented as an actor in Big Data ecosystem (see Fig. 2).

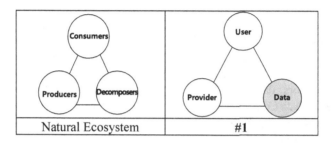

Fig. 2. Natural ecosystem and the ecosystem of Big Data by Ryan & Co Analysis 2013.

The second of the previous studies of interest, #2 is "An Analysis of Big Data Structure Based on the Ecological Perspective" by Jiyeon Cho, Taisiya Kim, Keon Chul Park and Bong Gyou Lee (2012). In this research paper, the actors of Big Data ecosystem were classified as User, Value producer and Big Data. Big Data was sub-divided into public and private data, which is generated by various industrial fields itself and re-processed data for the purpose as well as the relationship between each data. Users were classified as Big Data producer and Big Data distributor. In their research, through the comparison between the information ecosystem and knowledge ecosystem they suggest the similarities between their Big Data ecosystem (Fig. 3).

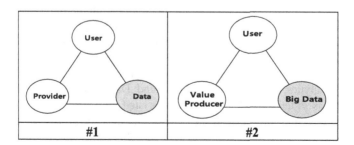

Fig. 3. The ecosystem of Big Data by Ryan & Co Analysis and by Jiyeon Cho, Taisiya Kim, Keon Chul Par and Bong Gyou Lee, An Analysis of Big Data Structure Based on the Ecological Perspective (2012).

The third historical study, #3 covers actors of Big Data ecosystem by the NIST (National Institute of Standards and Technology), Big Data Public Working Group (2013). The Big Data Working Group was established in June 2012 to develop a "Technology Roadmap." The roadmap is intended to determine appropriate "analytic techniques, technology infrastructure, and data usage" to support the secure adoption of Big Data. Defining Big Data will be an important part of this work. The study and development of Big Data depends on consensus in definitions, taxonomies, secure reference architectures, and a technology roadmap. More specifically, the aim is to create vendor-neutral technology and infrastructure agnostic deliverables. These attributes would enable Big Data stakeholders to pick-and-choose best analytics tools for their processing and visualization requirements on the most suitable computing platforms and clusters, while allowing value-added from Big Data service providers and flow of data between the stakeholders in a cohesive and secure manner.

According to the NIST Big Data Public Working Group (2013), the actor of Big Data was classified as Data Provider, Data Consumer, System Orchestrator, Big Data Application Provider and Big Data Framework Provider. The following is a look at the roles for each actors, data provider makes available data internal or external to the system and data. Data Consumer uses the output of the system. System orchestrator has duties of governance, requirements, monitoring and big data application support. Big data framework provider offers supporting resources (Fig. 4).

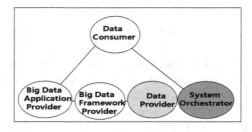

Fig. 4. Actors of Big data ecosystem by NIST Big Data Public Working Group (2013)

The final historical study, #4, the chart of the Big Data Ecosystem by Matt Turck and Shivon Zilis attempt to make sense of the rapidly evolving Big Data ecosystem back in June 2012. Initially, they were going to do this as an internal exercise to make sure they understand every part of the ecosystem. They suggested this as an "open source" project to benefit from people's thoughts and input. The chart of the Big Data Ecosystem took in December 2012, by Matt Turck and Shivon Zilis (2012), shows the following segments: Infrastructure, Analytics, Applications, Infrastructure, and Cross infrastructure/analytics, Open Source Project, and Data Sources (see Fig. 5). This classification was in terms of data analysis associated with Big Data business related companies. This is only described from utilizing the data for processing ends, which does not include the whole life cycle of ecosystem. This contributes to the literature by better contextualizing the Big Data ecosystem industry. Figure 5 explains the Big Data landscape, which shows the firms related to the service of Big Data. This landscape is the economic community that produces Big Data services.

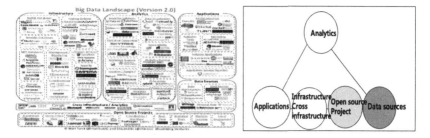

Fig. 5. The chart of the Big Data Ecosystem by Matt Turck and Shivon Zilis

3.2 Implications of the Research

In comparison to the natural ecosystem, the previous four studies of the Big Data ecosystem present data in a manner very different than that of decomposers in natural ecosystem. This shows that data has a uniquely important role as an actor in Big Data ecosystem. As we have seen from the first three previous research papers, consumers and providers appear as actors such as users and service providers. When we examined studies #3 and #4, there are additional agents: Application Provider and Framework Provider in #3 and Applications in #4. This also shows that the environment of analysis has a very important role in defining the Big Data ecosystem (see Table 1).

Based on the analysis of previous studies, broad categories for the Big Data ecosystem actors are derived as shown in the Fig. 6 below. By synthesizing the previous studies of Big Data ecosystem's actors, we suggested the actors of the core business of the big Data ecosystem. These core actors of Big Data ecosystem in this paper are service user, infra-provider and data holder. In order to explain the business ecosystem

Table 1. A Comparison of this Study and Historical Research

	Natural ecosystem	#1	#2	#3	#4	This research
Actors	Consumer	User	User	Data consumer	–	Service user
	Provider	Provider	Value producer	Big data application provider	Analytics applications	Service provider
				Big data framework provider	Infrastructure	
					Cross infrastructure	Infra provider
				System orchestrator	Open source project	
	Decomposer	Data	Big data	Data provider	Data sources	Data holder
Features	Basic concept of ecosystem	Suggest the data as an actor	Suggest provider as a value producer	Divide the provider as application, framework and orchestrator	Did not mention about consumer	Present the actors in big data business

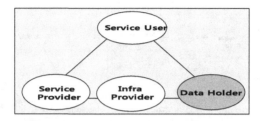

Fig. 6. Core components of Big Data ecosystem

of Big Data, we use the Moore's business ecosystem classification, core business, extended business and business ecosystem. In this study, the following findings are present about the business ecosystem of Big Data and actors of Big Data ecosystem. The following section details each actor.

3.3   Actors in the Ecosystem of Big Data

Obtained through the definition of Big Data ecosystems, this paper discusses key actors, data holders, service users, service providers and infrastructure providers to examine the role of the organic relationship between each of the actors. Data holders hold a large amount of data. The data may be subdivided into public data, private data (company or individual data retention) and social data. Social data is particularly rich in its inclusion of sensed data. Data holders stores and manage the data that are collected by services and business activities, either automatically or manually. These agents process raw data as well. Data holders collect, store and manage data components that are differentiated in other ecosystems. The most important ecological actor in Big Data business generates or creates new value from Big Data. Public data is data held by the national public authority and is processed in accordance with privacy regulations, which limits the potential new value creation. A typical example of public data is the nation's health care data held by the National Health Insurance Agency. Private data can be divided into two types, company holding data and personal data and an example of company holding data can be telecommunications company's data. Personal data is an email and etc. produced in accordance with personal activity. Social data are provided by social media for the purpose of individual or company promotional and marketing activities. This can be collected and utilized by interested businesses. Typical examples of social data are that which is collected and stored by Facebook and Twitter.

Service users include the consumption who use the resulting value via service and are classified as government and public agencies, private organizations (companies and organizations) and individuals according to the attributes of institutions. Service users appear on the market in the ecosystem as a consumer to purchase and use big data services. They perform as a user of services as well as producer of new data. Government and public institutions refer the service organization to improve the people's well-being and quality of life measures such as risk prediction and disasters preparedness programs. Companies and organizations in the private space may play a vital role profiting from Big Data. Facebook is a prime example of its service platform updating posts or pictures, videos, etc. that profits off of the personalized advertising. People can improve their life quality by using Big Data services. A prime example is SKT's T-map information service which shows the fastest route when driving.

The service provider creates value using Big Data and this can be classified with individual data collection, storage services, data processing services, data analysis services and visualization services. A service is not only providing S/W but also personnel, consulting services, and etc. The service provider in natural ecosystems acts as a producer, acting with infrastructure providers. It also generates a business value as an ecological actor.

Data collection and storage services may be defined as collecting various data and the service management securely. This service is required to collect and utilize public or non-public data. Data collections are classified into two types: manually collected data to serve a particular need and automatically collected data as a by-product of an existing business. Data processing services analyzes data to interpret the data for value creating insights. Visualization services enable data to become more intuitive by transforming it from its raw state to more familiar shapes and forms.

Infrastructure service providers manage technologies and infrastructures in order to use Big Data service and can be classified as H/W infrastructure, S/W infrastructure and the network infrastructure. Infrastructure providers play a leading role in activating the Big Data market through technological development and providing servers, storage, and etc. S/W infrastructure includes various S/W products: Hadoop, Cassandra, or Mongo. Network infrastructure includes any communication means, which support the data collection and storage process for Big Data. Until now, we examined the actors and roles of Big Data ecosystem which we defined in this paper.

The actors of the Big Data ecosystem may organically promote the Big Data industry through cooperation. To design supportive policy for Big Data industry development, we need to better understand the complexities of the Big Data ecosystem.

4 Business Ecosystem and the Ecosystem of Big Data

4.1 Business Ecosystem of Big Data

Continued interest in analyzing the industry in terms of business ecosystems is important. In this paper, Big Data ecosystem is framed under Moore's business eco-system theory. After classifying business ecosystem in terms of a core business, extended business and business ecosystem, we studied the actors in the core business of the Big Data ecosystem. In extended business, indirect suppliers are sensor equipment, devices, social networks, the Internet, etc. Indirect consumers are finance, transporta-tion, logistics, and etc. The business ecosystem of Big Data also includes governments, standardization bodies and other relevant organizations. Government agencies and other quasi-government regularly organizations belongs in the business ecosystem (see Fig. 7).

In this paper, the business ecosystem of Big Data is economic community which includes extended business. This community depends on environment, security,

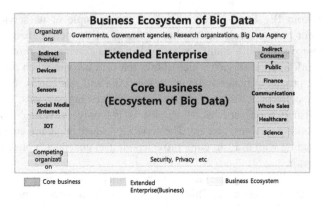

Fig. 7. Business ecosystem of Big Data

privacy, etc. as well as supporting the ecosystem by government agencies, educational institutions, research organizations and etc. Figure 7 depicts a new business ecosystem of Big Data. It shows that the core business is the ecosystem of Big Data. The following sections investigate the concepts and actors of Big Data ecosystem. The core business area of the business ecosystem is the ecosystem of Big Data defined the center and the rest is built in relation to the core.

4.2   Ecosystem of Big Data

The Big Data ecosystem is the core business of Big Data Business Ecosystem and broadly defined by previous studies. In this paper, the business ecosystem of big data is the economic community which includes extended business (enterprises) and depends on the environment, security, privacy, support of the ecosystem by government agencies, educational institutions, research organizations and etc. Figure 8 shows a new business ecosystem of Big Data, which contains the new ecosystem of Big Data as a core business. The concept of Big Data business ecosystem was established on understanding of organic relationship between actors of the Big Data ecosystem.

In this study, the business ecosystem of Big Data was explained based on Moore's business ecosystem. The ecosystem is broken down into three areas: core business, extended business and business ecosystem. From these areas, we identify the actors of Big Data ecosystem from a synthesis of the four previous researches. The business ecosystem of Big Data flows from an organic relationship that occurs in the ecosystem: a value creation process in the business ecosystem as well as characteristics of Big Data. In this study, the ecosystem of Big Data is an organic network of data holder, service users, service providers and infrastructure providers to corporate (co-operate) with each other in order to create new value.

Data holders collect, hold and dig large amounts of data. The service users are the value consumers using the service. Service provider creates value to the user of the data by providing software and manpower. Infrastructure provider is the actor that provides

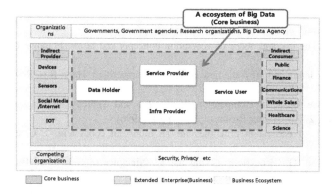

Fig. 8. Business ecosystem of Big Data

H/W, S/W, and network infrastructure. The business ecosystem of Big Data is the economic community including extended business and business-based environment, such as security, privacy and etc.

5 Conclusion and Discussion

Since 2010, the study of Big Data is growing rapidly because of its importance to the business market. In this paper, we contribute to the growing literature on the ecosystem of Big Data. This paper describes the industry and characterizes the actors that make up the ecosystem, fostering better understanding of the Big Data industry. This study may help inform policy and guide the development and support of the Big Data industry. This study organizes and analysis the business ecosystem of Big Data based on Moore's business ecosystem. The paper investigates the core business of the ecosystem of Big Data. The business ecosystem of Big Data depicts core business as ecosystem of Big Data, extended business that includes the core business, and the whole business ecosystem that includes the extended business. Through this characterization and subdivision of the Big Data ecosystem, the model develops a framework for discussing the key actors of the system. In this study, the ecosystem of Big Data is the organic network of data holder, service users, service providers and infrastructure providers to corporate each other in order to create new value. Data holders collect, hold and dig large amounts of data and service users have a role as the value consumer using the service. Service providers offer value to the user of data through software and man-power. Infrastructure providers develop H/W, S/W, and network infrastructure for big data. Business ecosystem of Big Data is the economic community including extended business and business-based environment, such as security, privacy and etc.

Through an understanding of the business ecosystem, this study can be used as research material and industrial development direction of government policy making. Constant research in Big Data area is needed.

Acknowledgements. This work was supported by National Information Society Agency in Korea.

References

Tansley, A.G.: The use and abuse of vegetation concepts and terms. Ecology **16**(3), 284–307 (1935)

Moore, J.F.: Predators and prey: the new ecology of competition. Harv. Bus. Rev. **71**(3), 75–83 (1993)

Moore, J.F.: The Death of Competition: Leadership and Strategy in the Age of Business Ecosystems, 297 p. Harper Business, New York (1996)

Korhonen, J.: Four ecosystem principles for an industrial ecosystem. J. Clean. Prod. **9**(3), 253–259 (2001)

Power, T., Jerjian, G.: Ecosystem: Living the 12 Principles of Networked Business, 392 p. Pearson Education Ltd., London (2001)

Iansiti, M., Levien, R.: The Keystone Advantage: What the New Dynamics of Business Ecosystems Mean for Strategy, Innovation, and Sustainability, 225 p. Harvard Business School Press, Boston (2004)

Moore, J.F.: Business ecosystem and view from the firm. Antitrust Bull. **51**(1), 31–75 (2006)

Mckinsey: Big data: The next frontier for innovation, competition (2011)

Cho, J., Kim, T., Park, K.C., Lee, B.G.: An analysis of big data structure based on the ecological perspective. Korea Soc. IT Serv. **11**(4), 277–294 (2012)

Gatner: Big Data: 12-dimensional Extreme Information Management (EIM) (2012a)

Gatner: 2012 10 Strategy Technologies. IDC, EMC (2012b)

Turck, M., Zilis, S.: The Chart of the Big Data Ecosystem (2012)

Korea Internet & Security Agency: Big Data ecosystem-based analysis of the future of society and industry outlook tasks entrusted Request for Proposals (2012)

National Information Society Agency: Big Data Era: market competition and the surrounding ecosystem strategy analysis. IT & Future Strategy Report (2012)

Ryan & Co Analysis: Future society forecast based on Big Data and Analysis Consulting of Industrial ecosystem (2013)

Grady, N., Balac, N., Lister, E.: Definition and Taxonomy Subgroup Presentation. NIST Big Data Public Working Group (2013)

Pigott, D.: Obituary: Arthur Roy Clapham, CBE, FRS (1904-1990). J. Ecol. **80**(2), 361–365 (1992)

Tansley, A.G.: Practical Plant Ecology (1923)

Peltoniemi, M., Vuori, E.: Business ecosystem as a tool for the conceptualization of the external diversity of an organization. In: Proceedings of the Complexity, Science and Society Conference, pp. 11–14 (2005)

Using Big Data Strategy for the Development
of the Communication Industry

Soonduck Yoo, Jungihl Kim$^{(\boxtimes)}$, and Kwangsun Ryu$^{(\boxtimes)}$

Hansei University, Gunpo, Republic of Korea
harry-66@hanmail.net, jungihlkim@naver.com,
master@how2sns.com

Abstract. This study is being discussed in order to strengthen and secure communications services revenues as a strategy for survival among fierce competition in the telecom services market. Telecommunications companies are experienced in connecting a variety of information, as well as providing various types of data storage. Due to the advent of Big Data era, analysis and utilization of data has emerged as an important issue. Therefore, telecommunications companies have to develop new service and achieve excellence for consumers by utilizing internal data they collect so they can increase profit. In the past, companies underwent internal utilization, but utilizing data sets in the external phase has recently begun. Ultimately, revenue should be obtained through the data and a variety of analyses by providing services directly to consumers. The carriers' revenue growth factor leads to a reduction in call charges for consumers. At the same time, Big Data can lead to market growth.

Keywords: Big data · Communication industry · Service market model · Data industry · Business model in telecommunication

1 Introduction

1.1 Telecommunications Service Market Trend

From 2011 to 2012, domestic (Korean) sales of telecommunications services increased from 53.879 trillion won in to 55.122 trillion won, a 2.1 % year-over-year growth. (see Table 1) In accordance with the global economic recession, in 2013 the domestic economic growth is forecast to slow-down. However, according to mobile operators, sales in Korea, the U.S., Japan, and Europe showed that Korean mobile communications industry climbed 4 % compared with the same period last year; in the United States and Japan, sales grew 3 % and 1 %, respectively. The United States and Japan are also dominant in the long-term evolution (LTE) domestic growth and penetration levels in the world, leading the market. According to them, availability of LTE servies is an important growth factor in the tellecommunication market.

According to another report, "Korea Communication Services & Equipment Market, 2013−2017 Forecast & Analysis," published by Korea IDC (www.idckorea. com), the telecom services market in Korea grew by 1.9 % year-over-year to 30.123 trillion won. It predicts a compound annual growth rate (CAGR) of 0.1 % growth over the next five years to approximately 30.243 trillion won in 2017, which is the largest

© Springer International Publishing Switzerland 2014
Y. Chen et al. (Eds.): WAIM 2014, LNCS 8597, pp. 349–359, 2014.
DOI: 10.1007/978-3-319-11538-2_32

Table 1. Telecom services market revenue Source: Korea Association for ICT Promotion (Unit: One hundred million Korean won)

Division	2006	2007	2008	2009	2010	2011	2012
Total amount of service sales	428,688	454,128	487,034	509,839	523,488	538,791	551,226
Wireline communications services	174,005	177,160	180,693	174,640	165,849	160,390	155,263
Wireless communication services	162,858	172,208	183,290	195,570	203,146	203,277	207,095
Equipment rental and resale recruit circuit. intermediation	17,742	19,945	21,694	17,821	13,210	14,547	12,607
Additional telecommunications services	32,034	35,788	40,940	47,950	54,593	60,583	65,785
Broadcasting telecommunications convergence services	42,049	49,027	60,417	73,858	86,690	99,994	110,476

increase up to now. As shown in Table 1, the Korea Communications Convergence area grew from 999.94 trillion won in 2011 to 1,104.76 trillion won in 2012, and reflects an increase of 10.5 % year-over-year growth in most of the areas [12].

The Korean telecom service market is growing, but the mobile market is saturated. Many companies has introduced data services to distinguis itself from conventional carrier services. In the case of mobile operators of high-speed communication networks, the development of smart phone technology has allowed them to store and process vast amounts of information. This started monetization of data services in the market.

Looking at research on companies that use data, the Korea Communications agency noted that the "ICT industry's 'golden egg' emerged as User Data," (2013) allowing them to create and actively utilize services; the movement is tending toward selling [2, 13]. Furthermore, the development direction of mobile operators led Bae Han Cheol and Oh Yun Su to review the contents of the previous studies, noting the "Strategy and Implications of the growth of global carriers" (2013): they argued that the mobile communication carriers are saturated due to new growth and diversification of services [11].

In this study, the carriers will create a profitable business plan for a new look at the data communication service utilizing Big Data and receive direction for the development of strategies presented. The immense advantage of the value of the data in respect to data availability will be discussed.

2   Big Data

2.1   Data Industry Market Outlook

Korea's DB market is growing with the increasing demand for the new DB system in order to take advantage of increasing data accumulation. The global growing trend of the DB industry ahead of a full-scale entry into a "Big Data" era is very optimistic. The market amounts to 13 trillion by 2015; the market is expected to be 14 trillion in 2017, according to an average annual growth of 5.6 %. Database business is estimated to continue to rapidly grow, at the level of approximately 5.8 % in 2017, compared to 2011 (see Table 2) [14, 15].

Table 2. DB industry market share trend in Korea *Source:* http://www.dbguide.net/

Division	2011	2012	2013	2014(P)	2015(P)	2016(P)	2017(P)	CAGR (11-17)
DB establish	42,374	45,120	46,564	48,566	50,752	52,736	54,573	4.30%
DB onsulting	2,437	2,630	2,839	2,926	3,073	3,251	3,387	5.60%
DB solution	15,800	17,467	20,936	24,014	27,641	30,240	33,735	13,5%
DB service	43,218	44,847	46,257	47,425	49,559	50,839	52,371	3.30%
SUM	103,829	110,064	116,596	122,931	131,025	137,067	144,067	5.60%

At this time in modern society, data is delivered to others via a communication network. Due to the development of networks and smart phone technology, the telecommunications companies hold a lot of data and its typical enterprises in the DB industry. They also serve as a DB creator in the DB market. Previously, telecommunication companies had been using their data as material for new business and internal service development, but since the advent of the Big Data era, they have been devoting more attention to providing external in addition to internal value creation. The following section will discuss how Big Data creates value.

2.2 Big Data

Data becomes more important when there are increased demand for the production and utilization of information [4, 17, 18]. As such, global consulting firms such as McKinsey and Gartner expect to take advantage of Big Data as a key source of national competitiveness. Big Data contains a variety of data that could be useful to a user [7, 16]. Data is generated and stored instantly, hence, needs for data management and analysis have increased accordingly. Due to recent technological development, rise in interest in Big Data has effected related technologies and tools (collection, storage, retrieval, sharing, analysis, visualization, and etc.) Organizations that hold data attempt to create business model by analyzing the data [3]. Data analysis is predominantly carried out through formalized data processing tools. Handling data in a way that ensures data security and privacy is not easy, especially in this early stage of the market [8].

It is almost impossible for companies that do not hold data to enter the Big Data market because of the difficulty of securing data. As a new market, the government is trying to put effort into Big Data-related business. Some large companies are trying to enter the market attracted to the potential business value of servicing Big Data. The telecommunications companies already hold a lot of data and have an advantage because this.

Cisco IBSG carried out twelve in-depth interviews with telecommunications business executives in order to find out whether they are ready for Big Data market. According to the survey results, 80 % of respondents showed that, for the next three years, Big Data is being recognized as their strategic priorities (Fig. 1) [9, 10].

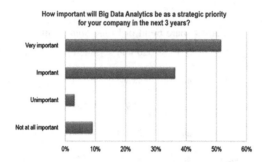

Fig. 1. Cisco IBSG poll question at Telco 2.0/STL Partners Conference, June 2012

2.3   Limitations of Open Data

Telecommunications companies have generated some information, such as financial and transportation information that can be utilized directly with their existing services [6]. Telecommunications companies have an interest in utilizing the data is because there is a need for more affordable Data Integration. Increase in data type, size, and complexity makes it hard for users to rapidly process data or conduct heavy data analysis. In terms of new business opportunities, data provides many opportunities for telecommunication companies. In order to use data effectively, the market is asking for open data, but telecommunications companies are finding it difficult to accomodate such needs [5]. Due to changes in the market environment for business growth and continued threats from potential competitors, telecommunications companies are very sensitive to profitability and are facing many threats in multilateral changes, not to mention customer privacy concerns. For example, due to credit card companies' recent privacy breaches, privacy breaches have emerged as a social problem in Korea. Therefore, safely managing organization's collected data is very important. Processing the data also comes at a high cost, and if revenue is not secured in accordance with sales, data processing can cause a great loss to a company. Thus, services directed to the public, as well as safe management and data processing by the organization that holds data, is considered the best practice.

3   Development of a Business Model in Telecommunication

3.1   Changes in Business Models of Telecommunications Companies

In the past, telecommunications companies' phone call revenue accounted for 90 % of their entire sales revenue [1]. However, registry rate for a new mobile phone has been stagnant and there has been downward pressure to voice call service fees. In addition, SMS (Short Message Service) revenue has sharply dropped in accordance with the development of smart phones that give often freen and unlimited acceses to internet text-messaging services. Thus, communication companies should put more effort into finding new services in order to create a profitable business. The direct carrier billing service is one of the additional profitable businesses for the domestic telecommunications companies; it appeared in the early 2000s and became an approximately 4 trillion won market in 2013. It is shown that service changes in accordance with the technical development.

Global telecommunication companies are trying to increase their productivity in the intense competition of the mobile market. Global carriers are trying to increase growth in global communications market and maximize profit of existing telecommunication business, such as wireless data. Softbank, NTT Do-Co Mo, and other Japanese wireless carriers' wireless data revenue has already grown to take up more than 50 % of total sales, and U.S. telecommunications' data revenues grew to 40 % of total sales (Fig. 5).

Korean carrier companies have focused on domestic market rather than expanding to global market. They tend to beef up their size by converging variety of services: telecommunication, internet, payment and billing, and etc. Korean carriers that have experienced limited growth in the domestic market can find new growth opportunities in the global convergence region through the introduction of Big Data, as well as the growth of the telecom sector, such as LTE. Thus, they are sensitive to the development of new technologies, and new channels for revenue should be considered very aggressively. Korean telecommunication companies effect their entrance into the business area through excavation of the convergence zone. This will prove to ensure the safety, growth, and profitability of Korea's telecom market from toll cuts, regulatory constraints, and many other internal constraints.

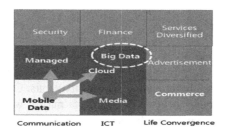

Fig. 2. Expansion strategy of the service operators in telecommunication market. Source: Digital ecosystem (www.digieco.co.kr)

3.2 Industrial Development Phase of Big Data Usage According to the Cisco IBSG

Some experts said that Big Data is "in the wake of a new leap" for telecommunication firms.

According to "Unlocking Value in the Fragmented World of Big Data Analytics" by IBSG (Internet Business Solutions Group), Big Data was classified in three waves of industry evolution (see Fig. 3). Today, most industries are squarely entrenched in Wave One. In this stage, data is compiled, but the impact of these reports on any decisions is

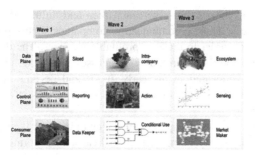

Fig. 3. Big data evolution

indirect. Their main worry is about the security and integrity of their data, and a vague feeling of potential data abuse prevails. In Wave Two, Data can move across individual silos within the organization and out into a common repository. Once there, the much broader set of collected data can be analyzed and shared for more trenchant decision making. By Wave Three, data is being collected, shared and processed across an entire ecosystem. Enterprises can leverage not only their own data, but also a myriad of third party data, including weather patterns, economic indicators, "anonymized" retail sales, social media traffic, and so forth. They are now standing in a Wave One situation; an enterprise-wide strategy for Big Data should be established. As can be seen in Cisco's view, utilizing the data of telecommunications companies appears to be very important, and data holders have seen the potential impact of the agency on the future market. Using this service will offer consumers many advantages (Fig. 2) [10].

4 Practices

Now, with the introduction of the Big Data market, telecommunications companies have the environment to take advantage of new technologies and holding information, as well as to make revenue by providing services to clients. Currently, telecommunications companies are trying to enter the Big Data market with a new business by utilizing the data. Telecommunication firms are in the stage of introducing new services directly to platform using their data. In early stage, they use data to meet their own corporate needs through Big Data analysis, and gradually they will attempt to provide external services from the inside.

4.1 To Take Advantage of Data for Internal Analysis

One international example of using Big Data for internal analysis is the "Precision Market Right," launched in 2012, which tracks the location information of mobile phone subscribers, apps, etc. This can replace existing market and research data, formally collected by direct aquisition, to accurately analyze the usage of information held by telecommunication companies. This service offers possibilities that were previously found difficult: to gain a new point of view of consumers. As a prime example, by analyzing the consumers who attended one of America's most famous events, the Super Bowl (Baltimore Ravens versus the San Francisco), it can be determined that 16 % of viewers were from Baltimore and 6 % from San Francisco. Since people from Baltimore were more actively engaged in this event, companies can adjust their marketing strategies accordingly.

4.2   To Provide Information in the Form of a Data Set to Outside Vendors

The UK carrier O2 services subscribers in cooperation with Starbucks: within a certain radius, Starbucks ships promotional discount coupons. This generates revenue through data collected in real time and sold to Starbucks. As a prime example in the Korean market, late-night bus routes in Seoul are determined by taking advantage of KT (Korea Telecom)'s data. This project uses the customers' communication information, provided by the carrier, to determine when and where users are most frequently in need of transportation.

Another example is communication data analysis used to estimate the number of participants that will attend local festivals. During events, data is collected from radio base stations in specific areas. Analyzing the usage pattern of call data in the entire area of the event over time helps to determine the number of participants for the festival and, therefore, festival organization (government) can estimate the supporting budget scale. The government budget support program has been able to operate efficiently through this resolution.

The commercial analytics service, "Geo vision," which is provided by SKT (one of Korea's telecommunication firms) is one data set selling model. This service uses data held by SKT in real-time to perform analysis of consumer patterns, behavior patterns and the local Information (see Table 3). People who want to open their store use this service to search the best location for their business and call data is core one to produce this analysis. Table 3 explains that what kind of analysis result can be shown in the service. The consumption patterns are described as (1) Businesses sector growth trends, (2) Total sales of commercial market (3) Sales trends by sector and 4) Analysis of consumer patterns. (1) Status of potential customers, (2) Residential population and (3) Changes in regional population influx are shown the behavior patterns in the certain area where they want to analyze. Local information, such as the main facilities and real estate price and trading information are also used for analysis (see Table 3).

Figure 4 depicts distribution of certain stores (product) in Seoul, Korea. This service helps to find the best location for their business as well as to make marketing plan for their product promotion.

According to the technology development, telecommunications companies have secured new revenue-based services to provide indirect benefits for consumers.

4.3   To Provide Services Directly to Customers by Utilizing Data

In this case, telecommunications companies give direct service to users by building services utilizing the data that they hold. Currently, telecommunications companies

Table 3. The analysis result from Geo vision

Consumption patterns	Behavior patterns	Local Information
1) Businesses sector growth trends 2) Total sales of commercial market 3) Sales trends by sector - Hourly turnover rate - Weekday, weekend sales ratio - Revenue share of loyal customers - The average amount spent per customer - Average purchase commercial customers: customer traffic 4) Analysis of consumer patterns - Analysis of consumption style	1) Status of potential customers: Location, area 2) Residential population - Population growth trend with housing type - Changes in house prices - Residential population, gender, age lifestyle changes 3) Changes in regional population influx - Scale influx of population - Hourly population trends - Style trends 4) Weekly resident population -Size of the population resides in weekly - Status of corporate employees 5) Analysis of behavior patterns	1) The main facilities details - Major Facilities status - Distribution of schools around the commercial area - Transportation facilities 2) Real estate price and trading information - Commercial Lease status - Residential status in downtown - Residential Real Estate price 3) Development of information - Pre-sale information - Development of transportation information

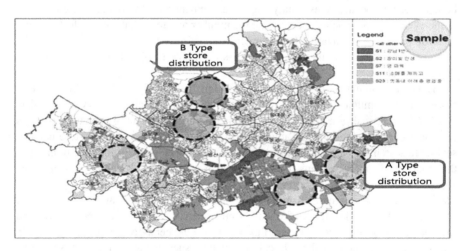

Fig. 4. Analysis of product (store) area by Geo vision

which have a lot of data have little ability to source the data to the outside because of security issues. For example, recent information leakage caused by a credit card company brought enormous damages to all parties involved. In order to solve such problems, direct services must be provided to the consumer via data management. With the advent of the era of Big Data, telecommunication companies need to provide various services and secure revenue by effectively utilizing their data.

5   Carrier Business Model Development Stage by Introduction of Big Data

In this study, the direction of carriers' revenue development using Big Data was classified in terms of utilization of leverage, both internally and externally. The purpose of internal utilization is to develop services from the data for their internal needs. Because of the issues and concerns about open data, opening up data to public has been difficult. In this situations, companies can generate revenue by enhancing services provided in addition to introducing basic Big Data services. From this, companies may be able to create new value from data which will set themselves more attractive to consumers than the competitors.

Companies can also attempt to enter the external services in order to take advantage of Big Data and profit from it. This involves establishing business model that is differentiated from services that are already provided internally. Corporate customers are often interested in such services. This is providing a service to outside companies by analyzing the information–such as call volume–within the privacy policy of the data user. A prime example of this is a one-time processing of the data itself. Another form of external utilization is to provide direct Big Data services to individual users and the enterprise by developing its own services.

When Big Data services market reaches maturity, private customers are connected to a specific site where they can shop for data and conduct data analysis in streamline process. During external utilizing, the important idea is that the data sets from the many services provided by telecommunications companies are sold to public and private companies in order to build a Big Data service. However, selling the data set as a one-off sales service presents some difficulties in generating sustainable revenue. When opening the processed data to the outside, open data that is different from the user's need is difficult to use. If disclosure of the data is done incorrectly, then there are liability issues for damage caused by open data. In the case of recent security breach by a credit card company, the data holding institution caused serious social problems. However, using the data as a service to individuals and businesses is advantageous in

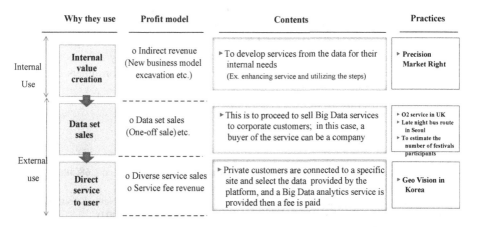

Fig. 5. Development direction of telecommunication data

terms of ensuring a variety of revenues. In addition, security, privacy, and other issues can be resolved through the direct and effective management of data.

In order to generate sustainable revenue, telecommunications companies must generate various Big Data services using the retained data. In order to create value from the user through each service, revenue should be obtained from service fees. In this case, the privacy-related measures, such as self-management and operation, can minimize the risk, and the burden of providing data directly to the outside can be reduced. Core companies in the field of data communication, if they have secured good revenue models, should be able to cut phone call costs (rather than continue to increase them) by providing stable communications service. According to Cisco, Wave One uses data internally, and Wave Two consists of the integrated management of data as a result of the utilization of outside-limiting purposes. Data sets on sale in this study can be classified into a similar category. Wave Three is the stage in which the management of services, as well as the production and sale of various types of data, is done, and it can be described as providing services directly through data, as presented in this study.

6   Conclusion

The possibility of the use of a variety of Big Data has evolved. Unlike the conventional method that predicts the future through sampling and statistical models, we live in an environment that creates profits when enterprises and institutions find correlations among new concepts through exhaustive research. As we are introduced to the era of Big Data, a variety of communication strategies should be explored by utilizing the data held by telecommunications companies, so we investigated development strategies in accordance with the utilization of Big Data. Data should be directly generating revenue to provide various services beyond selling. It cannot make long-term profits from data sets in the form of a one-off sales strategy. Data utilization can increase the revenue of the telecommunication companies, as well as provide a variety of benefits to the user by creating various services. Furthermore, utilizing a variety of data services for users can contribute to the activation of the Big Data market.

In order to find applicable services for telecommunication data, experts' opinions and ideas, excavation proposals, and overseas case references should be used. Telecommunications companies are secured by various revenue models through data and generate new revenue. To increase revenue through enhanced safety reduces the need for fee hikes to long-term customers; therefore, many benefits for consumers could be provided as well.

References

1. Antonelli, C., Baranes, E.: Communication & Strategies. Int. J. Digit. Econ. **68**(4th quarter), p. 67 (2007)
2. Fransman, M.: Innovation in the New ICT Ecosystem: How New Knowledge is Endogenously Created in the ICT Sector, University of Edinburgh (2007)
3. Big Data, NASSOM (2012)

4. The Guardian British media, Big data use cases presented everyday life, Korea Internet & Securiy Agency (2012)
5. Yoen, C.J.: An Analysis of Big Data Structure Based on the Ecological Perspective, Korea Society of IT Services (2012)
6. Hyun, C.G.: This applies to domestic and international examples of Big Data Implications, National IT Industry Promotion (2012)
7. Osika, C., Griffin, K.: Service Providers Are Just Getting Started with Big Data (2012)
8. White, C.: A Next-Generation Analytics Ecosystem for Big Data, BI Research (2012)
9. Gerhardt, B., Griffin, K., Klemann, R.: Unlocking Value in the Fragmented World of Big Data Analytics. How Information Infomediaries Will Create a New Data Ecosystem, Cisco Internet Business Solutions Group (IBSG) (2012)
10. Cisco IBSG poll question at Telco 2.0/STL Partners Conference (2012)
11. Bae, H.C., Oh, Y.S.: Suggestions strategic direction and growth of the global communication (2013)
12. Korea Communication Services & Equipment Market, 2013−2017 Forecast & Analysis (2013)
13. ICT industry's new golden egg emerged as user data, Korea Communications agency, vol. 67 (2013)
14. Big Data industry trends, KISTI Market Trend, Special Issue April (2013)
15. Big Data Market Analysis, KAIT (2013)
16. Big Data trend analysis and application strategies, DMC media (2013)
17. Case of Big Data Success Guide, IDG Group (2013)
18. Compher, D.: The Big Deal About Big Data (2013)

A Data Model for Object-Based Seamless Digital Map Using Digital Map 2.0

Hyeongsoo Kim[1], Hyun Woo Park[1], Wonwoo Cho[2], Incheon Paik[3],
and Keun Ho Ryu[1(✉)]

[1] Database/Bioinformatics Laboratory, Chungbuk National University,
Cheongju, South Korea
{hskim,hwpark,khryu}@dblab.chungbuk.ac.kr
[2] Geospatial Information Technology Co., Ltd., Seoul, South Korea
zzonu@nate.com
[3] School of Computer Science and Engineering, The University of Aizu,
Aizuwakamatsu, Japan
paikic@u-aizu.ac.jp

Abstract. Recently, a demand for a spatial information service as well as spatial data model that can manage spatial objects effectively has increased rapidly with the emergence and proliferation of the smart societies due to the introduction and diffusion of Ubiquitous, Intelligent Transport Systems (ITSs), Telematics and Location-based Systems (LBSs). Various spatial data models have been proposed to meet the demand. However, most of these models have some problems such as the construction of exceptional functions, inefficient data management and limited availability. Digital map 2.0, a tile-based data model, which is very popular in South Korea, suffers from the aforementioned problems. This paper tackles the limitations of tile-based digital maps by suggesting a data model for Object-based seamless maps. In essence, the proposed data model can be easily implemented (constructed) and applied to a variety of application domains by constructing object-based data model build on the Digital map 2.0. The proposed model was applied to a real world application scenario as a verification of the model's applicability to various systems and application domains.

Keywords: GIS · Digital map · Spatial data model · Object-based · Seamless

1 Introduction

A demand for a spatial information service has increased rapidly with the emergence and proliferation of the smart societies due to the introduction and diffusion of Ubiquitous, ITSs, Telematics and LBSs. In addition, the use of Geographic Information System (GIS) has increased in various applications such as navigation systems and Global Positioning Systems (GPSs) or Map services.

The management of the spatial data in the digital map is computationally expensive and time consuming. Thus, an efficient support system is crucially important to systematically manage, and smoothly support the spatial data. Consequently, several advanced counties have put considerable research efforts to manage spatial data efficiently [1, 2].

Y. Chen et al. (Eds.): WAIM 2014, LNCS 8597, pp. 360–371, 2014.
DOI: 10.1007/978-3-319-11538-2_33

Generally, management methods of digital map's spatial data are divided into tile-based and object-based. Tile-based data which is managed by tile can be easily implemented. However those have some problems related to storing the data, tracing and offering of correct history information. In contrast, object-based data which is managed directly by object unit is relatively complex. However the data has a good property which can enhance usefulness and solve problems about updating and history management of tile-based data [3–6].

Shapefile and GML are representative examples for spatial data models applied to various systems. However, the spatial data in these models are either managed by file or tile unit. Thus, these spatial data models need efficient data manipulation mechanisms for object insertion, modification and deletion. In South Korea, the most popular data model is Digital map 2.0, which is an example of a tile-based model. A demand for the object-based data model that guarantees efficient management and currency of spatial data has increased due to the fact that existing tile-based data model lags far behind satisfying its user's requirements.

In order to solve existing data model's problems, we proposed a suitable data model for Object-based seamless digital map by analyzing and using the data structures of the Digital map 2.0. In essence, we designed an efficient object-based data model that manages data using OSID. Furthermore, we have applied the proposed model to a real world application scenario as a verification of the model's applicability to various systems and application domains.

This paper is organized as follows. In Sect. 2, we review existing models and management methods of spatial data. We describe data modeling in Sect. 3. And we verify the applicability of the proposed data model in Sect. 4. Finally, we give our conclusions in Sect. 5.

2   Related Work

2.1   Review of Spatial Data Models

The spatial data models are divided into standard spatial data models and business spatial data models. ISO/TC211 spatial schema and XML-based spatial data models such as GML and CityGML are commonly used standard spatial data models. Shapefile is a widely used business spatial data model [5, 7].

ISO/TC211 is an international standard organization that standardizes various fields of digital geographic information. In ISO/TC211, there are many standards about spatial data. CityGML is an XML-based spatial data model for the storage, management and transmission of three-dimensional city model. This model is defined as OGC standard and registered in ISO 19136 [8, 9].

One of the business spatial models is Shapefile. This model is the most widely used spatial data model [10], and it includes three types of files *shp*, *dbf*, and *shx*. *Shp* file represents shape information of each geometric object and *dbf* file shows common attributes for objects. *shx* is an index file which is used to connect *shp* and *dbf* files. In South Korea, Digital map 2.0 is the most popular spatial data model which is an extension of Digital map 1.0. This model's structure is very similar to Shapefile and consists of *ngi* file and *nda* file that represent shape information and common attributes like Shapefile [7, 11, 12]. Therefore, many companies or institutions in South Korea use Shapefile based data model that follows the property regulations and spatial data models of Digital map 2.0.

2.2   Management Method of Spatial Data

Spatial data for digital map can be divided into tile-based data and object-based data. Tile-based data is managed by tile and can be easily implemented. Nevertheless those have many problems related to storing data, tracing and offering of correct history information, caused by the overhead of managing large number of spatial objects by the tile unit [4–6, 12]. In contrast, object-based data which is managed directly by object unit is relatively complex. However the data has a property that can solve problems related to updating and history management of tile-based spatial data. Hence this system is able to immediately update any change in the data and to supply users with the updated data without any delay [6, 7].

Recent literature proves sufficient studies about object-based spatial data model and Unique Feature Identifier (UFID) in South Korea [13, 14], Nevertheless, these studies have either used data that is hard to convert into object-based, or cannot be generalized to different application domains. In spite of the merits of the object-based spatial data, a system applicable to general uses has not been developed yet in Korea. In order to solve the aforementioned problems, we suggest a data model for object-based seamless digital map based on Digital map 2.0.

3   Designing Data Model for Object-Based Seamless Digital Map

3.1   Conceptual Modeling

In this process, we define conceptual model and schema from analyzing Digital map 2.0, regulating the range of objects and defining the concepts of class, and class group. Overall, the object-based seamless digital map is composed of two data models; one is the spatial data model that represents the shapes of objects in space, *i.e.* geometric model, and the other is a non-spatial data model that represents the geographic information. The spatial data model provides geometrical objects that express spatial information as coordinates from which points, lines, and surfaces can be composed. To raise compatibility we have designed spatial data model build on OpenGIS's Simple Feature Geometry model. It is not necessary to express the features of object-based seamless map using the all types which are defined in Simple Feature Geometry model [15]. Rather, these features can be expressed using simple classes. Therefore, in our design, although we have overlooked the three-dimensional aspects, we have considered the important elements to express the features of Digital map 2.0. Figure 1 shows (b) a geometrical model for object-based seamless digital map based on Digital map 2.0 and (a) OpenGIS's geometric model. The non-spatial data model of Object-based seamless digital map applies the data structures and classification system of Digital map 2.0. The data structure is divided into 2-themes of 8-ClassGroup each of which is comprised of 83 classes of objects, forming a hierarchical structure. This model can sufficiently express the characteristics of all objects as well as managing the features as one object. In this paper, we illustrate our proposed data model using Traffic Class-Group as typical case. Transportation is more complex than other themes within which update operations occur frequently. However, the proposed model can be applied easily to features of other themes.

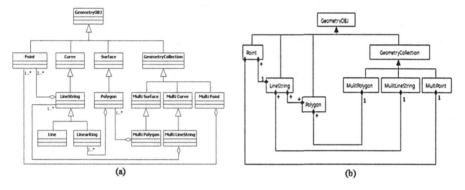

Fig. 1. (a) Geometric model of OpenGIS, (b) geometrical model for object-based seamless digital map

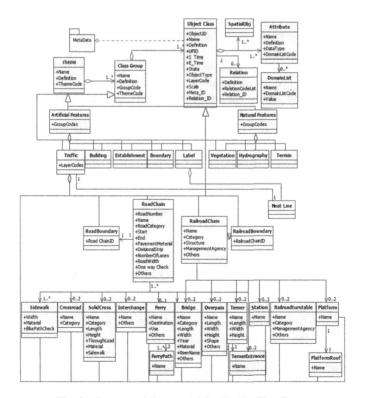

Fig. 2. Conceptual data model for Traffic ClassGroup

Object and Component Definition. It is the most important part to define features in conceptual modeling. Traffic *ClassGroup* is made of 18 features. Among these features, *RoadChain* and *RailroadChain* are important features and they form one huge traffic network. The remaining 16 features perform specified functions through correlations with *RoadChain* and *RailroadChain*. Positional correlation between these can be used to verify the accuracy and currency of the updated data. Figure 2 shows the overall conceptual data model for the Traffic *ClassGroup*.

Relationship Between Objects. Update operation in the digital map is the process in which existing objects disappear while new objects are created. The data inspection process needs to prevent errors that may occur during the update process. In the existing Digital map 2.0, it is difficult to process data inspection automatically because objects are stored independently. The proposed data model defines relationships between each object table in order to facilitate the detection of update related errors. Relationships between object tables can be categorized as follows

(1) *Subordinate relation between objects.* Some objects inherit all characteristics from other objects. In Traffic *ClassGroup*, *RoadBoundary* represents the boundary of *RoadChain*, and it has all attributes and boundary information of *RoadChain* object. To avoid duplicate data and integrity errors, all the attribute values of *RoadBoundary* object refer to attributes of *RoadChain* feature. This subordinate relation can be used to check errors during the update process. For example, when a new *RoadBoundary* object is inserted, *RoadChain* is inserted as shown in Fig. 3 (a). If *RoadChain* object is modified or deleted, *RoadBoundary* connected with *RoadChain* is also modified and deleted simultaneously. Figure 3(a) shows the subordinate relation between *RoadBoundary* and *RoadChain*. The subordinate relation is designed to store information of *RoadChain* that corresponds to *RoadBoundary* by adding the primary key of *RoadChain* as a foreign key in the *RoadBoundary* table. *RoadBoundary* object can refer to the necessary attributes via *RoadChain*'s primary key that is added to its own table.

(2) *Connection relation between objects.* Several objects of Traffic *ClassGroup* maintain a function that can connect each *RoadChain* or *RailroadChain*. The representative example is Bridge- an object that connects *RoadChains* when a river is blocking two *RoadChains*. That is, for the bridge to achieve its function, two *RoadChains* or *RailroadChains* should be connected. If there is no *Road-Chain* or *RailroadChain* connected with Bridge when one Bridge is created, the accuracy should be confirmed because this is a high possibility of input error. We defined dual relationship between Bridges and *RailroadChains* or *RoadChains*. Bridge object has a foreign key that refers to the primary key of *RoadChain* or *RailroadChain*. If there is no connection between objects, the system displays warning message as shown in Fig. 3(b).

(3) *Confluence Relation between objects.* When deleting a *RoadChain* that is a part of a *CrossRoad* (*i.e.* an intersection), the entries of *RoadChains* mainlined by the *CrossRoad* object should decrease by an equal amount of deleted *RoadChains* as shown in Fig. 4. Consequently, in a special case of two *RoadChains*, if one

RoachChain is deleted, its associated information should also be deleted from the *CrossRoad* object and hence the *CrossRoad* object itself will become obsolete and should be deleted. This case is illustrated in Fig. 4.

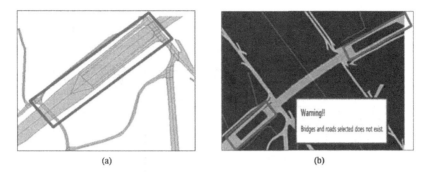

(a)   (b)

Fig. 3. (a) Subordinate relation between RoadChain and RoadBoundary, (b) Insertion error of Bridge object

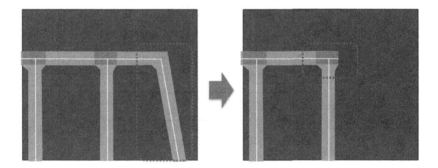

Fig. 4. Confluence relation between RoadChains and Crossroad by deletion

Seamless Expression and Management of History Data. We supplied Object-based Seamless Identifier (OSID) using the feature code of Digital map 2.0 to identify objects as shown in Fig. 5. OSID is classified as three elements code. The first 8 digits are reserved for the feature code; which is a combination of the 4 digit layer code of digital map 2.0, and 4 digits layer code of digital map ver.1.0. The second 7 digits represent a serial number that can handle more than nine million objects of equal type. Also, this grant system presents OSID structure by 16 digits including parity bit of 1 digit for error confirmation.

History management is an important property of the proposed data model. Existing data models are characterized by their complex management of the history data because they maintain two separate databases for the new and history data. To reduce this inconveniences, the proposed model integrated the two databases into a unified database. In other words, we applied S_time (Generation time of object – Start time) and E_time (Extinction time of object – End time) that was defined only in history database schema of the existing systems, to the all objects of the data model. As a result, all objects have their own survival cycle, and new objects have a specific value named "Until Now" as their E_Time value.

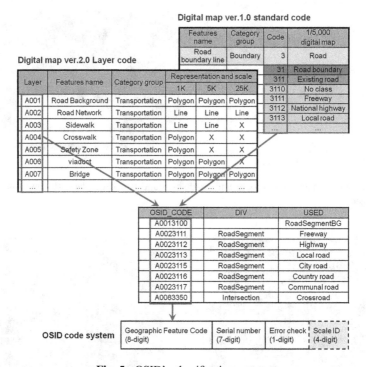

Fig. 5. OSID's classification system

3.2 Logical and Physical Modeling

We update the data model by applying normalization to the conceptual model. Namely, each relationship is redefined by removing unnecessary attributes and adding necessary attributes or relation tables through logical modeling. In 1 to 1 relationships or 1 to N relationships, a direct relation can be created between the associated tables. However, M to N relationships can be materialized by creating an intermediate table because a direct relation cannot be formed. Figure 6 shows Traffic *ClassGroup*'s logical model that have an intersection relation represented by M to N relationship.

We designed logical data model using actuality relation data model of Oracle DBMS. Logical data model is divided into Meta table schema and Entity table schema. Meta table schema stores information of theme or Group tables and forms hierarchy of classification system for the overall Object-based seamless digital map. Entity table schema stores actuality information of each object. Lastly, we formed final data model by assigning concrete data type about logical data schema and established an index for efficient management of the data at the physical model.

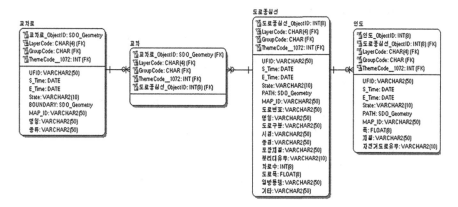

Fig. 6. Logical model for Intersection relation

4   Suitability Verification

4.1   Seamless Expression for an Object

Proposed data model identified logical object using OSID and all geographic object in real world has their unique code. Given there is a building named "43 building, Chungbuk National University" and its OSID code is "B00145624320416". This code was assigned when the "43 building, Chungbuk National University" object was inserted. If a user wants to search for this building from the database, he can issue the following query

```
Select *
From Building
Where OSID == 'B00145624320416'
```

When the query is executed, the system returns two objects in the result as shown in Fig. 7. The building object is divided into two physical objects and stored in the database. That is, the building object is cut into two different tiles which are "36706039" and "36706040". However, we assign a same OSID code to the physical objects in order to appear as one logical object.

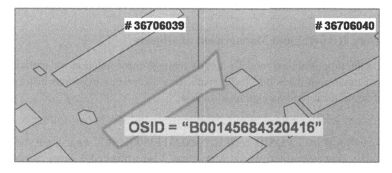

Fig. 7. Logical model for Intersection relation

4.2   Data Inspection for Real Time Updates

Data inspection is checking for errors that might occur due to updates. This model supports automatic checks for logical errors utilizing inter-table relationships. To verify the process of inspection, we considered an update case by deleting a *RoadChain* as shown in Fig. 8(a). Because there is M to N relationship between Crossroad and *RoadChain*, these two tables must connect through relation table called Crossing table. Whenever new *RoadChain* is connected to Crossroad, a new relation instance is created in crossing table. In Fig. 8(b), 3 relation instances exist on crossing table that connects one Crossroad and 3 *RoadChains* as shown.

For example, when one *RoadChain* whose serial number is 3 is deleted, the system would delete *RoadChain* object from database and then count the number of *Road-Chain* objects connected to Cross-road from relation of the data model. As a result, the number of relation instances in crossing table decreases by 2. However, from a conceptual point of view, Crossroad was created when more than 3 *RoadChains* were connected at one point. To put it simply, since one Crossroad object have to be connected to more than three *RoadChain* objects, the system would delete Crossroad object automatically at the same time from database as shown in Fig. 9.

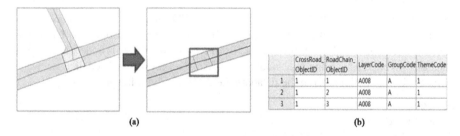

(a)   (b)

Fig. 8. Data inspection case by deleting a RoadChain

CrossRoad_ObjectID	RoadChain_ObjectID	LayerCode	GroupCode	ThemeCode
1				

Fig. 9. A result of data inspection case Fig. 9

4.3   History Retrieval and Management of Objects

To confirm the proposed data model's history manage-ment function, we explain by an example of tracing up-dates on "local road". If a user wants to search history information of some local road as explained bellow:

How many times No.594 local road was updated from 2005 to 2008?

If No.594 local road's OSID code is "A00231135001196", we can get the results through query processing and would know that it was updated 3 times as shown in Fig. 10. In addition, from the results, we can further investigate about the details of each update operation. As it is explained in Fig. 10, OSID is an important identifier to the search history information of an object. Thus, proposed model can trace the history for updated objects at any given time.

OSID	Data_type	Table_Name	Update_Method	Update_User	Update_Purpose	State	S_TIME	E_TIME
A00231135001196	LINE	RoadChain	Comparison	Hyeongsoo	Road repairing	Correction	2006-02-15	2006-10-31
A00231135001196	LINE	RoadChain	Comparison	Jin Hyoung	Road repairing	Correction	2006-10-31	2007-02-20
A00231135001196	LINE	RoadChain	Comparison	Sunggyu	Road repairing	Correction	2007-02-20	UC
A00231175000744	LINE	RoadChain	Comparison	Kildong	Road repairing	Correction	2007-04-16	UC
A00231175000736	LINE	RoadChain	Comparison	Sung Jik	Road repairing	Correction	2007-02-10	2007-04-21
A00231175000736	LINE	RoadChain	Comparison	Han Seok	Road repairing	Correction	2007-04-21	2007-05-12
A00231175000736	LINE	RoadChain	Comparison	Sang Yeob	Road repairing	Correction	2007-05-12	UC
A00231105000702	LINE	RoadBoundery	Comparison	Hyeongsoo	Road repairing	Correction	2007-10-13	UC
A00231105000704	LINE	RoadBoundery	Comparison	Sung Jik	Road repairing	Correction	2007-05-12	2008-05-11
A00231105000704	LINE	RoadBoundery	comparison	Sang Yeob	Road repairing	correction	2008-05-11	UC

* UC(Until Change) = NOW

First updating   Second updating   Third updating

Fig. 10. Example of history tracing

4.4 Retrieval of Multi-scale Object

Digital map 2.0 is generally divided into three scales 1:25,000, 1:5,000 and each of these representations for the object are different. For example, Chungbuk national university called "CBNU" is expressed as one Building object, and "Hakyeonsan" is not expressed at the scale of 1:25,000. On the other hand, "CBNU" is not expressed at the scale of 1:1,000. Instead of "CBNU", Specific building objects were included such as "Hakyeonsan", "Gymnasium" and "Library". That is, Specific building objects are merged to one building object like "CBNU" while moving from large-scale to small scale as shown in Fig. 11.

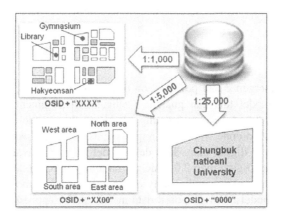

Fig. 11. Example of multi-scale expression for object

4.5 Comparison with Existing Data Models

We compared the proposed model with other main spatial data models such as Shap file, GML and Digital map 2.0. Table 1 shows the results of the comparison along with the characteristics of each spatial data model. First, All data models use tile-based management system except the National Geographic Framework Database and the proposed model. Especially, proposed data model can support all of the tile-based and object-based management methods.

Most of the existing data models did not support functions for history management because they were designed solely for the construction and expression of spatial data, while our data model can process history management promptly and easily. Our data model provides periodic and real time updates due to its object-based management.

Functional differences between each model are related to the difference of their file structure. Shapefile, GML and Digital map 2.0 have SHP, XML and NGI as file structures. Our data model supports systematic data manipulation operations. Namely, it supports efficient data insertion, deletion and modification. Identical objects are stored as independent objects and are given different scales in the existing systems. In contrast, our data model can express identical objects in multi-scale digital map using OSID's scaleID code.

Table 1. The results of Comparison with each spatial data model

Data model functions		Shapefile	Digital map 2.0	National geographic framework DB	Our model
Tile-based management		O	O	X	O
Object-based management		X	X	Difficult to manage	O
Periodic update		O	O	Difficult to manage	O
Real time update		X	X	Difficult to manage	O
Update data inspection		X	X	Basic operations	O
Geometric elements	0 ~ 2D	O	O	O	O
	3D	X	X	X	Δ
File system		SHP file	NGI file	DBMS	DBMS

5 Conclusion

Recently, by the emergence and proliferation of the smart societies, a demand for the object-based data model that can guarantee efficient management and currency of spatial data has increased because existing tile-based data model could not satisfy user's requirements. To cope with these requirements, we proposed a data model for Object-based seamless digital map. This data model improved availability using the data structures of the Digital map 2.0 which is famously used model in South Korea. Furthermore, we analyzed the applicability of the proposed data model and compared with existing data models. Our proposed data model can be applied to various spatial information services.

Acknowledgements. This research was supported by the MSIP(Ministry of Science, ICT and Future Planning), Korea, under the ITRC(Information Technology Research Center) support program(NIPA-2014-H0301-14-1022) supervised by the NIPA(National IT Industry Promotion Agency) and the National Research Foundation of Korea(NRF) grant funded by the Korea government(MEST) (No. 2008-0062611) and Basic Science Research Program through the National Research Foundation of Korea (NRF) funded by the Ministry of Science, ICT & Future Planning (No.2013R1A2A2A01068923).

References

1. Vangenot, C., Parent, C., Spaccapietra, S.: Modelling and manipulating multiple representations of spatial data. In: Richardson, D.E., van Oosterom, P. (eds.) Advances in Spatial Data Handling, pp. 81–93. Springer, Berlin (2002)
2. Illert, A.: Production of Digital Maps for the German Authoritative Topographic-Cartographic Information System ATKIS (2002)
3. Lee, Y.K., Lee, E.J., Ryu, K.H.: Design of multiversion index for history management of discretely changing spatio-temporal objects. In: the 30th KISS Fall Conference, vol. 30, pp. 184–186, KISS, Seoul, Korea (2003)
4. Kim, S.K.: Design of a feature-based spatial data management system. Master's Thesis, Chungbuk National University, Korea (2004)
5. Chi, J.H., Kim, S.K., Ryu, K.H., Kim, M.J.: Design of a feature-based spatial data management system for digital map. J. Korea Spat. Inf. Syst. Soc. **7**(3), 107–118 (2005)
6. Kim, H., Kim, S.Y., Lee, Y.K., Seo, S., Park, K.S., Ryu, K.H.: Spatial data model of feature-based digital map using UFID. J. Korea Spat. Inf. Syst. Soc. **11**(1), 71–78 (2009)
7. Lim, K.H., Jin, C.H., Kim, H., Li, X., Ryu, K.H.: A design of feature-based data model using digital map 2.0. J. Korea Soc Comput. Inf. **17**(7), 33–44 (2012)
8. OpenGIS Consortium: OpenGIS Geography Markup Language Encoding Standard. OpenGIS Stnadard, OGC 07-036 (2007)
9. ISO/TC211: GeographicInformation-Geography Markup Language(GML). ISO 19136 (2007)
10. ESRI: ESRI Shapefile Technical Description. An ESRI White Paper (1998)
11. Kim, S.Y., Kim, H., Seo, S., Ryu, K.H.: Design of spatiotemporal data model for managing history of digital map. In: the 32th KIPS Fall Conference, vol. 15, pp. 352–355, KIPS, Seoul, Korea (2008)
12. KICTEP: Technology Development of Digital Mapping for Next Generation. KICTEP R&D final report (2011)
13. Kim, J.H., Jung, D.H., Kim, B.G.: A study of methodology to grant UFID (Unique Feature IDentifier) of geographic features. J. Korea Open GIS Assoc. **5**(2), 23–31 (2003)
14. Jo, J.R., Shin, S.C., Kwon, C.O., Jin, H.C.: Design on the system and OSID structure for managing object-based seamless digital map. J. Korea Spat. Inf. Soc. **19**(3), 73–81 (2011)
15. OpenGIS Consortium: Simple Features Specification For SQL 1.1. OpenGIS Project Document, OGC 99-049 (1999)

Parallel Gene Clustering Using MapReduce

A.K.M. Tauhidul Islam, Chae-Gyun Lim, and Byeong-Soo Jeong[✉]

Department of Computer Engineering, Kyung Hee University,
1-Seocheon-dong, Yongin-si, Gyeonggi-do 446-701, Korea
{tauhid,rayote,jeong}@khu.ac.kr

Abstract. Data clustering has been considered as one of the most important techniques for unsupervised learning in diverse applications. Gene clustering is to find out groups of genes similarly expressed in large size of microarray data. Meanwhile, recent development of microarray technology generates a very large number of microarray data with low cost and handles more than 10,000 genes simultaneously in one chip. Thus, high performance computing of gene clustering has become increasingly important in microarray data analysis. In this paper, we propose a scalable parallel gene clustering method using the MapReudce programming model. The proposed method utilizes the *k-means* algorithm for identifying similar groups of genes. Experiment results show that the proposed method can offer good scalability with data size increases, and different numbers of nodes, and it can also provide effective clustering results against real microarray data.

Keywords: Microarray · Gene clustering · K-means algorithm · Map reduce

1 Introduction

Microarray technology has recently become a popular technique for bioinformatics, especially in clinical diagnosis, disease classification and finding gene regulation. This technique observes gene expressions of thousands of genes at one time and analyzes gene expression levels for clinical diagnosis and finding correlation between genes. For example, gene expression microarray data are widely used for identifying gene clusters which contain functionally related genes. With the development of microarray technology, data volumes generated by DNA chip are continuously increasing which would in turn require high performance computing. Efficient parallel gene clustering algorithm and implementation techniques are the key to meet the scalability and performance requirements entailed in microarray data analyses.

K-means algorithm is the most well-known and commonly used clustering method. Despite the emergence of different enhanced methods over the years, the early *K-means* clustering algorithm still remains as a popular approach on account of its simplicity and fast convergence time. It takes the input parameter,

© Springer International Publishing Switzerland 2014
Y. Chen et al. (Eds.): WAIM 2014, LNCS 8597, pp. 372–381, 2014.
DOI: 10.1007/978-3-319-11538-2_34

k and partitions a set of n objects into k clusters so that the resulting intra-cluster similarity is high whereas the inter-cluster similarity is low. In *K-means* algorithm, the most intensive computation is the calculation of distance functions. In each iteration of *K-means* algorithm, it would require a total of (nk) distance computations where n is the number of objects and k is the number of clusters being created. Since the distance computation between one object with the centers is irrelevant to the distance computations between different objects with centers, they can be executed in parallel though iterations must be executed serially.

On the other hand, high performance computing has become extremely important for analyzing large amount of biological data. MapReduce is an easy-to-use and general purpose parallel programming model that is suitable for large data analysis on a commodity hardware cluster. In MapReduce framework, users specify the computation in terms of a *map* and a *reduce* function and the underlying run time system automatically parallelizes the computation across large-scale clusters of machines, handles machine failures and schedules inter-machine communication to make efficient use of the network and file systems. Google and Hadoop both provide MapReduce runtimes with fault tolerance and dynamic flexibility support [4,16,17].

In this paper, we propose MapReduce based gene clustering method. Our method adapts *K-means* algorithm [16] in MapReduce framework which is implemented by Hadoop to make the clustering method applicable on large size of microarray data. By conducting comprehensive experimentation, we show scalability and accuracy of the proposed method.

The rest of the paper is organized as follows. In Sect. 2, we briefly examine existing works closely related with our work. Section 3 describes the proposed gene clustering approach and Sect. 4 shows experimental results and evaluates our parallel gene clustering algorithm with respect to scalability and accuracy. Finally, we conclude in Sect. 5.

2   Related Works

Microarray gene expression data consists of $N \times M$ matrix where N is the number of genes and M is the number of samples (or experimentations). One of the characteristics of gene expression data is that it is meaningful to cluster both genes and samples. On one hand, in gene-based clustering, co-expressed genes can be grouped in clusters based on their expressive patterns (genes are treated as objects while samples are features). On the other hand, in sample-based clustering, the samples can be partitioned into homogenous groups. The subspace clustering (biclustering) is to capture clusters formed by a subset of genes across a subset of samples.

Gene-based clustering techniques are usually used to group genes with similar expression patterns based on the organization of gene expression data. Several studies have been done on this kind of microarray data analysis [1–3,5]. The widely adopted clustering techniques for gene expression data include hierarchical clustering [1,7], self-organizing maps (SOM) [6], *K-means* clustering [3,5,6] and so on. Hierarchical clustering is one of the earliest used to cluster genes. It successively merges clusters until all elements belong to the same cluster from an initial partition. SOM is a method inspired by neural networks in the brain. It uses a competition and cooperation mechanism to achieve unsupervised learning. *K-means* is among the most competitive for gene expression data. It classifies genes to one of the k groups, determines the cluster membership by calculating the centroid for each group and assigns each object to the group with the closest centroid. Recently, new hierarchical clustering methods using genetic algorithms (GA) for the analysis of gene expression data are proposed [1,7]. They apply evolutionary techniques to build hierarchical clustering. In [1], they improve GA-based hierarchical clustering method by previously fitting several parameters based on some criteria.

On the other hand, handling large volume of expression data has been another concern of analyzing microarray data. MapReduce [8] developed by Google is a cloud technology which is amenable to processing and generating large size of datasets. For such reasons, many recently developed methods applied MapReduce in several data intensive applications. Reference [4] propose a way to handle clustering process for a very large multi-dimensional dataset. During the clustering process, they spot bottlenecks automatically and choose best strategy to minimize I/O and network cost. Zhao et al. [16], propose a way to extend *K-means* algorithm to execute it in parallel using MapReduce. In this work, we adopt MapReduce based parallel *K-means* for gene clustering. Our method validates intermediate gene clusters through gene function similarities.

3   Proposed Method

In this section, we describe the proposed MapReduce based gene clustering method in detail. First, we explain overall concept and also the principles hidden in each step of our method. Then, we describe each step thoroughly.

3.1   Method Outline

Figure 1 shows overall procedure of our method. In microarray experiments, samples are managed as rows and genes are as columns. However, they are presented as $N \times M$ matrix, where N is the number of genes and M is the number of experimental samples because the number of genes is much higher than the number of samples. We present an example microarray dataset in Table 1 for explaining the proposed method. We consider each gene's expressions g_i as feature vector rather than samples and execute *K-means* clustering in parallel using MapReduce to get correlated gene cluster. In the following descriptions, g_i is used to denote both gene-id and gene expressions vector. Once intermediate clusters G_c are generated, we verify biological relevance of the genes in a cluster. For this purpose, we consider interactions and ontology based data to reveal whether two statistically correlated genes are true positive or not. Moreover, the method increases scalability for large datasets.

Table 1. Sample microarray data

$Gene_{id}$	s_1	s_2	s_3	s_4	..	s_{m-2}	s_{m-1}	s_m
1	2	2	6	42	...	3	6	5
2	10	5	3	6	...	25	3	6
3	5	15	40	58	...	34	7	37
4	4	16	29	56	...	14	32	64
5	7	22	57	56	...	30	15	27
6	17	24	13	41	...	23	13	25
7	14	11	50	15	...	17	12	26
8	8	18	26	29	...	27	23	12
9	6	14	26	21	...	22	32	23
10	18	19	26	19	...	15	37	12
..	..	..	..	..	..	..	..	..
N	..	..	..	..	..	..	..	..

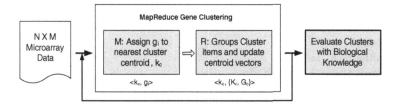

Fig. 1. Workflow of the proposed method

3.2 Parallel Gene Clustering

As discussed in existing parallel cluster methods [16,17], the tasks should be divided into parallel and sequential portions. Iterations of the clustering method are executed sequentially. However, the tasks into an iteration are done in parallel. Algorithm 1 gives detail descriptions of one iteration of the proposed method. Iterations of the proposed method begin by setting k-centroids as shared variables. In first iteration, the dataset is divided into k segments randomly and subsequently K-$centroids$ are measured. During next iteration, we get updated K-$centroids$ from preceding $Reduce$ tasks output. Considering sample dataset in Table 1 and $k=2$, we initially generate two segments of the data. The MapReduce driver class computes initial centroids from mean values of cluster members and set as globally shared variables. Here, $k_1 = (5.60, 12, 27, 43.60, 21.20, 12.60, 27.80)$ & $k_2 = (12.60, 17.20, 28.20, 25, 20.80, 23.40, 19.60)$. The dataset splits are then assigned to available worker nodes for Map task.

Map task inputs g_i expressions one at a time and determines most relevant cluster center for a g_i. We do this by calculating Euclidian distance of k_l and g_i; $Dist(k_l, g_i)$. For example, $Dist(k_1, g_1) = 1451.2$ and $Dist(k_2, g_1) = 1958$.

Algorithm 1. MR K-means Cluster

Data: Microarray Data (NXM); Broadcast Variables: K - centroids
Result: k gene clusters

1 **begin** *MR* Job:
2   **Map:**
    Fetches expression values of g_i in microarray data
    $Dist_c = \infty, k_c = -1$
    forall the k-centroids k_l **do**
3   **if** $Dist(k_l, g_i) < Dist_c$ **then**
4             $Dist_c = Dist(k_l, g_i)$;
5             $k_c = l$;
6   **end**
7   **end**
8   Emit $<k_c, g_i>$
9   **Combine:**
    Initialize K_{lc} and G_{lc} ;
10   **forall the** g_i of k_c in a local node **do**
11         $K_{lc}+ = g_i$;
12         $G_{lc} = Concatenate(G_{lc}, g_i)$;
13   **end**
14   Emit $<k_c, \{K_{lc}, G_{lc}\}>$
15   **Reduce:**
    Initialize K_r and G_c ;
16   **forall the** $<K,V>$ of k_c **do**
17         $K_r+ = k_{lc}$;
18         $G_c = Concatenate(G_c, G_{lc})$;
19   **end**
20   Emit $<k_c, \{K_r, G_c\}>$
21 **end**
22 Access reduce outputs from driver class to set as global variables

Hence g_1 remains to cluster k_1. However, $Dist(k_1, g_2) = 2640$ and $Dist(k_2, g_2) = 1770$ which requires reassignment of g_2 from k_1 to k_2. For g_1 and g_2, the *Map* task will emit $<k_1, g_1>$ and $<k_2, g_1>$ respectively. *Combine* task groups intermediate $<K, V>$ pair in a local node to reduce network traffic and workload of *Reduce* task. We make combine task slightly different from the reduce task so that updated k-centroids are calculated simultaneously. The intermediate values are concatenated and each g_i feature vectors are summed locally. Output of a *Combine* task will be $<k_c, \{K_{lc}, G_{lc}\}>$ where K_{lc} is the local summation of feature vectors and G_{lc} is the gene list in a local node. Finally, *Reduce* task emits complete list of genes of a cluster along with updated cluster centroids. The driver class loads updated cluster information for next iteration. In case, there is no update or the iterations completed, the final output would be put into HDFS.

3.3   Cluster Validation

After the clusters are generated from MapReduce execution, the genes in a cluster are filtered with publicly available domain knowledge. Biological sound correlations are more significant than only statistically correlated genes. We take interactions and gene functions data from NCBI [9] repository for human datasets and from SGD [10] for yeast datasets. Correlated genes interact directly among them and mentioned repositories document publicly available interaction knowledge. Hence direct interaction between two genes $DI(g_i, g_j)$, imply its significance. On the other hand, direct interactions are limited because of insufficient domain knowledge. Therefore, ontology based similarities are also important to find correlated genes. Gene Ontology (GO) [11] is the most widely used ontology knowledge base to find gene functions similarities. It provides a controlled vocabulary to provide gene functions characteristics regarding function, process and component through a graph structure. In short, we can measure correlated genes through gene function similarities, $GF(g_i, g_j)$. If two genes share specific GO terms, then the gene pair is considered to have biological significance and recommended for further experiments. For both $DI(g_i, g_j)$ and $GF(g_i, g_j)$, if there is any match then we considered the genes as true positive. Otherwise, their correlations are counted as false positive.

For example, let us consider g_1 and g_6 of Table 1 belong to cluster k_1. We verify the genes by $DI(g_1, g_6)$ and $GF(g_1, g_6)$. If there is any match from either of the function, we consider g_1 and g_6 as true positive and false positive otherwise. This way, we measure gene correlation accuracy of the clusters.

4   Experimental Results

We have experimented proposed method in a seven node cluster, each node having four cores. The memory size is 15 GB and storage capacity is 800 GB. The operating system is Ubuntu 11.10. For the *MapReduce* library, we use *Apache Hadoop*'s distribution of 1.1.0. One of the nodes is set as the master node. The remaining nodes are set as worker nodes. Each worker node has two slots of Map and Reduce. The HDFS block size is 32 MB and each block has three replications. We apply our proposed method to three publicly available microarray datasets and three synthetic datasets. Along with effective cluster generation, we examine various *MapReduce* metrics such as node scalability, data scalability and I/O costs. Since the sizes of real datasets were small, we validate our method's accuracy by using synthetically generated data sets. We generate three synthetic datasets from S. cerevasae [14] and Prostate Cancer [13] datasets and use those to exhibit data scalability in a *MapReduce* environment. For node scalability, we change the number of active nodes in the cluster and check execution time differences.

4.1   Datasets

Table 2 gives a brief description of the three real datasets used in our experiments. The Hughes et al. [14] dataset contains 300 microarray experiments and

Table 2. Description of real datasets

Datasets	Samples	Genes
Prostate cancer [13]	102	12600
Hughes et al. [14]	300	6325
Spellman et al. [15]	215	6316

6325 genes. Two other datasets, Prostate cancer [13] and Spellman et al. [15] contain 102 and 215 samples respectively. The class labels of the samples are not considered for our experiments.

Table 3. Description of synthetic datasets

Dataset	Samples	Genes	Size(MB)
HugesX5	1500	6325	51
ProstateX5	510	12600	37
ProstateX10	1020	12600	75

We generate three synthetic datasets from S. cerevasae [14] and Prostate Cancer [13] datasets. Though the generated datasets are not very large with regards to the MapReduce model, our intention is to observe the *MapReduce* scalability metrics on relatively large number of samples. Table 3 shows a brief description of the synthetic data.

4.2   Clusters Generation

To extract correlated genes from the datasets, we execute the proposed MapReduce algorithm in our lab-size distributed environment. A k-means implementation may require many iterations to converge. However, I/O costs are critical for MapReduce performance measure. Hence, we limit iterations heuristically to optimize results in different parameters. Along with this, k is required to define prior execution. Since, the proposed method focus on gene correlations rather than samples or class labels, the higher number of k is more suitable. For example, we found $12 \leq k \leq 18$ to be more suitable for experiment datasets.

4.3   Biological Relevance

We validate the extracted clusters with publicly available biological resources. Interactions and ontology based resources are taken from NCBI [9] and SGD [10] databases. NCBI-Gene resources are widely recognized for completeness and regular synchronization with major repositories such as BIND, GO, HGNC and EMBL. SGD provides most updated gene information for S. cerevasae. Therefore,

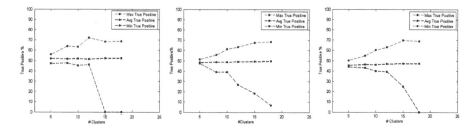

Fig. 2. Maximum, minimum and average true positive (%) among datasets over different #k, (a) Huges et al. (b) Spellman et al. and (c) Prostate cancer dataset

we take relevant data files from these repositories and integrate them into $MySql$ databases using XAMPP [12] package. Each gene (g_i) in a cluster is matched with other genes regarding gene functions and interactions. g_i and g_j will be considered a true-positive if they hold at least one match and false-positive otherwise. Figure 2 shows maximum, average and minimum true-positive (%) correlations achieved for each dataset on different cluster sizes. Figure 2(a) provides experimental result for Huges et al. [14] dataset. Maximum accuracy is achieved for $k = 12$. For a bigger #k, the maximum accuracy reduces slightly which indicates inherent clusters should be among 12–15 for the dataset. The minimum true positive curve reduces to zero as the cluster size increases. This refers to the fact that irrelevant genes are likely to be put in few clusters. However, average true positive correlations do not change over varying cluster sizes. Figure 2(b) and (c) show output for the Spellman et. al and Prostate cancer datasets respectively.

4.4 Hadoop Scalability

We measure effectiveness of the proposed distributed algorithm on two metrics: node scalability and data scalability. A *MapReduce* implementation has some overhead to initialize execution environment. Therefore, relatively large datasets are desirable for observing significant performance upgrades through parallelization. The publicly available microarray datasets are fairly small in size. Hence we utilize three synthetic datasets shown in Table 3 for scalability tests. Lab size clusters usually consist of small amount of memory and limited number of nodes. Therefore, increasing the number of nodes and memory significantly speeds up MR job execution. Publicly available cloud services are good candidates for such systems unless security is a major issue.

Figure 3(a) shows execution time of real datasets over varying number of nodes. Smaller datasets experience less speed up compared to relatively large dataset, i.e., Huges et. al. The two other datasets show almost similar execution time for four or more nodes. Despite their differences regarding size, sample number and gene number, they match closely regarding execution time. This is because, the network traffic cost and I/O cost compensate each other. In Fig. 3(b), the execution times of the synthetic datasets are shown. Each newly

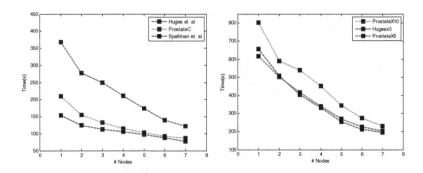

Fig. 3. Hadoop scalability on (a) Real and (b) Synthetic datasets

added synthetic sample is generated by averaging three randomly selected samples. As the number of nodes is decreased, the slope gets sharper over increasing data size. However, the synthetic datasets get more speed up than that of real datasets regarding their size. This supports the claimed effectiveness of MapReduce programming model.

5   Conclusion

In this paper, we proposed an efficient microarray gene clustering algorithm for distributed environment. The MapReduce based method extracts correlated genes rather than existing sample clustering methods. The method generated encouraging biologically relevant results for relatively big k-value than class labels due to large number of loosely connected genes. For implementation, we develop an efficient *MRkMeans* algorithm in open source *Hadoop* implementation. We experiment three real datasets for accuracy measures and three synthetically generated datasets for scalability measures. Extensive experiment results verify the effectiveness of our method. In future, We shall incorporate more domain knowledge in cluster generation phase to improve accuracy of the generated correlations.

Acknowledgment. This work was supported by Basic Science Research Program through the National Research Foundation of Korea (NRF) funded by the Ministry of Education, Science and Technology (NRF-2013R1A1A2006236).

References

1. Castekkanos-Garzon, J.A., Diaz, F.: An evolutionary computationary model applied cluster analysis of DNA microarray data. Expert Syst. Appl. **40**, 2575–2951 (2013)
2. Yi, G., Sze, S.-H., Thon, M.R.: Indenifying clustering functionally related genes in genomes. Bioinformatics **23**(9), 1053–1060 (2007)

3. Zhihua, D., Wang, Y., Ji, Z.: PK-means: A new algorithm for gene clustering. Comput. Biol. Chem. **32**, 243–247 (2008)
4. Cordeiro, R.L.F., Traina, C. Jr., Traina, A.J.M., Lopez, J., Kang, U., Faloutsos, C.: Clustering very large multi-dimensional datasets with MapReduce. In: International Conference on Knowledge and Data Discovery (2011)
5. Hartigan, J.A., Wong, M.A.: A K-means clustering algorithm. Appl. Stat. **28**, 126–130 (1979)
6. Lam, Y.K., Tsang, P.W.M.: eXploratory K-means: a new simple and efficient algorithm for gene clustering. Appl. Soft Comput. **12**, 1149–1157 (2012)
7. Greene, W.A.: Unsupervised hierarchical clustering via genetic algorithm. In: Congress on Evolutionary Computation, pp. 998–1005 (2003)
8. Dean, J., Ghemawat, S.: Mapreduce: simplified data processing on large clusters. Commun. ACM **51**(1), 107–113 (2008)
9. National Center for Biotechnology Information. http://www.ncbi.nlm.nih.gov. Accessed 03 Feb 2014
10. The Saccharomyces Genome Database(SGD). http://www.yeastgenome.org. Accessed 03 Feb 2014
11. The Gene Ontology project(GO). http://www.geneontology.org/. Accessed 02 March 2014
12. The XAMPP open source package. http://www.apachefriends.org/en/xampp.html. Accessed 02 March 2014
13. Singh, D., Febbo, P.G., Ross, K., Jackson, D.G., Manola, J., Ladd, C., Tamayo, P., Renshaw, A.A., D'Amico, A.V., Richie, J.P.: Gene expression correlates of clinical prostate cancer behavior. Cancer cell **1**(2), 203–209 (2002)
14. Hughes, T.R., Marton, M.J., Jones, A.R., Roberts, C.J., Stoughton, R., Armour, C.D., Bennett, H.A., Coffey, E., Dai, H., He, Y.D., et al.: Functional discovery via a compendium of expression profiles. Cell **102**(1), 109–126 (2000). Elsevier
15. Spellman, P.T., Sherlock, G., Zhang, M.Q., Iyer, V.R., Anders, K., Eisen, M.B., Brown, P.O., Botstein, D., Futcher, B.: Comprehensive identification of cell cycle-regulated genes of the yeast saccharomyces cerevisiae by microarray hybridization. Mol. Biol. Cell, Am. Soc. Cell Biol. **9**(12), 3273–3297 (1998)
16. Zhao, W., Ma, H., He, Q.: Parallel K-means clustering based on mapReduce. In: Jaatun, M.G., Zhao, G., Rong, C. (eds.) CloudCom 2009. LNCS, vol. 5931, pp. 674–679. Springer, Heidelberg (2009)
17. Sun, Z.: A parallel clustering method study based on mapReduce. In: 1st International Workshop on Cloud Computing and Information Security, Atlantis Press (2013)

An Efficient Patent Storing Mechanism Based on SQLite on Hadoop Platform

Xuhua Rui, Baul Kim, and Dugki Min$^{(\boxtimes)}$

Department of Computer, Information and Communication Engineering,
Konkuk University, Seoul, South Korea
{abealasd, bangwol, dkmin}@konkuk.ac.kr

Abstract. This paper describes a patent storage and analysis system based on Hadoop and SQLite. This system is proposed to solve data loading overhead in Map Phase of patent analysis applications. System proposed in this paper utilizes SQLite data structure as HDFS block container to enhance patent documents storage and query performance. In addition, a hierarchical index-based patent data processing approach has been implemented in this system to support patent analysis applications. This paper presents a data storing mechanism which enhances query jobs by at least 2x faster than original Hadoop. And our mechanism decreases 90 % data loading overhead in map phase.

Keywords: Hadoop · SQLite · Patent analysis · Big data

1 Introduction

In recent decades, data generated by human beings has reached an unbelievable magnitude. As Dick Costolo (CEO of Tweeter) mentioned, more than 400 million tweets are published every day [1]. Facebook users upload more than 300 million pictures every day [2]. Thus we have to face the truth that we are in the era of information explosion. Patent data is one of those highly growing data. Patents represent the current state of technology and trend of human society. In recent years, there have been a tremendous growth in research and development in science and technology, patent documents become the important input data source for R&D. However, as time goes, the number of patent documents grows inconceivably. In 2000, 175979 patents have been granted by USPTO. But in 2012, the number has drastically increased to 276788. During the past decade, more than 2.5 million patents have been granted by USPTO. As patent contains 90 to 95 percent of the world's science and technology information [3], patent analysis becomes an efficient approach to discover knowledge. Patent data also describes competitive intelligence of companies [4]. Therefore to learn knowledge or search information from such a huge number of patent documents becomes a challenge nowadays.

In early 2004, the most popular web searching service provider, Google, published a solution to manage huge number of data and to utilize those explosive increase of data [5]. They introduced a program paradigm called MapReduce which works as a core part in Google. They introduced their solution for processing entire web data. MapReduce is a programming paradigm for making parallel programs simpler than ever. Hadoop is

© Springer International Publishing Switzerland 2014
Y. Chen et al. (Eds.): WAIM 2014, LNCS 8597, pp. 382–392, 2014.
DOI: 10.1007/978-3-319-11538-2_35

the most famous open-source MapReduce framework currently. Hadoop provides a scalable, reliable distributed computing environment. Being a successful open source project, Hadoop attracts more and more developers towards it. Although Hadoop has been proved as a successful big data processing platform, it is still not sufficient for all kinds of data types [6]. As patents are designed to conserve professional knowledge from human society, most of patents have been well formatted. Most patent based researchers are considering only part of entire patent documents [7].

In this paper we propose a mechanism utilizing SQLite data structure for storing and analyzing patent data on Hadoop platform efficiently. To consider compatibility, we implemented our mechanism without changing any existing part of current Hadoop. Details will be introduced later. In Sect. 2, we illustrate the motivation of our work. And Sect. 3 introduces the architecture design of our system and Sect. 4 gives detail implementation issues and our solutions. Sections 5 and 6 shows experimental result of our system. And finally we conclude our work in Sect. 7.

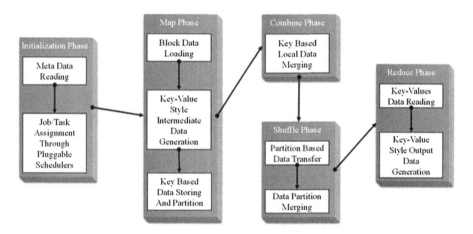

Fig. 1. Traditional Hadoop MapReduce processing flow

2   Motivation

As Fig. 1 shows, traditional Hadoop MapReduce processing flow contains 5 phases. Initialization phase includes job configuration, target data analysis, task assignment and job launching steps. Map phase includes block data loading, map function launching, intermediate data generation and intermediate sorting and partition steps. Combine phase is an additional phase which includes local data merging work steps. Shuffle phase includes intermediate data transferring, intermediate data merging and intermediate data sorting steps. Reduce phase includes reduce function launching and result data outputting steps.

Current researchers mostly focus on enhancing MapReduce performance by optimizing combine phase, shuffle phase and reduce phase [8–10]. However most patent analysis applications spend a lot of time on querying and filtering work [7] which can be done within map phase. Therefore we propose a Hadoop based system which enhances patent data query performance by utilizing SQLite block format.

3   Overview Architecture

Our proposed patent storage and analysis system is built based on Hadoop 1.0.3. Considering the compatibility, we didn't change any part of the original Hadoop. However, in order to achieve our goal we added and extended Hadoop modules.

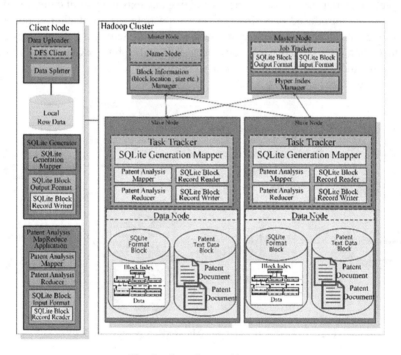

Fig. 2. Overview architecture

As Fig. 2 shows, there are 2 parts in our system. A client node represents a user who utilizing this system for patent analysis. Hadoop cluster represents our backside data processing cluster. A client node contains a data uploader which is based on the default Hadoop DFS client with a modified data splitter, a SQLite generator which is a MapReduce program converting original patent text data blocks to SQLite format blocks, and a customized MapReduce application which provides functionality for users to build patent analysis application.

Our system provides a SQLite block record reader which is extended from Hadoop basic block record reader to support data reading functionality for blocks in SQLite format. The SQLite block input format does a data splitting job for data which will be processed in map phase. Following the default Hadoop design, we provide a SQLite block output format to control output data behavior of our system. The SQLite block record writer does the real data writing jobs to Hadoop distributed file system (HDFS). Different from original Hadoop, we designed a hyper index manager to maintain a hyper index which is designed to enhance data query performance. In our system, hyper index contains a table which stores frequently queried data and blocks storing them. Since we mainly consider to store data in SQLite format, the default low level block management mechanism is not sufficient. In order to keep the compatibility, we choose to add a block information manager which contains a copy of block location, block size and other block information out of Name Node. This manager synchronizes block information from Name Node and updates checksums of SQLite blocks when they have been changed (ex. index generation).

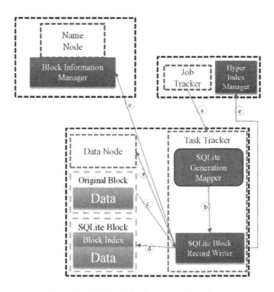

Fig. 3. SQLite block generation flow

SQLite block generation steps are shown in Fig. 3:

(a) Job Tracker assigns a SQLite Generation Mapper to a specific Task Tracker to convert an Original Block to a SQLite Block.
(b) The SQLite Generation Mapper confirms information of the target Original Block and extract data from it to SQLite Block Record Writer.
(c) Once SQLite Block Record Writer receiving a data, it writes it into local SQLite Block.
(d) After writing all data into SQLite Block, SQLite Block Record Writer sends information of new SQLite Block to Data Node, Block Information Manager and Hyper Index Manager.

Figure 4 shows how our system assigns map tasks for processing data stored in SQLite Blocks:

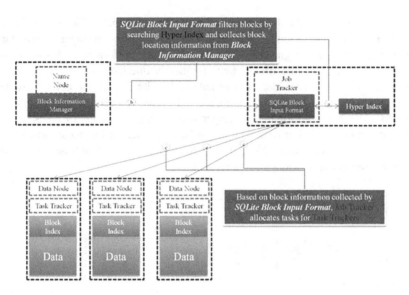

Fig. 4. Map task assignment flow for SQLite block

(a) When Job Tracker receives a MapReduce application, it utilizes SQLite Block Input Format to confirm data size, data range and data location. SQLite Block Input Format look up hyper index to confirm if all blocks should be processed.

(b) SQLite Block Input Format requests to Block Information Manager to receive block size, block location and data range of blocks. And inform this data to Job Tracker.

(c) Job Tracker assigns map tasks based on data collected in previous steps to Task Trackers.

Figure 5 exhibits how our system loads data from SQLite Block to Mapper:

(a) When Task Tracker receives a map task assignment, it creates a JVM to launch the map task.

(b) Task Tracker utilizes SQLite Block Record Reader to receive block information from local data node.

(c) Then SQLite Block Record Reader reads Block Index to confirm the location of target data.

(d) Finally SQLite Block Record Reader read data which needs for the mapper and send them to map function.

4   Implementation

We implement a data uploader based on original Hadoop DFSClient with a modified data splitter. The Hadoop data splitting approach only consider data block size. However patent data consist of a large amount of small documents. Our modified data splitter combines small size of huge amount patent documents as one block, and add

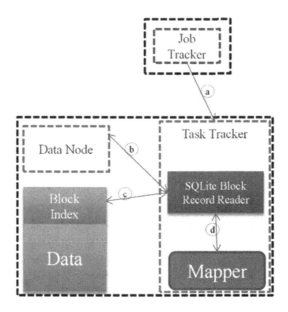

Fig. 5. SQLite block data loading flow

separation symbol between 2 patent documents. We implement SQLite generator as a MapReduce application. It converts patent block data in text format to SQLite format in a distributed way. Essentially, our SQLite generator only has map phase, because we implement our idea in a block to block mapping way. Each block conversion task is one isolated map task. To support SQLite operations, we utilize [11] in our SQLite block record reader and SQLite block record writer. In our system, the low level SQLite operations are done by SQL statements. We support a parameter to receive user customizing SQL statements, so user can define their own data selection method for each SQLite blocks. Since we consider to generate indexes for each block, data of each block should be isolated. In addition, size of indexes are not predictable before we generate indexes for blocks. Therefore we implement a block information manager to synchronize updated SQLite block information to Name Node and inform SQLite block location information to hyper index manager.

The implementation of block information manager is the most challenging work for us. Because in our original idea, we prefer not to modify Hadoop source code to keep the compatibility. However traditional block management approach of Hadoop is not sufficient for our scenario. And Hadoop does not support APIs for manually receiving and updating block information because of security issues. Finally we do the implement work in a tricky approach. We implement the block information manager as the same security level as Name Node so that Name Node accepts requests from our block information manager. In previous section, we introduce a hyper index which is used to enhance data query performance. In our system, hyper index only effects when processed data is stored in indexed SQLite blocks. In our patent analysis system, hyper index keeps frequently requested and identifiable index key with its block location. Currently we store hyper index in local SQLite database.

5   Experiment Environment

5.1   System Configuration

We deploy our proposed system on a 14-node cluster which is connected through a Cisco SF200E-24 switch. Detailed description of our experimental cluster has been given in Table 1.

Table 1. Specification of experimental cluster

	Storage	Memory	CPU	# of Machines
Master	98 GB	2 GB	Intel(R) Core(TM)2 Duo CPU E4500 @ 2.20 GHz	1
Slave Type A	123 GB	2 GB	Intel(R) Core(TM)2 Duo CPU E4500 @ 2.20 GHz	2
Slave Type B	184 GB	1 GB	Intel(R) Core(TM)2 Duo CPU E8200 @ 2.66 GHz	4
Slave Type C	123 GB	3 GB	Intel(R) Core(TM)2 Duo CPU E8400 @ 3.00 GHz	7

In our experiment, we deploy Name Node, Job Tracker, Block Information Manager and Hyper Index Manager in a same machine which is called master node. The rest machines are configured as Data Nodes and Task Trackers which are called slave nodes. We set number of replications as '2' instead of the default value '3' to increase hard disk utilization. Hadoop authentication function has been set as 'false' and number of default reduce tasks has been set as '13' to match slave node number in our system. Table 2 describes configuration files we changed and value of core parameters we used in our system.

Table 2. Hadoop cluster configuration

Configuration file	Property	
	Name	Value
hdfs-core.xml	dfs.replication	2
core-site.xml	hadoop.security.authentication	false
	mapred.reduce.tasks	13

To increase computation locality in our system, we use the fair scheduler instead of the default FIFO scheduler. Table 3 shows parameters we defined for fair scheduler in our system.

Table 3. Fair scheduler configuration

Property	
Name	Value
mapred.jobtracker.taskScheduler	org.apache.hadoop.mapred.FairScheduler
mapred.fairscheduler.locality.delay	120000

5.2 Experimental Dataset

Patent data in our experiment are downloaded from Google's USPTO full patent text data repository [12]. Our experiment considers patent documents granted from 1976 to 2000 including 2437840 patents. Original patent data are stored in text format. We first upload all patent data to HDFS. When we upload them we combine all patent documents as one entire file and add separation symbol between two continued patent documents.

Secondly, we launch a patent analysis MapReduce application which extracts and lemmatizes nouns, adverbs, and adjectives from each patent documents. Via this application, each word frequency is extracted from different patent documents. We treat words extracted by this application as feature words for those patent documents. Thirdly, we convert those feature word blocks to SQLite format block. We define four columns (patent id, feature word, count of each feature word and count of all feature words in one patent document) to contain data in SQLite format. In addition, we generate index for feature word column. Finally, we have three different block formats (original text format, SQLite format and indexed SQLite format). In our experiment, we test feature word query performance on this three block formats.

6   Experimental Result and Analysis

To compare query performance of three block formats, we do a query performance test. We launch a searching MapReduce job on three block formats to find patents which contain "computer" as feature word. In this section we will discuss our experimental results.

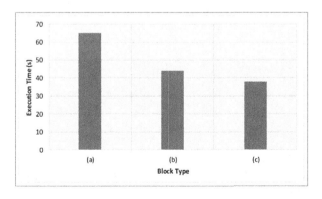

Fig. 6. Application execution time (a) Original text block (b) SQLite block (c) Indexed SQLite Block

Figure 6 shows total execution time of our test application running on each block format. Application launching on original text format blocks costs the longest time. Obviously, SQLite block format enhances query performance in MapReduce computing paradigm. However execution time of SQLite block format and indexed SQLite block format shows an inconspicuous difference. Because both of them cost a lot of time on non-data-processing works (ex. task assignment, JVM creation).

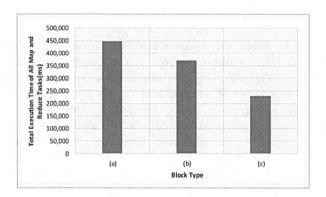

Fig. 7. Total execution time of all map and reduce tasks (a) Original text block (b) SQLite block (c) Indexed SQLite block

Figure 7 shows total execution time of all map and reduce tasks. Comparing indexed SQLite block format with others, we can find that query on indexed SQLite block format costs fewer amount of time. Because map tasks running on indexed SQLite block format only processed rows which contains "computer" as a feature word. Significantly, SQLite block format also shows better performance than original text block format. Because SQLite format maintains data in a B + tree data structure, which has already been optimized for query [13].

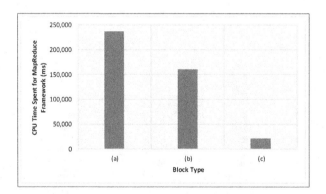

Fig. 8. CPU time spent for MapReduce framework (a) Original text block (b) SQLite block (c) Indexed SQLite block

Figure 8 demonstrates total CPU time spent for MapReduce framework (ex. input data to map tasks). Comparing all three applications, only block data loading part is different. Since we generated index for feature word column, our system reduce 90 % CPU time spent for MapReduce framework.

7 Conclusions

This paper presents a Hadoop based patent storage and analysis system. Our system utilizes SQLite data structure as block format to contain patent data and optimize patent analysis performance especially patent query performance. Since we utilize SQLite format as our block container, our system support partial SQL benefits such as indexing. Our proposed system saves CPU utilization for loading data into mappers. As our experiment exhibits, we save CPU time spent for MapReduce framework. The saving is significant as demonstrated by previous experiments. And by utilizing index we load only necessary data to mappers which avoids data loading overhead in Map Phase.

Most benefits of our system are effected by utilizing B + tree and B- tree of SQLite data structure. Currently we are implementing a hierarchical index-based MapReduce framework for patent data analysis. System presented in this paper is a prototype implementation of Hadoop and SQLite for patent documents storage and query.

Acknowledgement. This work (Grants No. C0141582) was supported by Business for Cooperative R&D between Industry, Academy, and Research Institute funded Korea Small and Medium Business Administration in 2014.

References

1. CNET News Internet & Media Twitter hits 400 million tweets per day, mostly. http://news.cnet.com/8301-1023_3-57448388-93/twitter-hits-400-million-tweets-per-day-mostly-mobile/
2. Facebook is collecting your data - 500 terabytes a day. http://gigaom.com/2012/08/22/facebook-is-collecting-your-data-500-terabytes-a-day/
3. Huang, L., Yuan, Y., Zhao, Z.: A study on the application of data mining in the patent information analysis for company. In: Second International Workshop on Education Technology and Computer Science (ETCS), Wuhan, vol. 1 (2010)
4. Karki, M.M.S.: Patent citation analysis: a policy analysis tool. World Patent Inf. **19**(4), 269–272 (1997)
5. Dean, J., Ghemawat, S.: MapReduce: simplified data processing on large clusters. Commun. ACM **51**(1), 107–113 (2008)
6. Buck, J.B., Watkins, N., LeFevre, J., Ioannidou, K., Maltzahn, C., Polyzotis, N., Brandt, S.: SciHadoop: array-based query processing in hadoop. In: Proceedings of 2011 International Conference for High Performance Computing, Networking, Storage and Analysis, Seattle, p. 66 (2011)
7. Tseng, Y.-H., Lin, C.-J., Parker, D.S.: Text mining techniques for patent analysis. Inf. Proc. Manage. **43**(5), 1216–1247 (2007)

8. Yang, H.-C., Dasdan, A., Hsiao, R.-L., Parker, D. S.: Map-reduce-merge: simplified relational data processing on large clusters. In: ACM SIGMOD International Conference on Management of Data, Beijing, pp. 1029−1040 (2007)
9. Herodotou, H., Babu, S.: Profiling, what-if analysis, and cost-based optimization of MapReduce programs. In: Very Large Database Endowment, vol. 4, pp. 1111−1122 (2011)
10. Lin, M., Zhang, L., Wierman, A., Tan, J.: Joint optimization of overlapping phases in MapReduce. Perform. Eval. **70**(10), 720–735 (2013)
11. Java Database Connector for SQLite. https://bitbucket.org/xerial/sqlite-jdbc
12. USPTO Bulk Downloads: Patent Grant Full Text. http://www.google.com/googlebooks/uspto-patents-grants-text.html
13. The SQLite Database File Format. https://www.sqlite.org/fileformat.html

Considering Rating as Probability Distribution of Attitude in Recommender System

Xiangyu Zhao, Zhendong Niu, Wentao Wang, Ke Niu,
and Wu Yuan[⊠]

School of Computer Science and Technology,
Beijing Institute of Technology, Beijing, China
yuanwu@bit.edu.cn

Abstract. Recommender systems play increasingly significant roles in solving the information explosion problem. Generally, the user ratings are treated as ground truth of their tastes, and used as index for later predict unknown ratings. However, researchers have found that users are inconsistent in giving their feedbacks, which can be considered as rating noise. Some researchers focus on improving recommendation quality by de-noising user feedbacks. In this paper, we try to improve recommendation quality in a different way. The rating inconsistency is considered as an inherent characteristic of user feedbacks. User rating is described by the probability distribution of user attitude instead of the exact attitude towards the current item. According to it, we propose a recommendation approach based on conventional user-based collaborative filtering using the Manhattan Distance to measure user similarities. Experiments on MovieLens dataset show the effectiveness of the proposed approach on both accuracy and diversity.

Keywords: Recommender systems · Collaborative filtering · Rating inconsistency

1 Introduction

In the current age of information overload, it is becoming increasingly hard for people to find relevant content. Recommender systems have been introduced to help people in retrieving potentially useful information in a huge set of choices [12,16].

Conventional recommendation methods are based on users' rating values. These rating values are considered as indications of users' preference level towards the rated items. Recommender systems estimate the ratings of items that have not been rated by the target user based on the rating history, and recommend top-N items with highest predicted ratings. However, some researchers have found that there is inconsistency when users are giving their ratings with experiments which are using rate-rerate procedures [2,3]. This inconsistency is considered as rating noise.

© Springer International Publishing Switzerland 2014
Y. Chen et al. (Eds.): WAIM 2014, LNCS 8597, pp. 393–402, 2014.
DOI: 10.1007/978-3-319-11538-2_36

There have been some studies on rating noise of recommender systems. Hill et al. [8], Cosley et al. [6], and Amatriain et al. [2] try to quantify the rating noise between different rating trails. O'Mahony et al. [14] and Amatriain et al. [3] use de-noising methods to improve recommendation quality.

In this paper, we try to solve the problem in a different way. In conventional ideas, a rating value which a user rates on an item in treated as ground truth of the user's taste. Different from them, according to the inconsistency among user feedbacks, a rating value is considered as a probability distribution that the current user's attitude towards the item. Based on it, the distance between two ratings can be computed by the expected distance between their probability distributions. By modeling users with Vector Space Model (VSM), we can use the distance between vectors to measure the distance, and then the similarity between users. Using the calculated user similarities, recommender systems can search nearest neighbors for individual users, and generate recommendations by the neighbors' opinion. This is our proposed recommendation approach. The experiments in Sect. 5 using MovieLens dataset demonstrate its effectiveness in top-N recommendation task.

The remainder of the paper is organized as follows. Section 2 reviews related work, including conventional recommendation approaches and recent works on rating noise. Section 3 analyzes rating noise, and proposes our method to deal with it. Section 4 introduces the evaluation metrics used in our experiments. Experiments are carried out on MovieLens dataset in Sect. 5 to compare the proposed approach with existing ones. Finally, Sect. 6 concludes the paper by summarizing the contributions and future work.

2   Related Work

Recommendation techniques have been studied for several years. Various recommendation algorithms have been developed, including content-based filtering (CB) and collaborative filtering (CF). CF is a popular one, since it is not necessary to analyze the content of the candidate items and uses collective intelligence instead. Furthermore, it can be easily adapted from one domain to another. CF algorithms can be divided into two classes: memory-based algorithms and model-based ones [1,4].

Memory-based algorithms are heuristic methods that make rating predictions based on the entire collection of items previously rated by users [4,5,11,12]. These algorithms require all ratings, items, and users to be stored in memory. They are based on a basic assumption that people who agreed in the past tend to agree again in the future. For example, in user-based collaborative filtering (UserCF), the level of agreement can be measured by similarity between users. There are several functions to measure the similarity. In this paper, MD between users with VSM is chosen to measure user similarities. Based on the similarity calculation, a set $N(u)$ of the nearest neighbors of user u is obtained. After that, ratings for unknown items are predicted by the adjusted weighted sum of known ones, and items with high predicted values are recommended to users [7].

Model-based CF is another kind of typical CF methods. Model-based algorithms use the collection of ratings to learn a model, typically using some statistics machine-learning methods, which is then used to make rating prediction. Examples of such techniques include Bayesian clustering [4], matrix factorization [15], and probabilistic Latent Semantic Analysis [9]. These systems periodically create a summary of rating patterns offline.

Conventional ideas consider a rating value which a user rates on an item as ground truth of the user's taste. However, there is rating noise in recommender systems. Amatriain *et al.* have found that there are inconsistency among user feedbacks [2]. They tackle the problem of analyzing and characterizing the noise in user feedback through ratings of movies. They quantify the noise in user ratings that is due to inconsistencies, and analyze how factors such as item sorting and time of rating affect this noise. In addition, some other studies, e.g., Hill *et al.* [8] and Cosley *et al.* [6], also quantify the rating noise between different rating trails. O'Mahony *et al.* [14] classify noise in RS into natural and malicious. The natural noise refers to the definition of user generated noise provided in their paper, while the malicious noise refers to the noise that is deliberately introduced in a system in order to bias the results. To deal with them, they propose a de-noising algorithm in which the ratings labeled as noise are discarded. Focusing on the natural noise, Amatriain *et al.* [3] propose a recommendation algorithm to de-noise existing datasets by means of re-rating: i.e. by asking users to rate previously rated items again. Furthermore, they find that asking users to re-rate items is more beneficial than asking users to rate unseen items. This indicates that rating noise is important for recommendation quality.

To the best of our knowledge, the previous researches on recommender systems that explicitly address the problem of inconsistencies in user ratings are focusing on either quantifying rating noise or using de-noising methods to improve recommendation quality. Different from them, we solve the problem with a novel perspective. In our opinions, rating inconsistency can be considered as an inherent characteristic of user feedbacks. Therefore, a rating can be considered as a probability distribution of the user's taste instead of the exact attitude towards the current item. The details of our ideas will be proposed in the next section.

3   Method

As mentioned above, there is rating noise in recommender systems. Users are inconsistent in giving their feedback. The inconsistency among user ratings is that a user may rate an item with a value in one trial and rate the same item with a different value in another. As a result, if recommender systems commonly rely on user ratings to compute their predictions, the inconsistencies in these ratings will have an impact on the quality of the recommendations.

The main goal of this paper is to reduce this impact. In our opinion, because of the inconsistencies of ratings, a rating value rated by a user on an item does not indicate the exact attitude towards the item. It is just a probability that his/her attitude towards the item is the rating value, and there are other possibilities by which the user's attitude is other value. The idea can be formalized as:

Definition 1. *If a user rates an item with value v_0, his/her true attitude towards this item is a probability distribution (V_{v_0}), which can be written as:*

$$V_{v_0} = [v_1, v_2, \ldots, v_n] \tag{1}$$

$$v_k = P(k|v_0), k \in (1, n) \tag{2}$$

where n is the rating scale of current data, $P(k|v_0)$ is the conditional probability of user attitudes that a user's attitude towards an item is k, if he/she rates the item with value v_0. In addition, because of the definition of conditional probability, the summation of the probability distribution for a certain value should be one. Therefore, it is necessary to meet a condition as:

$$\sum_{k \in (1,n)} P(k|l) = 1, l \in (1, n) \tag{3}$$

According to the definition, the distance between two ratings is not the numerical distance between their values, but the expected distance between their probability distributions. The distance between rating value s and t can be calculated as:

$$dis(s, t) = \sum_{k \in [1,n]} \sum_{l \in [1,n]} P(k|s) \cdot P(l|t) \cdot |k - l| \tag{4}$$

VSM is used to model users. Therefore, a user can be considered as a vector of the probability distribution of ratings, which can be written as:

$$U(u) = [u_1, u_2, \ldots u_m] \tag{5}$$

$$u_i = \begin{cases} V_{v_0}{}^T & , i \in I(u) \\ 0 & , i \notin I(u) \end{cases}, i \in [1, m] \tag{6}$$

where $I(u)$ is the item set rated by user u, v_0 is the rating value which user u rate item i with, m is the number of items, 0 represents a zero vector.

Using this user model, we can measure users' distance by computing Manhattan distance (MD) between user vectors. MD is the sum of the absolute differences of each element in user vectors. It can be written as:

$$MD(u, a) = \sum_{i \in I(u,a)} dis(u_i, a_i) \tag{7}$$

where $I(u, a)$ represents the item set co-rated by user u and user a, u_i and a_i are the value which user u and a rate item i with, respectively.

MD is the total distance between users' ratings. Higher value of MD indicates larger differences between two users' preferences. However, this value cannot be used to measure user similarities directly. It is because that the value of MD increases when the size of item set co-rated by both users increases, which actually indicates large similarity between users. Therefore, the number of co-rated items is used to adjust MD value. The adjusted value is then used to measure user similarity. The similarity function between user u and a can be written as:

$$sim(u,a) = \frac{1}{1 + \frac{MD(u,a)}{|I(u,a)|}} \qquad (8)$$

where $|I(u,a)|$ is the size of $I(u,a)$.

Base on this similarity calculation, a set $N(u)$ of the nearest neighbors of user u is obtained. After that, ratings for unknown items are predicted by adjusted weighted sum of known ones, and items with high predicted values are recommended to users [7]. The predicted rating of item i by user u can be calculated as:

$$R^*(u,i) = \overline{R(u)} + \frac{\sum_{a \in N(u)} sim(u,a) \cdot (R(a,i) - \overline{R(a)})}{\sum_{a \in N(u)} |sim(u,a)|} \qquad (9)$$

where $R(a,i)$ is the rating value that user a rates item i, and $\overline{R(u)}$ represents the average rating of user u.

Unfortunately, it is difficult to obtain the conditional probability of user attitudes. In recommender systems, all we can get are the explicit ratings from users. The exact attitude of users is unobserved. We use an approximate method to estimate the conditional probability of user attitudes. According to the test-retest idea, the consistency and inconsistency of ratings can be calculated between the different trials of the same item set. The consistency is the probability that a user always rates an item with the same value. This can be used to estimate the conditional probability that the rating value is the same as the exact attitude. The inconsistency is the probability that a user rates different value on a same item in different trails. The probability by which if users gave a rating of l in one trial, they give a different rating k in another is used to estimate the conditional probability $P(k|l)$. These kinds of probability are the transition probability of user ratings. They are used to estimate the conditional probability of user attitudes.

4   Evaluation Metrics

This paper focuses on top-N recommendation task, where a recommender system is trying to pick the best N items for people [16]. The Normalized Discounted Cumulative Gain (NDCG) [10] metric is a popular metric for evaluating the relevance of top-N results in information retrieval where the documents are assigned graded rather than binary relevance judgments. As the rating values can indicate the levels of users' preferences on items in recommender systems, the NDCG metric is suitable for evaluating recommendation quality in the top-N recommendation task. In this paper, we use NDCG as our main evaluation metric. It is an accuracy metric, which can be written as:

$$DCG@N(u) = \sum_{p=1}^{N} \frac{2^{R(u,p)} - 1}{\log(1 + p)} \qquad (10)$$

$$NDCG@N = \frac{1}{|U|} \sum_{u \in U} \frac{DCG@N(u)}{IDCG@N(u)} \qquad (11)$$

where $R(u, p)$ is the rating value of user u rating the item at the p-th position of the recommendation list, DCG@N(u) represents the Discounted Cumulative Gain value at the N-th position for user u, IDCG is the maximum possible DCG which is used for normalizing the NDCG value. However, in recommender systems, the recommended items are always unknown ones (the target user has not rated the items yet). Therefore, some researchers consider the recommendation problem as a ranking problem, and use NDCG to evaluate the algorithms while ranking the items in the test set [13]. In this paper, we recognize it as NDCG+ metric and consider it to be a comparative evaluation metric. The main difference between these two metrics are NDCG+ ranks the items contained in the test set, whereas NDCG ranks all the possible items[1] for each user.

In addition, 1-call at top-N recommendations is used as another accuracy metric. It reflects the ratio of users who have at least one relevant item in their top-N recommendation lists [15].

Furthermore, a number of recent studies have found that beyond accuracy there are other quality factors, which are also important to users, e.g., diversity and novelty [1]. Coverage is one of the most popular diversity metrics. It measures the percentage of items that an algorithm is able to recommend to users in the system. Denoting the total number of distinct items in top-N places of recommendation lists for all users as N_d, the N-dependent coverage is defined as:

$$COV(N) = \frac{N_d}{N} \qquad (12)$$

In addition, the coverage of recommendations in the long tail of the item set is also a significant evaluation metric, which indicates the novelty of recommendations to a certain degree. It can be written as:

$$CIL(N) = \frac{NL_d}{N} \qquad (13)$$

where NL_d represents the intersection of N_d and the long tail item set[2].

In summary, 5 evaluation metrics are used to evaluate our proposed approach. NDCG and 1-call are used to evaluate the top-N recommendation quality. COV is used for evaluate the diversity of recommendations, whereas CIL is mainly for evaluating novelty. NDCG+ is a metric to evaluate the ranking prediction quality.

---

[1] All items except the ones that the current user has rated in the training set.

[2] The whole item set is split into two subsets, the head set and the long tail one, according to the popularity of items. In this paper, the long tail set contains 80 % items with the least popularity according to the "20–80" rule.

5 Experiment

5.1 Experiment Setup

The proposed recommendation approach is evaluated on two datasets from MovieLens[3]. The first MovieLens dataset, denoted as ML1, contains 100 thousand ratings which are assigned by 943 users on 1682 movies. The second one, denoted as ML2, contains 1 million ratings from 6040 users on 3900 movies. Collected ratings from both datasets are in a 1-to-5 star scale.

We use 5-fold cross validation for the evaluation. Starting from the initial data set, five distinct splits of training and test data are generated. For each data split, 80 % of the original set is included in the training data and 20 % of it is included in the test data. Users' rating history in the training set is used to generate recommendations according to different algorithms. The test set is then used to evaluate the recommendation results.

The proposed approach is based on conventional UserCF. Its improvement is on the similarity calculation with MD on probability distribution (MD_PD). The approximate conditional distribution of user attitudes is estimated by the researches of Amatriain et al. [2,3] in which they use re-rating to quality rating consistency, and study the statistics of the user rating transition probability. In order to demonstrate our proposed idea, two baseline approaches are used as comparative ones. They both are based on conventional UserCF with Pearson Correlation (Pearson) and classical MD (C_MD) to measure user similarity, respectively. The size of nearest neighbors for all the three approaches is 50, which is a typical value in UserCF.

5.2 Experiment Results and Analysis

In this subsection, we present a performance comparison of both accuracy and diversity between our proposal and the benchmark approaches. Tables 1 and 2 illustrates the results of the proposed approach and the comparative ones in top-N recommendation task on ML1 and ML2, respectively. For each approach, we report the NDCG values at the $1st$, $3rd$ and $5th$ positions in the recommendation list, 1-call, COV and CIL at the $5th$ position, compared with NDCG+ at the $5th$ position. The bold cell indicates the best result for that metric.

Table 1. Performance of different approaches on ML1

	NDCG			1-call	COV	CIL	NDCG+
	1	3	5				
Pearson	2.07 %	2.45 %	2.91 %	15.5 %	102	54	83 %
C_MD	3.57 %	3.57 %	3.60 %	16.6 %	106	57	88 %
MD_PD	**3.79 %**	**3.65 %**	**3.87 %**	**18.2 %**	**107**	**58**	**89 %**

---

[3] http://www.grouplens.org/node/73

Table 2. Performance of different approaches on ML2

	NDCG			1-call	COV	CIL	NDCG+
	1	3	5				
Pearson	2.25 %	2.27 %	2.42 %	14.8 %	221	102	88 %
C_MD	2.00 %	2.27 %	2.38 %	14.7 %	264	137	**93 %**
MD_PD	**2.42 %**	**2.37 %**	**2.50 %**	**15.3 %**	**278**	**147**	93 %

As shown in Table 1, our proposed approach outperforms the others on all evaluation metrics on ML1. The results demonstrate the effectiveness of MD_PD. In terms of NDCG metrics, MD_PD gains the best performances, followed by C_MD. This indicates their capability of recommending relevance items, and verifies the effectiveness of both considering rating as probability distribution and using MD to measure user similarities. C_MD improves Pearson on 1-call about 7 %, and MD_PD further improves C_MD about 10 %. This indicates the contribution to provide valuable recommendations at top-N positions. It raises the chance that users receive at least one relevant recommendation among just a few top-ranked items. Furthermore, MD_PD gains the best COV and CIL performances, which illustrates that MD_PD can improve the baseline ones on diversity and novelty as well as on accuracy. NDCG+ is a comparative evaluation metric, and we only consider the value at the $5th$ position. On this metric, MD_PD still performs best. This indicates its capability of ranking items.

For the performances on ML2 dataset, which is much larger than ML1, similar observations can be found, as shown in Table 2. MD_PD still outperforms other approaches on both accuracy and diversity metrics. However, while the dataset changes from ML1 to ML2, C_MD does not outperform Pearson especially on accuracy metrics. This shows that the advantage of MD is reducing when the size of dataset increases. In this case, the effectiveness of MD_PD on ML2 is mainly because of considering rating as probability distribution.

The above experiments are offline ones, which are based on the statistical results of Amatriain *et al.* [2,3]. Therefore, a global probability distribution is used to measure the probability distribution of user attitudes. However, different users always have different habits to give their feedbacks. As a result, a much personalized approach may further improve the recommendation quality. Fortunately, our approach is suitable for personalizing. According to (4), the expected distance between different users rating on a same item can use a personalized probability distribution of the current user's attitudes instead of the global one. It can be written as:

$$dis(u_i, a_i) = \sum_{k\in[1,n]} \sum_{l\in[1,n]} P_u(k|u_i) \cdot P_a(l|a_i) \cdot |k - l| \qquad (14)$$

where P_u and P_a are the personalized probability distribution of the attitude of user u and a, respectively. In our subsequent researches, we will evaluate the personalized approach on our online experiment.

6 Conclusions

Recommender systems have made significant progress in recent years. Most studies consider user ratings as their exact attitude towards items. However, some researchers have found that users are inconsistent while giving their feedbacks, which can be considered as rating noise. In this paper, we consider the rating inconsistency as an inherent characteristic of user feedbacks, and use the probability distribution of user attitudes to describe user ratings instead of the exact attitude towards the current item. According to it, we propose a similarity function based on the Manhattan Distance between user vectors. The similarity function is used in convention UserCF to improve the top-N recommendation performance. Experiments on MovieLens dataset show the effectiveness of the proposal on both accuracy and diversity.

The experiments demonstrate our idea in offline experiment, and we will try to further demonstrate it in online experiments. In addition, the proposed approach uses global probability distribution to measure user ratings. We will find a much personalized way in our future work.

Acknowledgements. This work is supported by the National Natural Science Foundation of China (Project Nos. 61250010).

References

1. Adomavicius, G., Kwon, Y.: Improving aggregate recommendation diversity using ranking-based techniques. IEEE Trans. Knowl. Data Eng. **24**(5), 896–911 (2012)
2. Amatriain, X., Pujol, J.M., Oliver, N.: I like it.. I like it not: evaluating user ratings noise in recommender systems. In: Houben, G.-J., McCalla, G., Pianesi, F., Zancanaro, M. (eds.) UMAP 2009. LNCS, vol. 5535, pp. 247–258. Springer, Heidelberg (2009)
3. Amatriain, X., Pujol, J., Tintarev, N., Oliver, N.: Rate it again: increasing recommendation accuracy by user re-rating. In: Proceedings of the Third ACM Conference on Recommender Systems, pp. 173–180. ACM (2009)
4. Breese, J., Heckerman, D., Kadie, C.: Empirical analysis of predictive algorithms for collaborative filtering. In: Proceedings of the Fourteenth Conference on Uncertainty in Artificial Intelligence, pp. 43–52. Morgan Kaufmann Publishers Inc. (1998)
5. Chen, W., Niu, Z., Zhao, X., Li, Y.: A hybrid recommendation algorithm adapted in e-learning environments. World Wide Web, pp. 1–14 (2012)
6. Cosley, D., Lam, S., Albert, I., Konstan, J., Riedl, J.: Is seeing believing? How recommender system interfaces affect users opinions. In: Proceedings of the SIGCHI Conference on Human Factors in Computing Systems, pp. 585–592. ACM (2003)
7. Delgado, J., Ishii, N.: Memory-based weighted majority prediction. In: ACM SIGIR'99 Workshop on Recommender Systems. Citeseer (1999)
8. Hill, W., Stead, L., Rosenstein, M., Furnas, G.: Recommending and evaluating choices in a virtual community of use. In: Proceedings of the SIGCHI Conference on Human Factors in Computing Systems, pp. 194–201. ACM Press/Addison-Wesley Publishing Co. (1995)

9. Hofmann, T.: Collaborative filtering via gaussian probabilistic latent semantic analysis. In: Proceedings of the 26th Annual International ACM SIGIR Conference on Research and Development in Information Retrieval, SIGIR'03, pp. 259–266. ACM, New York (2003)
10. Järvelin, K., Kekäläinen, J.: Cumulated gain-based evaluation of IR techniques. ACM Trans. Inf. Syst. (TOIS) **20**(4), 422–446 (2002)
11. Koren, Y.: Factorization meets the neighborhood: a multifaceted collaborative filtering model. In: Proceeding of the 14th ACM SIGKDD International Conference on Knowledge Discovery and Data Mining, pp. 426–434. ACM (2008)
12. Linden, G., Smith, B., York, J.: Amazon.com recommendations: item-to-item collaborative filtering. IEEE Internet Comput. **7**(1), 76–80 (2003)
13. Liu, N.N., Zhao, M., Yang, Q.: Probabilistic latent preference analysis for collaborative filtering. In: Proceedings of the 18th ACM Conference on Information and Knowledge Management, pp. 759–766. ACM (2009)
14. O'Mahony, M., Hurley, N., Silvestre, G.: Detecting noise in recommender system databases. In: Proceedings of the 11th International Conference on Intelligent User Interfaces, pp. 109–115. ACM (2006)
15. Shi, Y., Karatzoglou, A., Baltrunas, L., Larson, M., Oliver, N., Hanjalic, A.: Climf: learning to maximize reciprocal rank with collaborative less-is-more filtering. In: Proceedings of the Sixth ACM Conference on Recommender Systems, pp. 139–146. ACM (2012)
16. Zhao, X., Niu, Z., Chen, W.: Opinion-based collaborative filtering to solve popularity bias in recommender systems. In: Decker, H., Lhotská, L., Link, S., Basl, J., Tjoa, A.M. (eds.) DEXA 2013, Part II. LNCS, vol. 8056, pp. 426–433. Springer, Heidelberg (2013)

On Computing Similarity in Academic Literature Data: Methods and Evaluation

Masoud Reyhani Hamedani$^{(\boxtimes)}$ and Sang-Wook Kim

Department of Computer Software, Hanyang Uiversity, Seoul, Korea
{masoud,wook}@agape.hanyang.ac.kr

Abstract. Similarity computation for academic literature data is one of the interesting topics that have been discussed recently in information retrieval and data mining. Consequently, a variety of methods has been proposed to compute the similarity of scientific papers. In this paper, we present various similarity methods and evaluate their effectiveness via extensive experiments on a real-world dataset of scientific papers.

1 Introduction

Scientific papers are one of the primary sources to share information and knowledge among researchers. Consequently, computing the similarity in the academic literature data is one of the interesting topics that have been discussed recently in information retrieval and data mining. A scientific paper contains two related aspects: content and citations. The content is the main aspect of a paper that demonstrates its context. On the other hand, citations selected carefully by the authors are a set of references to the related and authoritative papers. In other words, the main purpose of inserting citations in a paper is improving its content in order to make it more understandable for readers.

In the literature, a variety of methods have been proposed to compute the similarity of scientific papers. However, each of these methods applies two aspects of a scientific paper in the similarity computation differently. The link-based similarity measures such as SimRank [7], rvs-SimRank [19], and P-Rank [19] consider only the citation relationships between papers to compute the similarity and totally neglect the content. On the contrary, the text-based similarity measures such as Cosine [10], Dice Coefficient [10], BM25 [3, 11], and Kullback-Leibler Distance (KLD) [1, 16] consider only the content of papers to compute the similarity and ignore citations. CEBC [4], Keyword-Extension [18], [12], and [15] (here, we call them combination methods) consider content and citations together and utilize both of them in the similarity computation.

In this paper, we present an overview of the various similarity methods have been proposed to compute the similarity of scientific papers. Also, we analyze and compare the effectiveness of these methods by performing extensive experiments on a real-world dataset of scientific papers.

The rest of this paper is organized as follows. Section 2 describes the link-based similarity measures. Section 3 explains the text-based similarity measures. Section 4 presents the combination methods. Section 5 explains our dataset and analyzes experimental results. Section 6 summarizes and concludes the paper.

© Springer International Publishing Switzerland 2014
Y. Chen et al. (Eds.): WAIM 2014, LNCS 8597, pp. 403–412, 2014.
DOI: 10.1007/978-3-319-11538-2_37

2 Link-Based Similarity Measures

The link-based similarity measures compute the similarity of scientific papers by exploiting a citation graph where nodes specify papers and edges represent the citation relationship between a pair of papers. In the literature, there are various types of link-based similarity measures such as Coupling [9], Co-citation [14], Amsler [2], SimRank [7], rvs-SimRank [19], and P-Rank [19]. SimRank, rvs-SimRank, and P-Rank are the recursive version of Co-citation, Coupling, and Amsler, respectively, that significantly improve their accuracy [19]. SimRank considers the in-links recursively and neglects the out-links, and similarity between two papers is computed *only* based on the papers that cite them. rvs-SimRank computes the similarity between papers only based on the out-links recursively, so the similarity between two papers is computed *only* based on the papers cited by them. P-Rank considers both in-links and out-links recursively, so the similarity between a pair of papers is computed based on the papers that cite them and are cited by them. Furthermore, other link-based similarity measures are a special case of the P-Rank unified formulation as follows:

$$R_0(p, q) = \begin{cases} 0 & \text{if } (p \neq q) \\ 1 & \text{if } (p = q) \end{cases}$$

$$R_{k+1}(p, q) = \lambda \times \frac{C}{|I(p)||I(q)|} \sum_{i=1}^{|I(p)|} \sum_{j=1}^{|I(q)|} R_k\big(I_i(p), I_j(q)\big) + (1 - \lambda)$$
$$\times \frac{C}{|O(p)||O(q)|} \sum_{i=1}^{|O(p)|} \sum_{j=1}^{|O(q)|} R_k\big(O_i(p), O_j(q)\big), \tag{1}$$

where $R_k(p, q)$ denotes the P-Rank score between papers p and q in the iteration k, $I(p)$ is a set of papers that cite p, $O(p)$ is a set of papers cited by p, λ ($0 \leq \lambda \leq 1$) is a relative weight, and C ($0 \leq C \leq 1$), is a damping factor. If $\lambda = 1$, Eq. (1) is reduced to SimRank, and if $\lambda = 0$, it is reduced to rvs-SimRank.

The link-based similarity measures only consider the citation relationships between scientific papers in similarity computation and absolutely neglect the content of papers.

3 Text-Based Similarity Measures

The text-based similarity measures consider a paper as a group of terms and two papers are regarded more similar if they have more terms in common. In the literature, there are various types of text-based similarity measures. Cosine [10] and Dice Coefficient [10] are based on the vector space model [13]. In the vector space model, every paper is represented as an n-dimensional vector and each dimension corresponds to a term of the paper. The value of a dimension defines the weight of the term [17]. The weight is calculated as TF-IDF, the product of term frequency (TF) and inverse document frequency (IDF). TF indicates the number of times that the term appears in a paper, whereas IDF indicates the importance of the term in a dataset and is calculated usually by taking logarithm of dividing the total number of papers by the number of those papers that contain that term. The high TF-IDF value for a term in a paper denotes that the term is not common among so many papers in the dataset and at the same time, has high repetition in the paper. Therefore, this term is an important feature of the paper to be used in similarity computation.

Cosine computes the similarity of a pair of papers as the value of the angle between their corresponding vectors as follows:

$$Cosine(p,q) = \frac{\sum_{t\in(p\cap q)}(w_{t,p}.w_{t,q})}{\sqrt{\sum_{t\in p}(w_{t,p})^2 \sum_{t\in q}(w_{t,q})^2}} \tag{2}$$

where $w_{t,p}$ is the weight of term t in paper p. Also, Dice Coefficient computes the similarity of a pair of papers by considering their vectors as follows:

$$Dice(p,q) = \frac{2 \times \sum_{t\in(p\cap q)}(w_{t,p}.w_{t,q})}{\sum_{t}(w_{t,p})^2 + \sum_{t\in q}(w_{t,q})^2} \tag{3}$$

Kullback-Leibler Distance (KLD) [1, 16] and BM25 [3, 11] are based on the probabilistic models [5]. In probabilistic models, every term appears in a paper according to a probability. In other words, a term belongs to a paper with a specific probability value. The basic assumption is that terms are distributed differently within relevant and irrelevant papers. KLD considers a paper as a probability distribution over all the terms in the dataset, and the *distance* between two papers is computed according to these probabilities as follows:

$$KLD(Q,R) = \sum_{t\in V}(P(t,Q) - P(t,R))\log\frac{P(t,Q)}{P(t,R)} \tag{4}$$

where V is a term set, Q is a query paper, and $P(t, R)$ is the probability value of appearing term t in paper R. If term t does not appear in paper R, $P(t, R) = 0$, we should smooth paper R as follows:

$$P(t,R) = \begin{cases} \beta * P(t|R) & \textit{if } t \textit{ occures in } R \\ \varepsilon & \textit{otherwise} \end{cases} \tag{5}$$

$$P(t|R) = \frac{tf(t,R)}{\sum_{x\in R} tf(x,R)} \tag{6}$$

$$\beta = 1 - \sum_{t\in V, t\notin R} \varepsilon \tag{7}$$

where $tf(t, R)$ is the frequency of term t in paper R and ε is a smoothing value that can be set to any value less than the minimum existing value of $P(t|R)$. β is a normalization coefficient.

BM25 utilizes its own weighting scheme and puts more emphasis on the term frequency and the paper length to compute the similarity of a pair of papers as follows:

$$BM25(p,q) = \sum_{t \in q} idf_t.\alpha_{t,p}.\beta_{t,q} \tag{8}$$

$$\alpha_{t,p} = \frac{(K_1 + 1)tf_{t,p}}{K_1\left((1-b) + b.\frac{|p|}{avg(L)}\right) + tf_{t,p}} \tag{9}$$

$$\beta_{t,q} = \frac{(K_3 + 1)tf_{t,q}}{K_3 + tf_{t,q}} \tag{10}$$

idf_t is the inverse document frequency of term t, $|p|$ is the size of paper p, $avg(L)$ is the average size of the dataset, and $tf_{t,\ p}$ is the term frequency of term t in paper p.

The text-based similarity measures only take into account the content of scientific papers in similarity computation and absolutely ignore the citation relationship between them.

4   Combination Methods

The combination methods consider the content and citations *together* and utilize both of them in the similarity computation. In Ref. [12], the similarity of scientific papers is computed by combining the text-based and link-based similarity measures. The intuition behind this method is that the text-based and link-based similarity measures consider only one aspect of scientific papers, content or citations, respectively; however, both content and citations are valuable in computing the similarity. In this method, the similarity between a pair of papers is computed by applying a text-based and a link-based similarity measure separately. Then, these similarity scores are combined into a single value according to a weighted linear combination as follows:

$$S(p,q) = w_1 TS(p,q) + w_2 LS(p,q) \tag{11}$$

where $TS(p,q)$ and $LS(p,q)$ are similarity scores between papers p and q computed by a text-based and a link-based similarity measure, respectively. The optimal values of w_1 and w_2 are indicated by utilizing SVM^{rank} [8], which is based on the support vector machine (SVM).

In CEBC [4], Keyword-Extension [18], and Ref. [15], the content of a paper is enriched by adding additional terms from those papers that have a direct citation relationship with it. The main intuition behind these methods is that contents of two papers involved in a citation relationship are somehow related and similar. Thus, the missing terms in the available content of a paper, which are missed due to the limitation of crawling and parsing can be taken from neighbor papers in the citation graph [18]. Also, the major reason to insert citations in a scientific paper is improving its content. In other words, without citations the context of a paper cannot be entirely interpreted [4, 15]. Therefore, the accuracy of similarity computation will be improved if the content of a paper being enriched with the content of those papers that have citation relationship whit it. In all these methods, a paper is represented as an n-dimensional vector. Therefore, any similarity measures for vectors (such as Cosine, Dice, BM25, and KLD) can be applied for similarity computation.

CEBC constructs a bibliographic context, BC, for every paper p, as a set of the papers directly connected to paper p in the citation graph. Then, the content enrichment for paper p is done as follows:

$$EW(p,t) = \alpha OW(p,t) + (1 - \alpha) \sum_{q \in BC(p)} \frac{OW(q,t)}{|BC(p)|} \qquad (12)$$

where t is a term in the term set, $EW(p, t)$ is the enriched weight of term t in paper p, $OW(p, t)$ is the original weight of term t in paper p, $BC(p)$ is the BC set of paper p, q is a paper in $BC(p)$, and $0 \leq \alpha \leq 1$ is the important factor for combination. $OW(q, t)$ is divided by the size of BC in order to normalize the effect of $OW(q, t)$ in the content enrichment.

In Keyword-Extension, the content enrichment is done without applying any weighting scheme and normalization. In other words, the original weight of term t in paper p is combined simply with the weights of term t in direct neighbor papers in the citation graph. However, Keyword-Extension considers only the title and abstract to compute the similarity of scientific papers with the weighting ratio 0.3 and 0.7 for terms in the title and the abstract, respectively.

Reference [15], proposes a method for scholarly papers recommendation. Every paper in the publication set of a user is enriched by adding additional terms from direct neighbor papers in the citation graph. The cosine similarity value between two papers is used in the content enrichment as an important factor to combine the weight of terms. Then, the user profile is constructed as an n-dimensional vector based on the papers in the publication set of the user. Finally, recommendation is performed by comparing the user profile with candidate papers by applying Cosine as the similarity measure. In this method, if paper p has n direct neighbors in the citation graph, the content enrichment for paper p is done as follows:

$$EW(p,t) = OW(p,t) + \sum_{i=1}^{n} Cosine(p,p_i)OW(p_i,t) \qquad (13)$$

where t is a term in the term set, p_i is a direct neighbor of paper p in the citation graph, and $Cosine(p, p_i)$ is the cosine similarity value between papers p and p_i.

5   Evaluation

In this section, we explain our real-world dataset and represent our experimental results. In order to evaluate the effectiveness of link-based similarity measures, text-based similarity measures, and combination methods in computing the similarity of scientific papers, we applied them on a real-world dataset of scientific papers and analyzed their accuracy. All our experiments were performed on an Intel machine equipped with four 2.67 GHz i5 CPU, 12 GB RAM, and 64-bit Fedora Core 17 operating system. All methods were implemented in Java based on Open JDK 1.7.0 by using Eclipse IDE 4.2.0.

5.1   Dataset

We crawled information of 1,071,973 papers from DBLP[1] and obtained their citation information from MS Academic Search[2]. To make a precise ground truth for our dataset, we used a famous data mining textbook [6] where relevant papers are categorized by the experts (i.e., authors) according to the research topics in the bibliographic section of each chapter. We constructed eleven ground truth sets by selecting eleven research topics from different chapters: data preprocessing, association rules, classification, clustering, data stream mining, link mining, graph mining, data cubes, spatial databases, data warehouses, and web mining. Due to the copyright issue, we only have the title, abstract, and citation information for each paper in the dataset. As shown in [18], the lack of the body does not have negative effect on the accuracy of similarity computation.

5.2   Measures

In order to evaluate carefully the effectiveness of different methods in computing the similarity of scientific papers, we use MAP, precision at top 10 results (P@10), and recall at top 10 results (R@10) [10] as evaluation measures. We use all papers in a ground truth set as a separate query and compute MAP, P@10, and R@10 for every single ground truth set. Then, we calculate the average value of MAP, P@10, and R@10 over all eleven ground truth sets.

5.3   Results and Analyses

Figure 1 shows the accuracy of SimRank, rvs-SimRank, and P-Rank on different iterations from 1 to 5 in terms of MAP, P@10, and R@10. We set the damping factor $C = 0.8$ for all measures and the relative weight $\lambda = 0.5$ for P-Rank same as [19]. SimRank, rvs-SimRank, and P-Rank show their best accuracy in iteration 1, 2, and 5, respectively. However, P-Rank with five iterations largely outperforms SimRank and rvs-SimRank. The reason is that unlike SimRank and rvs-SimRank, P-Rank considers both in-links and out-links in the citation graph to compute the similarity.

For text-based similarity measures, in order to calculate the weight of a term in a paper, we utilized the TF-IDF value for Cosine and Dice, the TF value for KLD and the BM25 weight for BM25. As a result, we constructed three different vectors to represent

---

[1] http://www.informatik.uni-trier.de

[2] http://academic.research.microsoft.com/

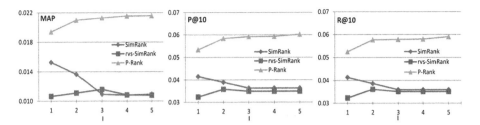

Fig. 1. Accuracy of SimRank, rvs-SimRank, and P-Rank

every single paper in our dataset. For BM25, we set parameters $K_1 = 1.2$, $b = 0.75$, and $K_3 = 2$ by following [3]. Figure 2 demonstrates the accuracy of Cosine, Dice, BM25, and KLD in terms of MAP, P@10, and R@10. In opposition to what has been stated in [19], the text-based similarity measures outperform link-based similarity measures. The reason is that they consider the content of papers in similarity computation, and the content represents the context of a paper more accurately than citations. Dice outperforms other text-based similarity measures. The accuracy of BM25 is *slightly* better than Cosine. Therefore, similarity measures based on the vector space model are more appropriate than those are based on the probabilistic models to compute the similarity of scientific papers.

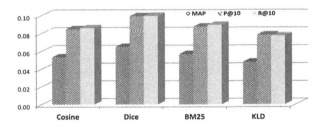

Fig. 2. Accuracy of Cosine, Dice, BM25, and KLD

Figure 3 represents the accuracy of combination method in Ref. [12] where the similarity of scientific papers is computed by combining the link-based and text-based similarity measures. In consistency with Ref. [12], we combined P-Rank (with five iterations) with Cosine, Dice, BM25, and KLD separately. In Fig. 3, we illustrate each combination as P-Rank+A where A denotes the text-based similarity measure. This combination method totally outperforms P-Rank and the text-based similarity measure that is combined with, in terms of MAP, P@10, and R@10.

As an example, P-Rank+Cosine outperforms the accuracy of both P-Rank and Cosine significantly because the text-based and link-based similarity measures only consider a single aspect of scientific papers in similarity computation; however, this method considers both content and citations together to compute the similarity.

In order to evaluate the accuracy of CEBC, we performed the content enrichment for every paper in the dataset based on TF-IDF values for Cosine and Dice, TF values

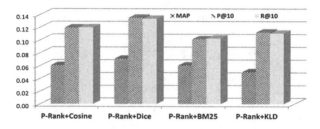

Fig. 3. Accuracy of combining P-Rank with text-based similarity measures

for KLD, and BM25 weights for BM25. In CEBC implementation, we set $\alpha = 0.5$ according to [4]. Figure 4 shows the accuracy of CEBC with Cosine, Dice, BM25, and KLD in terms of MAP, P@10 and R@10. Same as the combination method in Ref. [12], CEBC totally outperforms the text-based and link-based similarity measures. However, CEBC(Dice), CEBC(BM25), and CEBC(KLD) outperform P-Rank+Dice, P-Rank+BM25, and P-Rank+KLD in terms of MAP, P@10, and R@10, respectively; however, CEBC(Cosine) outperforms P-Rank+Cosine *only* in term of MAP. The reason is that the method in Ref. [12] computes the similarity of scientific papers based on content and citations separately and simply combines these two similarity scores into a single value. However, CEBC enriches the content of a paper by using the content of those papers have direct citation relationships with it. In other words, CEBC utilizes one aspect, citations, to enrich the other aspect, content.

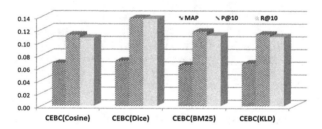

Fig. 4. Accuracy of CEBC with Cosine, Dice, BM25, and KLD

Figure 5 represents the accuracy of Keyword-Extension with Cosine, Dice, BM25, and KLD in terms of MAP, P@10, and R@10. Same as CEBC, we performed the content enrichment for every paper in our dataset based on TF-IDF values for Cosine and Dice, TF values for KLD, and BM25 weights for BM25. Also, we regarded the weighting ratio 0.7 and 0.3 for every term in the title and the abstract, respectively. Keyword-Extension absolutely outperforms the text-based and link-based similarity measures.

Figure 6 shows the accuracy of the combination method in Ref. [15] (here, we call it CW) with Cosine, Dice, BM25, and KLD in terms of MAP, P@10, and R@10. Same as CEBC and Keyword-Extension, we performed the content enrichment for every paper in our dataset based on TF-IDF values for Cosine and Dice, TF values for KLD, and BM25 weights for BM25. In the implementation, we utilized a same similarity

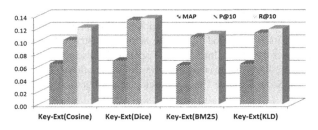

Fig. 5. Accuracy of Keyword-Extension with Cosine, Dice, BM25, and KLD

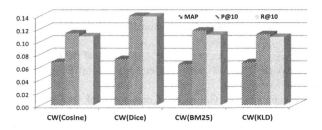

Fig. 6. Accuracy of CW with Cosine, Dice, BM25, and KLD

measure in the content enrichment and similarity computation (e.g., for applying CW with KLD to compute the similarity of scientific papers, we use KLD in the content enrichment). CW totally outperforms the text-based and link-based similarity measures.

In summary, the combination methods significantly improve the accuracy of the text-based similarity measures and totally outperform the link-based similarity measures in terms of MAP, P@10, and R@10 because they consider both content and citations together in computing the similarity of scientific papers. However, it is *difficult* to choose one of the combination methods as the superior one. The only evidence is that the combination method in Ref. [12] has slightly worse accuracy than other ones. Also, there is not any significant difference between the accuracy of CEBC, Keyword-Extension and the combination method in Ref. [15].

6   Conclusions

In this paper, we introduced eleven existing methods for computing similarity in the academic literature data and explained their working mechanism. Also, we evaluated the effectiveness of these methods by performing extensive experiments on a real-world dataset of scientific papers. According to our experimental results, the combination methods, which consider both content and citations, are more appropriate for computing similarity in the academic literature data than the text-based and link-based similarity measures. The combination methods absolutely outperform the link-based similarity measures. Also, they significantly improve the accuracy of the text-based similarity measures more than 40 % in average.

Acknowledgments. This work was supported by (1) Business for Cooperative R&D between Industry, Academy, and Research Institute funded Korea Small and Medium Business Administration in 2013 (Grants No. C0006278), (2) the MSIP (Ministry of Science, ICT, and Future Planning), Korea, under the ITRC (Information Technology Research Center) support program (NIPA-2013-H0301-13-4009) supervised by the NIPA (National IT Industry Promotion Agency), and (3) the Seoul Creative Human Development Program (HM120006).

References

1. Aktolga, A., Ros, I., Assogba, Y.: Detecting outlier sections in US congressional legislation. In: ACM SIGIR, pp. 235–244 (2011)
2. Amsler, R.: Application of citation-based automatic classification. Technical report, Texas University (1972)
3. Barrón-Cedeño, A., Eiselt, A., Rosso, P.: Monolingual text similarity measures: a comparison of models over Wikipedia articles revisions. In: European Conference on IR Research, pp. 305–319 (2003)
4. Chiki, N., Rothenburger, B., Gilles, N.: Combining link and content information for scientific topics discovery. In: ICTAI, pp. 211–214 (2008)
5. Fuhr, N.: Probabilistic models in information retrieval. Comput. J. **35**(3), 243–255 (1992)
6. Han, J., Kamber, M.: Data Mining: Concepts and Techniques, 2nd edn. Morgan Kaufmann, San Francisco (2006)
7. Jeh, J., Widom, J.: SimRank: a measure of structural-context similarity. In: ACM SIGKDD, pp. 538–543 (2002)
8. Joachims, T.: Optimizing search engines using clickthrough data. In: ACM SIGKDD, pp. 133–142 (2002)
9. Kessler, M.: Bibliographic coupling between scientific papers. Am. Doc. J. **14**(1), 10–25 (1963)
10. Frakes, W.B., Baeza-Yates, R.: Information Retrieval: Data Structures and Algorithms. Prentice-Hall, Englewood Cliffs (1992)
11. Lv, Y., Zhai, C.: When documents are very long, BM25 fails. In: ACM SIGIR, pp. 1103–1104 (2011)
12. Reyhani Hamedani, M., Lee, S., Kim, S.: On combining text-based and link-based similarity measures for scientific papers. In: ACM RACS, pp. 111–115 (2013)
13. Salton, G., Lesk, M.E.: Computer evaluation of indexing and text processing. ACM J. **15**(1), 8–36 (1968)
14. Small, H.: Co-citation in the scientific literature: a new measure of the relationship between two documents. J. Am. Soc. Inf. Sci. **24**(4), 265–269 (1973)
15. Sugiyama, K., Kan, M.: Scholarly paper recommendation via user's recent research interests. In: JCDL, pp. 29–38 (2010)
16. Tan, B., Shen, X., Zhai, C.: Mining long-term search history to improve search accuracy. In: ACM SIGKDD, pp. 718–723 (2006)
17. Yates, R.B., Neto, B.R.: Modern Information Retrieval. Addison Wesley, Boston (1999)
18. Yoon, S., Kim, S., Kim, J.: On computing text-based similarity in scientific literature. In: WWW, pp. 169–170 (2011)
19. Zhao, P., Han, H., Yizhou, S.: P-Rank: a comprehensive structural similarity measure over information networks. In: CIKM, pp. 553–562 (2009)

Ranking People by Integrating Professional Capability, Sociability and Time Decaying Effect

Hongguo Yang$^{(\boxtimes)}$, Derong Shen$^{(\boxtimes)}$, Yue Kou, Tiezheng Nie,
and Ge Yu

College of Information Science and Engineering,
Northeastern University, Shenyang, China
yanghongguo896@gmail.com,
{shenderong,kouyue,nietiezheng,yuge}@ise.neu.edu.cn

Abstract. In this paper, we propose a model which is designed for the ranking of people who are of a same working domain. Our model measures their ranks comprehensively. It calculates their rank-scores from two aspects: one aspect is about their sociability, which can be represented by the links connected to them from other related people, the other one is about the achievements completed by them, e.g., the achievements of a scientist can be measured by the quality of books and papers published by him/her. Our model also takes the decaying effect of time into consideration, because achievements would probably be forgotten by people or might become obsolete after a long time period, consequently the related rank-scores would be decayed along with time. Our experiments showed that our rank list is time sensitive and more meaningful in the situation when we want to get people ranked according to how much people are recognized and remembered by others at a specific time point.

Keywords: Time · Ranking · Person · Influence

1 Introduction

People Ranking is very meaningful in the situation when we need to find out those top ranked persons who are more important than others. For example, People Ranking can be utilized by search engines. Some search engines have launched a project called Cubic knowledge, which aims to return information about a person as many as possible. It would take a decent amount of computational and storage cost if a search engine tends to get cubic knowledge for all the person entities on web. At the mean time, it's of little meaning to gather the cubic knowledge for those who are common people and rarely be searched by users. It's better for search engines to find out those popular people which worthy their particular treatments. Here comes the People Ranking problem, the solution of which would tell them who are those popular persons. Besides, web users are mainly interested in those people who are popular in recent years. In order to get the recently popular persons, it's better to take the time decaying

© Springer International Publishing Switzerland 2014
Y. Chen et al. (Eds.): WAIM 2014, LNCS 8597, pp. 413–425, 2014.
DOI: 10.1007/978-3-319-11538-2_38

effect into consideration, as people are forgetful and a past famous person might not be as famous as past due to the time decaying effect. People Ranking can also be employed in the academic domain. We can use it to get researchers ranked and find experts in an interested research domain. Likewise, in show business domain, we can rank those performers and get a celebrity list. People Ranking can also be beneficial for web users to evaluate the rank of a commodity (e.g., book) according to the rank of its manufacturer (e.g., writer) returned by People Ranking algorithms. Furthermore, People Ranking can also be tailored to rank the employees in an organization for appropriate and effective task assignments and staff appointments.

With some existing ranking algorithms, you might also be able to get people ranked, but none of these algorithms take the decaying effect of time into consideration, which would result in selecting a person who was pretty famous before but has already retreated from public life for many years. Second, they are not able to return a rank list especially for a specific time deadline, for example, who were those influential persons in 1920 for a certain domain. Furthermore, unlike Fig. 1, they don't distinguish the achievement links (reflecting professional *capability*) from social links (*sociability*) and treat them with a same strategy by ignoring their different significance to a person's rank.

The ranking model proposed in this paper should possess the following functionalities. First, it should be able to return a rank list for a specific time deadline. Second, it can reflect the time decaying effect in the final rank list. Third, the ranking model should not only be able to reflect a person's *sociability* but also their professional *capability*. The ranking model proposed in this paper is denoted as *Influence* ranking model, as such a ranking model mainly measures the influential power of people.

In summary, the contributions of this paper are as followings. First, the *Influence* ranking model is proposed in this paper, which treats people's social links and achievement links differently and then combines them together. Second, time factor are taken into consideration in this paper. Third, an optimization method is provided for efficient calculation. Finally, the experiments demonstrate that, from the point of comparing people's *Influence* at a specific time point, our model gets a more meaningful rank list than other algorithms.

The remainder of this paper is organized as follows. Related works are discussed in Sect. 2. An overview of *Influence* ranking model is introduced in Sect. 3. The achievements and the decaying effect of time are measured in Sects. 4 and 5 respectively. Experimental results are given in Sect. 6. Lastly, we draw a conclusion in Sect. 7.

2   Related Works

Initially, some ranking algorithms [1,2] and optimizations [3–5] are developed for web-page ranking. Later, some entity ranking algorithms [6–9] are proposed, but none of these take time factor into consideration. Paper [10] aims to rank location entities while this paper is specifically for ranking Person Entities. Time factor is dealt with in the papers [11–13]. The algorithm proposed in paper [11] exploits time feedback into ranking algorithm for web-pages, the fundamental method of which is totally different from that of this paper which is for ranking person entities. Paper [12] measures page rank for a specific user through the time length the user spend on, while this paper uses time length for measuring the decaying effect of time on the rank score of people. Paper [13] mentioned the obsolescence of knowledge, but the specific evaluation method is not proposed and the general method for ranking is total different from that of this paper.

3   Overview of Influence Ranking Model

As we have mentioned in the introduction section, *Influence* ranking model combines time, social and achievement factors all together. Before deriving our final *Influence* ranking model, we first introduce some basic concepts–*Sociability*, *Capability* and so on based on these factors. For the clear presentation of them, we employ the following graph Fig. 1 for help. We also utilize the symbols listed in the Table 1 for easy presentation.

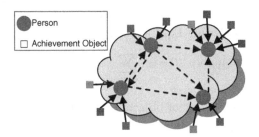

Fig. 1. Graph model used in our paper

Sociability. The dashed edges in Fig. 1 represent the relationships between people, based on which we can get people ranked through a variant of PageRank [1] algorithm. This paper employs the concept *Sociability* to denote such a ranking model, which is formulated by the following Eq. 1 to calculate the *Sociability* of person p_i at the time point t.

$$\text{PersnSocia}(p_i, t) = \delta * \frac{1}{N} + (1 - \delta) * \sum_{p_k \in R(p_i, t)} \frac{\text{PersnSocia}(p_k, t)}{\text{OL}(p_k, t)}. \qquad (1)$$

Where N is the total number of candidate persons for ranking, $R(p_i, t)$ is the set of persons that have linked to person p_i before the time point t, $\text{OL}(p_k, t)$ is the number of outbound links of person p_k before the time point t.

Table 1. Overview of symbols

Symbol	Meaning
PersnSocia(p_i, t)	The sociability score of person p_i at time t
Capab(p_i, t)	The capability score of person p_i at time t
CapabRetn(p_i, t)	The retained capability score of person p_i at time t
PersnEmin(p_i, t)	The eminence score of person p_i at time t
PersnInfl(p_i, t)	The influence score of person p_i at time t
Imprt(ob_i)	The importance of object ob_i at time t
$Retn(ob_j, t)$	The proportion of the importance of ob_i retained at time t
Wght(c)	The weight assigned for category c
$ob_i.popFig$	The popularity figure of object ob_i
b_i	The basic fading amount of object ob_i

Capability. The squares in Fig. 1 represent the achievement objects, such as an article, a book and so on achieved by someone. This paper employs the concept professional *Capability* to denote the ranking model that only considers the links between objects and people (solid edges in Fig. 1). The *Capability* ranking model is formulated by the following Eq. 2 as the rank score of people at time t.

$$\text{Capab}(p_i, t) = \sum_{ob_j \in OB(p_i, t)} \text{Imprt}(ob_j). \tag{2}$$

Where $OB(p_i, t)$ is the set of objects that link to person p_i before time t. Imprt(ob_j) denotes the importance score of object ob_j, which is examined in Sect. 4.

Eminence. *Sociability* and *Capability* are the two baseline ranking models which calculate people's rank score by paying attention to only one aspect of a person's characteristics. This paper employs the concept *Eminence* to denote the ranking model that combines the *Sociability* and *Capability* ranking models together. This *Eminence* model is formulated by following Eq. 3.

$$\text{PersnEmin}(p_i, t) = \delta * \text{Capab}(p_i, t) + (1 - \delta) * \sum_{p_k \in R(p_i, t)} \frac{\text{PersnEmin}(p_k, t)}{\text{OL}(p_k, t)}. \tag{3}$$

The *Eminence* model doesn't take the time decaying effect into consideration, while the *Influence* model proposed in this paper aims to take the time decaying effect into account.

CapabilityRetained. In this paper, we also take the time decaying effect into consideration. A past sensational person might probably become less well-known after a long period of time, as his/her achievements which was done many years ago would be forgotten by people gradually. This paper employs the concept *CapabilityRetained* to denote the ranking model that considers the time decaying effect as well as the *Capability* ranking model. The *retained* Capability is formulated by the following Eq. 4.

$$\text{CapabRetn}(p_i, t) = \sum_{ob_j \in OB(p_i, t)} \text{Imprt}(ob_j) * Retn(ob_j, t) \,. \tag{4}$$

Where $Retn(ob_j, t)$ measures the time decaying effect on object ob_j, which is a value between 0 and 1 and represents the proportion of the importance of object ob_i retained at time t. Its specific calculation method is elaborated in Sect. 5.

Influence. Having introduced the above concepts, here comes our *Influence* ranking model, which combines the *Sociability* and *CapabilityRetained* together. This paper employs *Influence* to denote such a ranking model, as it generally reflects how much a person is eminent as well as remembered by other persons, i.e. how much influential a person is. Our *Influence* model is formulated by the following Eq. 5.

$$\text{PersnInfl}(p_i, t) = \delta * \text{CapabRetn}(p_i, t) + (1 - \delta) * \sum_{p_k \in R(p_i, t)} \frac{\text{PersnInfl}(p_k, t)}{\text{OL}(p_k, t)} \,. \tag{5}$$

With this formula, we not only consider a person's own achievements and time decaying effect, but also consider his/her sociability comprehensively.

One thing we should notice is that, as we take time factor into account, we deem the graph of Fig. 1 as a dynamically changing graph along with time. Old persons, achievement objects are fading out of our graph, while new persons are included in, due to the decaying effect of time.

4   The Importance of Objects

As we have mentioned in Sect. 3, this section aims to evaluate the importance of objects. Objects can be of different categories, which are illustrated by different colors in Fig. 1. Different categories of objects (e.g., book and article) are usually of different importance. We should assign different weights for them. The weight for a category c is denoted as $\text{Wght}(c)$ which is examined in Sect. 4.1. Inside a category, the level of objects is denoted as $\text{Level}(ob_i)$ which is measured in Sect. 4.2. The final importance of object ob_i is as the following Equation:

$$\text{Imprt}(ob_i) = \text{Level}(ob_i) * \text{Wght}(C(ob_i)) \,. \tag{6}$$

Where $C(ob_i)$ returns the category of object ob_i.

4.1 The Assignment of Category Weights

The number of categories is often very small in each domain. So it is feasible for us to define their category weights preliminarily by domain experts. Table 2 exhibits the weights designated by domain experts for categories in show business domain. c_i represents the i-th category. A category may have several sub-categories, which can have sub-categories further. c_i has $|c_i|$ sub-categories, and w_i is the weight for c_i. The weights assigned by domain experts satisfy the equation $\sum\limits_{i, F(c_i)=c_k} w_i = 1, 0 \leq w_i \leq 1$. Where $F(c_i)$ is the father category of c_i.

Table 2. Weights designated by domain experts

Categories in Show-business($c_0, |c_0| = 2, w_0 = 1$)			
TV Serial(c_1)($|c_1| = 2, w_1 = 0.55$)		Film(c_2)($|c_2| = 2, w_1 = 0.45$)	
Eastern(c_3)($|c_3| = 2, w_3 = 0.6$)	Western(c_4)	Eastern(c_5)	Western(c_6)
China(c_7)	Others(c_8)	($w_4 = 0.4$)	($w_5 = 0.5$)	($w_6 = 0.5$)
($w_7 = 0.55$)	($w_8 = 0.45$)		

The weights assigned by domain experts are conditional weights under their father category. Their absolute weights (w_i') are computed by the following Eq. 7:

$$w_i' = |c_k| * w_i * w_k', F(c_i) = c_k .\tag{7}$$

The Eq. 7 has the following two properties:

Property 1: Given any two sub-categories of a same father-category, the ratio between their conditional weights remains unchanged between their absolute weights, i.e.

$$\forall i, j, \text{ if } F(c_i) = F(c_j), \text{then } w_i' : w_j' = w_i : w_j .$$

Property 2: The average of the absolute weights for all the sub-categories is equal to the absolute weight for their father-category, i.e.

$$\frac{\sum\limits_{i, F(c_i)=c_k} w_i'}{|c_k|} = w_k' .$$

Property 1 and Property 2 indicates that the absolute weights derived from Eq. 7, remain their comparative status among its sibling categories as well as inherit the status of their father. We deem Eq. 7 as an appropriate method for determining the absolute weights for categories.

4.2   The Level of Object Under a Category

Until now, we've discussed how to attain the weights for categories, now it's time to measure the objects' comparative level under a same category. An object's level would be demonstrated by its popularity, i.e. the number of users that appreciate the object. We implicitly assume that if an object is clicked or viewed by a user on the web, the user casts a vote for this object. We utilize the number of votes as an object's popularity-figure ($popFig$) to measure its level. For example, in show business domain, we can use the view times of a movie as its $popFig$ to generally represent the movie's acceptance by people.

The level of an object is computed by the following Eq. 8.

$$\text{Level}(ob_i) = \frac{ob_i.popFig}{\sum\limits_{ob_k \in C(ob_i)} ob_k.popFig} . \tag{8}$$

Where ob_k is under the same category as ob_i, i.e. $C(ob_i)$.

5   The Decaying Effect of Time

Time has a decaying effect on the importance of an object. For example, a thesis which was produced ten years ago is usually not as useful as a one that was produced yesterday, as people are forgetful and fickle and object would probably become obsolete over a long time period. We exploit the forgetting curve model [14,15] proposed by Hermann Ebbinghaus to measure the time decaying effect.

5.1   The Object Importance Retention

The forgetting curve aims to depict the decline pattern in memory over time. In the year 1885, Hermann Ebbinghaus extrapolated the hypothesis that the forgetting law of memory follows an exponential nature [14]. The following formula can roughly describe it:

$$R = e^{-\frac{t}{s}} . \tag{9}$$

t refers to the time length from the start time that a thing is memorized, s refers to the strength of memory, and R refers to the percentage of memory that is still retained at time point t. People are always forgetting along with time. The longer t becomes, the fewer retention will be kept. As we can see, it is hard for people of stronger memory to forget something. The stronger memory s is, the bigger the retention will be.

We apply this forgetting curve model to measure the decaying effect of time on the importance of an object. Here we take the importance score of an object as (the strength of memory), as the more superior an object is, the harder it will be forgotten by people, and take the object's age as t, as the longer after the object was born, the easier for people to forget them. We calculate the *retention* of an object's importance at the time t by the following Eq. 10:

$$Retn(ob_i, t) = \begin{cases} e^{-k * \frac{t-t_b}{Imprt(ob_i)}}, t \geq t_b \\ 0, t < t_b \end{cases}.$$

(10)

Where k is a coefficient for tuning the final value, t_b is the born time of object ob_i. Equation 10 satisfies the following Property 3.

Property 3: The influence retention score of any object is between 0 and 1, and is decreasing with time i.e.

$$\forall t, i, \; 0 \leq Retn(ob_i, t) \leq 1;$$
$$\forall t_i < t_j, \; Retn(ob_i, t_i) \geq Retn(ob_i, t_j).$$

From the Eq. 10, this can be easily proved. According to property 3, we can deem the retention of an object as the proportion of the users who still remember it at time point t.

5.2 The Optimization of Retention Calculation

The retention function $Retn(ob_i, t)$ is a continuous function about variable t. It's not applicable and necessary to compute and store all the retention scores for any time point t. We could compute and update the retentions periodically. First, we need to determine the length of time *period* between two adjacent evaluating time points. We could set the *period* as one day, one week, or one year according to the specific situation. Then, we can tune our retention calculation by the following Eq. 11:

$$Retn(ob_i, t) = \begin{cases} e^{-k * \frac{\lfloor \frac{t-t_b}{period} \rfloor * period}{Imprt(ob_i)}}, t \geq t_b \\ 0, t < t_b \end{cases}$$

(11)

We use the following formula to define the basic decaying amount b_i for object ob_i. b_i can be deemed as the proportion of users that still remember ob_i after one period length.

$$b_i = e^{-k * \frac{period}{Imprt(ob_i)}}.$$

(12)

$$Set \; n(t) = \left\lfloor t - t_b / period \right\rfloor, n(t) = 0, 1, 2, \cdots \cdots$$

(13)

With Eqs. 12 and 13, we can rewrite Eq. 11 as follows:

$$Retn(ob_i, t) = \begin{cases} b_i^{n(t)} = Retn(ob_i, t - period) * b_i, t \geq t_b \\ 0, t < t_b \end{cases}$$

(14)

With Eq. 14, we can see that ob_i's retention at time t can be calculated by multiplying the retention at time $t - period$ with b_i when time $t >= t_b + period$. For example, (suppose the *period* is one month), then we can easily get this month's retention by multiplying the retention of the preceding month with b_i. b_i is irrelevant to variable t, so we can pre-compute and save it in our database.

Influence Ranking Algorithm. Until now, all the necessary components have been examined for the calculation of our *Influence* rank scores. The following algorithm is for computing the *Influence* rank list.

Algorithm 1: Influence Ranking

Input   : the time deadline t, graph data G, conditional weights W designated
           by domain experts, δ assigned for professional capability, k assigned
           for tuning the decaying speed of time
Output: the influence rank list $RL(t, G, W)$

1 Calculate the absolute weights for categories with Eq. 7
2 Calculate the level of objects under their category with Eq. 8
3 Calculate the importance of objects with Eq. 6
4 Calculate the basic decaying amount with Eq. 12
5 Calculate the *retention* of objects with Eq. 14
6 Calculate the *CapabilityRetained* with Eq. 4
7 Calculate the influence rank scores with Eq. 5
8 Rank the influence scores and get the final *Influence* rank list $RL(t, G, W)$

---

6 Experiments

We use two datasets–*AA* (actor and actress) and *DBLP* in our experiment. Dataset *AA* is extracted from the websites of douban, pps and hao123 with the extraction software-DeiXTo. It is used to rank the influence of actors and actresses in show-business domain. In order to scale down the volumes of DBLP, we only focus on ranking the researchers in database and data mining domain.

In this section, we compare the rank score of *Influence* (Eq. 5) with that of *Eminence* (Eq. 3) and *Sociability* (Eq. 1) on datasets *AA* and *DBLP*. Besides, we also compare the *Influence* and *Eminence* lists on *AA* with an existing celebrity list-*Forbes List of Chinese Celebrities*.

Our experiments showed the following advantages of *Influence*:

- Sinking and Rising Quicker: A capable person's *Influence* rank could sink down comparatively quicker than with *Eminence* if he/she has retreated from working domain for a long time, while a new rising star could rise up more easily if he/she has achieved a lot in recent years (Sect. 6.1).
- Comprehensive Consideration: Taking the person's self achievements into consideration makes the ranking results more convincing and meaningful than just *Sociability* (Sect. 6.2).
- Versatility: By tuning the parameter δ in Eq. 5 and k in Eq. 10, we can easily change the *Influence* rank score into *Sociability* or *Eminence*, which are pretty meaningful for some specific situations (Sect. 6.3).

6.1   Sinking and Rising Quicker

We first demonstrate the Sinking Quicker aspect of *Influence Ranking Model*. We choose two performers on AA and two researchers on DBLP as an example to explain this. The two performers were pretty famous twenty years ago and passed away from us ten years ago. The other two researchers have made a great many achievements and are still publishing valuable articles on database but not as many as used to be.

From Figs. 2 and 3, we can see that their *Eminence* ranks remains in comparatively high rank these years, while their *Influence* ranks has already sunk down. This is because we included the time decaying factor in the *Influence* rank calculation. It is fine to use the *Eminence* method when we just want to check a person's actual professional capacity and sociability, but if we want to know who are more influential than others, we'd better use the *Influence* method, which would filter out those who were once very successful but have already began fading out of work domain for a long time.

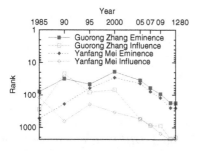

Fig. 2. Zhang and Mei

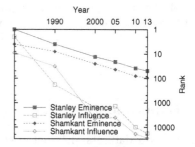

Fig. 3. Stanley and Shamkant

With the same reason, the promising persons who have done many achievements recently, their *Influence* rank would rise comparatively easier than *Eminence*, which are illustrated in Figs. 4 and 5.

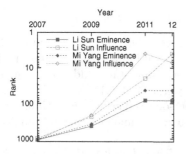

Fig. 4. Li Sun and Mi Yang

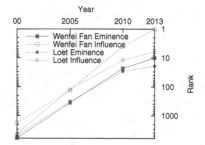

Fig. 5. Fan and Loet Leydesdorff

With *Influence*, we can also get a dynamic list, as anyone's *Influence* rank score would sink down to the bottom one day and save spaces for those newly rising stars.

6.2   Comprehensive Consideration

Sociability ranking model only considers the relationship links between people. While *Influence* not only checks a person's *Sociability* but also his/her professional *Capability*. It evaluates people more comprehensively.

In Tables 3(b) and 4, we can see that the *Sociability* of some famous persons are comparatively low. This is because *Sociability* measures people partially by just considering the relationships between them, and as a result, sinks those hard-working persons who take little time for socialization. In Table 3(a), we compared the ranks of *Influence* with that of *Forbes*. We can see that the *Influence* ranks of those, who have contributed a lot these years, are higher than that of *Forbes*. *Forbes* gives lower ranks for some rising star.

6.3   Versatility

The damping factor δ of Eq. 5 denotes the weight assigned for achievements while $1 - \delta$ for relationship. From Fig. 6, we can see that with the δ becoming smaller, You Ge's rank becoming higher while Wei Zhao's becoming lower. Here we can infer that You Ge is better connected but is less productive than Wei Zhao until

Table 3. Rank of 2012 in show business

(a) Compare Influence, Forbes, Eminence

Star Name	Forbes	Influence	Eminence
Mi Yang	13	7	44
Fucheng Guo	21	144	15
Chaowei Liang	27	25	8
Runfa Zhou	44	352	26
Kun Chen	49	42	17
Zhang Wen	71	5	34

(b) Compare Infuluence, Sociability

Star Name	Influence	Sociability
Xun Zhou	1	127
Zhang Wen	5	58
Mi Yang	7	28
Tianle Gu	9	94
Qiusheng Huang	98	23
Rongguang Yu	105	7

Table 4. Ranks of 2013 on DBLP

Researcher	Influence	Eminence	Sociability
Wenfei Fan	1	10	327
Loet Leydesdorff	6	19	252
Robert L. Glass	15	52	431
Stanley Y. W. Su	17786	49	14
Shamkant B. Navathe	27481	95	427
Michel E. Adiba	31001	246	348

2012. When $\delta = 1$, we get a *CapabilityRetained* (Eq. 4) rank list; when $\delta = 0$, we get an approximate *Sociability* rank list.

The coefficient k of Eq. 10 is for tuning the final retention proportion. When $k = 0$, no decaying effect is put on rank score, thus we get an *Eminence* rank list; when k becomes bigger, the retention would become smaller, i.e. an object's importance score would decay quicker. From Fig. 7, we can see that when k becomes bigger ($k = 0.005$), Robert's rank become higher while Stanley become lower, thus we can infer that Most of Robert's achievements have been done recently while that of Stanley were done some years ago. When k continues becoming enough bigger ($k = 0.02$), we can see that Stanley's rank is promoted a lot. In Table 4 we can see that Stanley's *Sociability* is comparatively high. When k is pretty bigger, the final rank would mainly reflect a person's sociability, which can be inferred from Eq. 10 too as a person's achievements score would almost become zero when they are decaying pretty quickly.

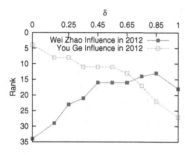

Fig. 6. Different δ of Eq. 5

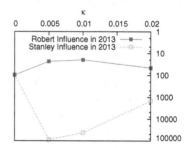

Fig. 7. Different k of Eq. 10

With our *Influence* model, we can freely set the parameter δ and k according to application needs. If we want their achievements decays quicker, we can assign k a bigger value and vice versa. If we want to put more emphasis on a person's achievements, we can assign a bigger value to δ and vice versa. In a word, the *Influence* model proposed in this paper is pretty versatile.

7   Conclusion

In this paper, we proposed an *Influence* ranking model which not only measures the sociability of people but also their professional capability. Besides, we also measures the time decaying effects on the achievements of people in our model. Our experiments demonstrate that, from the point of comparing people's influence, our *Influence* ranking model gets a more meaningful rank list. Our model is also pretty versatile for different purposes through tuning the value of parameters. The model proposed in our paper can be exploited in many different applications, such as search engines entity ranking, organization staff grading, experts finding and so on. The further works will focus on two aspects: first we will expand the time decaying effects on people's relationship links; then we will take the links between achievements into consideration.

Acknowledgement. This research was supported by the National Basic Research 973 Program of China under Grant No. 2012CB316201, the National Natural Science Foundation of China under Grant Nos. 61033007, 61003060 and the Fundamental Research Funds for the Central Universities under Grant Nos. N100704001, 130404015.

References

1. Page, L., Brin, S., Motwani, R., Winograd, T.: The pagerank citation ranking: bringing order to the web (1999)
2. Jeh, G., Widom, J.: Scaling personalized web search. In: Proceedings of the 12th International Conference on World Wide Web, pp. 271–279. ACM (2003)
3. Dubey, H., Roy, B.: An improved page rank algorithm based on optimized normalization technique. Int. J. Comput. Sci. Inf. Technol. **2**(5), 2183–2188 (2011)
4. Berkhin, P.: Bookmark-coloring algorithm for personalized pagerank computing. Internet Math. **3**(1), 41–62 (2006)
5. Fujiwara, Y., Nakatsuji, M., Shiokawa, H., Mishima, T., Onizuka, M.: Efficient ad-hoc search for personalized pagerank. In: Proceedings of the 2013 International Conference on Management of Data, pp. 445–456. ACM (2013)
6. Nie, Z., Zhang, Y., Wen, J.R., Ma, W.Y.: Object-level ranking: bringing order to web objects. In: Proceedings of the 14th International Conference on World Wide Web, pp. 567–574. ACM (2005)
7. Balmin, A., Hristidis, V., Papakonstantinou, Y.: Objectrank: authority-based keyword search in databases. In: Proceedings of the Thirtieth International Conference on Very Large Data Bases-Volume 30, VLDB Endowment, pp. 564–575 (2004)
8. Rode, H., Serdyukov, P., Hiemstra, D., Zaragoza, H.: Entity ranking on graphs: studies on expert finding (2007)
9. Chakrabarti, S.: Dynamic personalized pagerank in entity-relation graphs. In: Proceedings of the 16th International Conference on World Wide Web, pp. 571–580. ACM (2007)
10. Venetis, P., Gonzalez, H., Jensen, C.S., Halevy, A.: Hyper-local, directions-based ranking of places. Proc. VLDB Endow. **4**(5), 290–301 (2011)
11. Qi, Hc, Huang, Dc, Zhevg, Y.: An improved pagerank algorithm with time feedbacking. J. Zhejiang Univ. Technol. **33**(3), 272 (2005)
12. Guo, Y.Z., Ramamohanarao, K., Park, L.A.: Personalized pagerank for web page prediction based on access time-length and frequency. In: IEEE/WIC/ACM International Conference on Web Intelligence, pp. 687–690. IEEE (2007)
13. Yang, K.W., Huh, S.Y.: Automatic expert identification using a text categorization technique in knowledge management systems. Expert Syst. Appl. **34**(2), 1445–1455 (2008)
14. Wikipedia: Forgetting curve. http://en.wikipedia.org/wiki/Forgetting_curve
15. Loftus, G.R.: Evaluating forgetting curves. J. Exp. Psychol. Learn. Mem. Cogn. **11**(2), 397 (1985)

Author Index

Printed in the United States
By Bookmasters